Mesoscale Phenomena
in Fluid Systems

ACS SYMPOSIUM SERIES **861**

Mesoscale Phenomena in Fluid Systems

Fiona Case, Editor
Colgate Palmolive Company

Paschalis Alexandridis, Editor
The State University of New York

Sponsored by the
**ACS Divisions of Physical Chemistry and
Colloid and Surface Chemistry**

American Chemical Society, Washington, DC

Library of Congress Cataloging-in-Publication Data

American Chemical Society. Meeting (224th : 2002 : Boston, Mass.)

  Mesoscale phenomena in fluid systems / Fiona Case, editor, Paschalis Alexandridis, editor.

      p. cm.—(ACS symposium series ; 861)

  "Sponsored by the ACS Divisions of Physical Chemistry and Colloid and Surface Chemistry."

  Includes bibliographical references and index.

  ISBN 0–8412–3867–7

    1. Soft condensed matter—Measurement—Congresses. 2. Surface chemistry—Congresses. 3. Nanotechnology—Congresses. 4. Fluid dynamic measurements—Congresses.

    I. Case, Fiona, 1965- II. Alexzndridis, Paschalis. III. American Chemical Society. Division of Physical Chemistry. IV. American Chemical Society. Division of Colloid and Surface Chemistry. V. Title. VI. Series.

QC173.458.S62A44   2003
530.4'13—dc21   2003052337

The paper used in this publication meets the minimum requirements of American National Standard for Information Sciences—Permanence of Paper for Printed Library Materials, ANSI Z39.48–1984.

PRINTED IN THE UNITED STATES OF AMERICA

Foreword

The ACS Symposium Series was first published in 1974 to provide a mechanism for publishing symposia quickly in book form. The purpose of the series is to publish timely, comprehensive books developed from ACS sponsored symposia based on current scientific research. Occasionally, books are developed from symposia sponsored by other organizations when the topic is of keen interest to the chemistry audience.

Before agreeing to publish a book, the proposed table of contents is reviewed for appropriate and comprehensive coverage and for interest to the audience. Some papers may be excluded to better focus the book; others may be added to provide comprehensiveness. When appropriate, overview or introductory chapters are added. Drafts of chapters are peer-reviewed prior to final acceptance or rejection, and manuscripts are prepared in camera-ready format.

As a rule, only original research papers and original review papers are included in the volumes. Verbatim reproductions of previously published papers are not accepted.

ACS Books Department

Contents

Preface...xi

Characterizing Mesoscale Structure and Phenomena

1. **Atomic Force Microscopy Study of Trisiloxane Surfactant Aggregate Structures at the Solid–Liquid Interface**..............................2
   J. Dong, G. Mao, and R. M. Hill

2. **Cryogenic Temperature Transmission Electron Microscopy Study of Self-Aggregation of Sodium Lithocholate Single-Molecular Walled Nanotubes**...17
   Y. Talmon, J. Schmidt, and P. Terech

3. **Characterizing Mesoscale Structure and Phenomena in Fluids Using NMR**..27
   Peter Stilbs

4. **Characterization down to Nanometers: Light Scattering from Proteins and Micelles**..44
   Ulf Nobbmann

5. **Small-Angle Scattering Characterization of Block Copolymer Micelles and Lyotropic Liquid Crystals**...............................60
   Paschalis Alexandridis

6. **X-ray Scattering Studies of Long-Chain Alkanol Monolayers at the Water–Hexane Interface**..81
   Mark L. Schlossman and Aleksey M. Tikhonov

7. **The Structure and Composition of Mixed Surfactants at Interfaces and in Micelles**..96
   J. Penfold, R. K. Thomas, E. Staples, and I. Tucker

8. Particle Characterization by Scattering Methods
    in Systems Containing Different Types of Aggregates.
    Aggregation of an Amphiphilic Poly(paraphenylene)
    in Micellar Surfactant Solutions...117
    Tobias Fütterer, Thomas Hellweg, and Gerhard H. Findenegg

9. Aggregation and Gelation in Colloidal Suspensions: Time-
    Resolved Light and Neutron Scattering Experiments.........................143
    Peter Schurtenberger, Hugo Bissig, Luis Rojas, Ronny Vavrin,
    Anna Stradner, Sara Romer, Frank Scheffold,
    and Veronique Trappe

10. Aging of Soft Glassy Materials Probed by Rheology
    and Light Scattering...161
    Eugene Pashkovski, Luca Cipelletti, Suliana Manley,
    and David Weitz

11. Nanoscale versus Macroscale Friction in Polymers and
    Small-Molecule Liquids: Anthracene Rotation in PIB
    and PDMS...177
    Mark M. Somoza and Mark A. Berg

12. Sculpting Nanoscale Liquid Interfaces...191
    Richard C. Bell, Hanfu Wang, Martin J. Iedema,
    and James P. Cowin

Predicting Mesoscale Structure and Phenomena

13. Modeling Structure and Properties in Mesoscale Fluid Systems.......206
    Peter V. Coveney

14. Mesoscale Simulations: Industrial Applications...............................227
    Simon McGrother, Lam Yeng Ming, and Gerhard Goldbeck-Wood

15. Application of Dissipative Particle Dynamics to Materials
    Physics Problems in Polymer and Surfactant Science.......................242
    Massimo G. Noro, Frederico Meneghini, and Patrick B. Warren

16. Dynamics of Phase Separation in Polymeric Systems.......................258
    G. J. A. Sevink, A. V. Zvelindovsky, and J. G. E. M. Fraaije

17. Simulations of Polymer Solutions: A Field-Theoretic Approach........279
    Alfredo Alexander-Katz, André G. Moreira,
    and Glenn H. Fredrickson

18. Modeling Mechanical Properties of Resins Prepared
by Sol–Gel Chemistry..290
Stelian Grigoras

19. Grand Canonical Monte Carlo Simulations of Equilibrium
Polymers and Networks...298
James T. Kindt

20. Study of the Effects of Added Salts on Micellization
of Cetyltrimethylammonium Bromide Surfactant.............313
B. Lin, S. Mohanty, A. V. McCormick, and H. T. Davis

Applications of Mesoscale Phenomena

21. Formation of Multilamellar Vesicles from Ethylene Oxide–
1,2-Butylene Oxide Diblock Copolymers............................328
J. Keith Harris, Gene D. Rose, and Merlin L. Bruening

22. Cubosome Formation via Dilution: Kinetic Effects
and Consumer Product Implications....................................346
Patrick T. Spicer

23. The Importance of Mesoscopic Structures in the Development
of Advanced Materials...360
Cristina U. Thomas, Gregg Caldwell, Richard B. Ross,
Sanat Mohanty, and Miriam Freedman

24. Impact of Mesoscale Structure and Phase Behavior
on Rheology and Performance in Superwetting Cleaners.................376
Guy Broze and Fiona Case

25. Emulsions or Microemulsions? Phase Diagrams and Their
Importance for Optimal Formulations..................................394
Ingegärd Johansson and O Boen Ho

Indexes

Author Index...415

Subject Index..417

Preface

Mesoscale is a term often used to describe structures and behaviors that occur at length scales between nanometers (10^{-9}m) and 0.1 micrometers (10^{-7}m). As the name implies, mesoscale structures are intermediate in size. They are larger than the molecular or atomistic scale, which makes them inaccessible to many molecular characterization methods, and prohibitively expensive to model using atomistic-based simulation techniques (such as molecular mechanics). Yet because of their nanometer to micrometer structure they cannot be fully characterized by their bulk behavior or modeled using continuum models. These types of materials have also been called soft condensed matter and complex or nanostructured fluids.

This intermediate length scale is challenging to study, yet phenomena that occur at this scale determine the properties of many scientifically and commercially important materials. The target audience we had in mind as we developed the original American Chemical Society (ACS) Symposia and this book is an industrial research scientist (or member of an industrially focused academic group) working on materials whose performance depends on mesoscale phenomena. This could be a personal care product such as a liquid hand soap or shampoo, a cosmetic product such as a moisturizer or make-up, or a food. It could be paint or ink. It could be the microencapsulation or delivery system for a drug or an emulsion polymerization. We hope that this book will be a valuable resource, an encyclopedia of available techniques and applications, for the scientists faced with the challenge of controlling and designing these types of materials. However, we also recommend this work as a reference of complementary approaches to researchers focused on one or two of the techniques described. Mesoscale phenomena also operate on length scales critical for biological materials, and many of the techniques described in this book may be of interest to the life science community.

The book is divided into three sections. In the first and largest section, experimental techniques to characterize mesoscale structure are described by some of the leading experts. The second section focuses on mesoscale modeling and simulation methods. Because of the challenge of experimental characterization, particularly for complex mixtures, these are especially valuable for this class of materials. The final section of the book includes industrial case studies or applications examples showing how an understanding of mesoscale structure, properties, and phase behavior can be used in materials design.

The range of applications, characterization methods, and modeling techniques is broad. We expect that most readers will find at least one method that is new to them. Each of the contributors hopes that you will find the methods and techniques described in their chapter particularly useful in your research, and in addition to describing our latest research we have provided careful introductions placing our work in context, and key references for further reading.

We thank the ACS Divisions of Physical Chemistry and the Colloid and Surface Chemistry as well as the ACS Petroleum Fund for their support of the *Mesoscale Phenomena in Fluid Systems* symposium at the 224th ACS National Meeting in Boston in 2002. We also thank the ACS Division of Colloid and Surface Science for their support of the related *Nanoscale Organization by Self-Assembly* symposium at the 2002 Colloid and Surface Science Symposium, Ann Arbor Michigan, from which some of the contributions were taken. We thank the large international group of scientists who were willing to provide technical reviews of the chapters. We also thank Accelrys Inc. (www.accelrys.com) for their financial support that allowed color graphics in the modeling and simulation section, many of which were generated using their software packages. Fiona particularly thanks her husband, Martin Case, her parents Ian and Ann McCraw, and her colleagues in the Innovation and Strategy group at Colgate Palmolive for their invaluable assistance with proofreading. She also thanks the Colgate Palmolive Company for the opportunity to work with these fascinating materials, and for the time to edit this book.

Fiona Case
Innovation and Strategy Group
Colgate Palmolive Research
909 River Road
Piscataway, NJ 08855
fhcase@hotmail.com

Paschalis Alexandridis
Department of Chemical Engineering
University at Buffalo
State University of New York
Buffalo, NY 14260–4200
palexand@eng.buffalo.edu

Characterizing Mesoscale Structure and Phenomena

Chapter 1

Atomic Force Microscopy Study of Trisiloxane Surfactant Aggregate Structures at the Solid–Liquid Interface

J. Dong[1], G. Mao[1,*], and R. M. Hill[2]

[1]Department of Chemical Engineering and Materials Science,
Wayne State University, Detroit, MI 48202
[2]Dow Corning Corporation, 2200 West Salzburg Road, Midland, MI 48686

The aggregate structure of a nonionic trisiloxane surfactant at the liquid/solid interface was studied by Atomic Force Microscopy (AFM) as a function of substrate surface energy. The hydrophobicity of oxidized silicon wafer was gradually increased by increasing the amount of n-octadecyltrichlorosilane (OTS) monolayer coverage. AFM soft-contact imaging and force measurements captured the variation in surfactant aggregate structures from spherical micelles to elongated micelles, defected monolayer, and continuous monolayer with increasing surface hydrophobicity. The aggregate structural evolution at the solid/liquid interface resembles the microstructural sequence of the surfactant in bulk solution at room temperature. It is speculated that the hydrophobic attraction between the surfactant and the solid surface induces two-dimensional analogues of the surfactant bulk microstructures. The hydrophobic attraction affects the geometric packing parameter in a similar fashion as the increase in surfactant concentration.

Introduction

The wetting, surface activity, and phase behavior of silicone surfactants have been studied extensively because of their wide use in anti-foaming, enhanced wetting, corrosion protection, bactericides, and cosmetic formulations *(1)*. Specifically a group of trisiloxane surfactants have excited much research interests because they promote rapid spreading of aqueous solution on hydrophobic surfaces. Various causes were said to contribute to the superspreading capability of the trisiloxane surfactants, which include the molecular structure, solution turbidity, and formation of lamellar- or bicontinuous-like structure near the interfaces *(2)*.

Surfactant phase behavior and microstructure in bulk solution are well understood *(3,4)*. However only recently significant progress has been made in the understanding of the aggregate structure at the solid/liquid interface. It is possible to visualize the surfactant aggregate structure directly at the solid/liquid interface by soft-contact AFM *(5)*. Soft-contact AFM is performed by floating the AFM tip above the adsorbed surfactant layer with the image force set below the breakthrough force to the underlying substrate. Soft-contact imaging together with direct force-versus-distance measurements make it possible to define the aggregate morphology at the nanometer scale. Nonionic oligo(oxyethylene) n-alkyl ethers, i.e. C_iE_j surfactants, were found to form an epitaxial layer of hemicylindrical hemimicelles on highly oriented pyrolytic graphite (HOPG) *(6)*. In the absence of epitaxy, the aggregate structure of C_iE_j surfactants was found to depend on temperature *(7)* and surface chemistry *(8,9,10,11)*. Combined with data from other techniques such as ellipsometry, fluorescence, and neutron reflection, the adsorption behavior of C_iE_j surfactants on hydrophilic surfaces versus on hydrophobic surfaces is summarized as following *(12)*: 1) on hydrophilic surfaces, C_iE_j surfactants form surface micelles or flat bilayers depending on the geometric packing parameter; and 2) on hydrophobic surfaces, C_iE_j surfactants form flat monolayers similar to those at the air/water interface. In one study where the substrate hydrophobicity was gradually changed by the mixed monolayers of thiols, no intermediate structures between micelles/bilayers and monolayers were observed *(10)*. Here we search for additional surfactant aggregate structures by studying a surfactant with unusually rich phase behavior - nonionic trisiloxane surfactant $((CH_3)_3SiO)_2Si(CH_3)(CH_2)_3(OCH_2)_{12}OH$ or in short $M(D'E_{12}OH)M$ *(13,14)*. The AFM images and force curves presented here will show that new types of aggregate structures were formed by $M(D'E_{12}OH)M$ at the solid/liquid interface. The AFM study was carried out in a fixed concentration of $M(D'E_{12}OH)M$, 2 × Critical Aggregation Concentration (CAC), on surfaces with gradual variation of hydrophobicity. The surface hydrophobicity was gradually increased by increasing the surface coverage of n-octadecyltrichlorosilane (OTS) monolayer on oxidized silicon. The surface hydrophobicity was characterized by its contact angle with water. We present a possible link between the surfactant surface aggregate structural evolution and its bulk microstructural sequence.

Experimental Section

Materials

M(D'E$_{12}$OH)M was used as received from Dow Corning Corp. with 95% purity. n-Octadecyltrichlorosilane (C$_{18}$H$_{37}$SiCl$_3$) or OTS was purchased from United Technologies and distilled just before use. Water was deionized to 18 MΩ-cm resistivity (Nanopure system, Barnstead). One-sided polished N type silicon (111) wafer was purchased from Wafer World with resistivity of 50 to 75 Ω-cm. The silicon substrate was oxidized following the "RCA" procedures used in the integrated circuit manufacturing *(15)*. The freshly oxidized substrate was completely wettable by water with a contact angle close to 0°.

Substrate Surface Modification

Vapor phase deposition of OTS on oxidized silicon was carried out in a desiccator. The substrate was placed 2 to 3 cm above OTS liquid droplets. Vacuum was pulled until the OTS liquid started to boil at approximately 1.3 kPa. The deposition time was changed from 15 to 90 min in order to obtain substrates with water contact angles between 20° and 107°. Subsequently the substrate was annealed at 150 °C for 2 hr. Only freshly prepared substrates were used in the experiments. All substrates were imaged by AFM prior to injection of the surfactant solution. All substrates showed flat and featureless surfaces with mean surface roughness between 3 and 8 Å over a scan area of 500 × 500 nm^2. No domain structure was imaged on any OTS-modified substrates.

Characterization Methods

The contact angle was measured by an NRL contact angle goniometer (Model 100, Rame-Hart) in the laboratory atmosphere. A water droplet of 20 μl was placed on the substrate and contact angles were read on both sides of the droplet. Five droplets were placed at various spots near the center of the substrate, and contact angles were averaged with an error of ± 3°.

AFM imaging and force measurements were conducted in a liquid cell using Nanoscope IIIa AFM (Digital Instruments). An E-scanner with maximum scan area of 16 × 16 μm² was used. The z scale of the scanner was calibrated with Ultra-Sharp Calibration TGZ02 set (step height 100 nm, Silicon-MDT). Silicon nitride (Si_3N_4) integral tips (NP type) were used with a nominal tip radius of 20 to 40 nm. In order to compare our results, we present data that were obtained by the same cantilever tip. The spring constant of the cantilever was calibrated using the deflection method against a reference cantilever (Park Scientific Instruments) of known spring constant (0.157 N/m) *(16)*. The calibrated value 0.17 ± 0.05 N/m was used in all force plots. The force calibration curve was converted to the force-versus-separation plot following a standard procedure *(7)*. AFM images of the surfactant layer were obtained in the soft-contact mode. The scan rate was between 3 and 12 Hz. Height images were captured with feedback gains between 3 and 5. Deflection images were captured with feedback gains less than 1. Height images were flattened in order to remove background slopes. No other filtering procedures were performed on these images. The tip to substrate velocity was fixed at 0.1 μm/s in force calibration. We found that the force curves became independent of approach speed below 0.5 μm/s for the nonionic surfactant systems. The temperature was maintained at 22 ± 1 °C.

Results

The concentration of M(D'E$_{12}$OH)M was fixed at 2 × CAC (5.4 × 10^{-4} mol/l). Deflection images obtained by contact mode and amplitude images obtained by tapping mode are shown here unless specified. All oxidized silicon substrates with or without OTS coverage appeared to be smooth and featureless under AFM at the nanometer scale. A typical surface roughness Ra was measured to be 5 Å over a scan size of 500 × 500 nm². There is no evidence for domain formation by OTS at any surface coverage according to the AFM results.

Figure 1a and 1b are the soft-contact AFM image and force-versus-separation curve between an AFM tip and oxidized silicon in the M(D'E$_{12}$OH)M solution. The oxidize silicon was completely wettable by a water droplet. On oxidized silicon, M(D'E$_{12}$OH)M molecules aggregated into small globular particles with unit size between 6 to 9 nm. The diameter was determined manually in the AFM sectional analysis by measuring the center-to-center distance between neighboring particles. The particle size is identical to the size of M(D'E$_{12}$OH)M micelles in solution below 10 %w/w, which is 7.0 nm *(13)*. The force curve showed a repulsion starting at 9.1 nm. The repulsion reached a local maximum of 0.9 nN at 3.9 nm. Force decreased and then increased rapidly

a

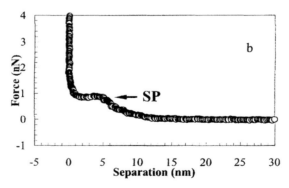

b

Figure 1. AFM image and force-versus-distance curve on oxidized silicon wafer in M(D'E$_{12}$OH)M solution. (a) AFM image showing globular particles. The contact angle is written at the upper right corner. (b) Force-versus-distance curve with the approximate set point force marked as SP.

upon further approach. The force curve lacked a clear gap or jump-in event. The jump-in event is associated with a sudden removal of surfactant layer from the interacting zone. The absence of this abrupt event implies continuous compression and/or removal of surfactant layer. Previous AFM studies of C_iE_j surfactants on silica have associated the continuous type of force curves with surface micelles and the discontinuous type with bilayers *(9,7)*. The wide spatial distribution of surfactant molecules in micelles interacting with substrate functional groups may cause the surfactant molecules to be removed at various compression distances. It was concluded that M(D'E_{12}OH)M formed spherical micelles on oxidized silicon.

The structure of M(D'E_{12}OH)M on OTS-modified oxidized silicon substrate with 20° water contact angle is represented by Figure 2a and 2b. Bundles of elongated rod-like features were observed with width between 6 and 10 nm and length between 30 and 80 nm. The force curve identified as Force Curve 1 showed a continuous repulsion that started at 4.5 nm. No jump-in was observed. The range of the repulsion was in between monolayer thickness and micelle diameter *(13)*. However, the measured force curves varied widely on this surface because of the surface heterogeneity. Force Curve 1 represented those that gave longer range of repulsions with the tip contacting mostly micelle-covered area. Force Curve 2 represented those that were positioned in between micelles. All the force curves should be treated as averages between the rods and that of the background. On a substrate with 40° water contact angle, rod-like features with lower aspect ratio (between 8 and 15 nm in width and between 25 and 35 nm in length) but higher coverage (roughly 8 and 4 rods per 100 nm^2 on 40° and 20° surface respectively) were observed as shown in Figure 3a. The corresponding force curve is shown in Figure 3b. The individual rods co-existed with bigger aggregates of 40 to 80 nm in size. The force curve showed a continuous repulsion starting at 9.5 nm. OTS did not form anisotropic domains on silica and therefore cannot account for the elongated rod formation. The rods cannot be induced by AFM scanning because they appeared at random angles, at the lowest allowable set point force, and only on relatively hydrophilic surfaces. However, the rods gradually rotated to be perpendicular to the fast scan direction with repeated scanning. Therefore the alignment of rods was likely an artifact of the AFM scanning. The elongated features on 20° and 40° surfaces suggest that M(D'E_{12}OH)M formed cylindrical (or less likely hemicylindrical) micelles on the relatively hydrophilic surfaces.

The structure of M(D'E_{12}OH)M aggregates on an OTS-modified substrate with 80° water contact angle is represented by Figure 4. Figure 4a shows a smooth film with randomly distributed defects as indicated by the arrows. The height image (Figure 4b) indicates that the defects were holes. The depth of the holes was less than 1.5 nm as determined in the AFM sectional height profile of Figure 4c. The defects were unchanged during repeated scanning and with

8

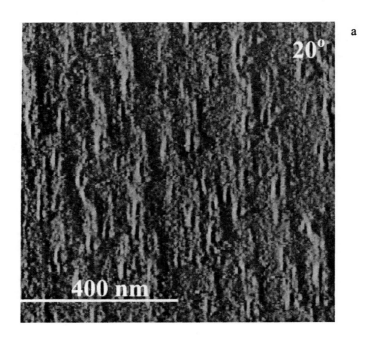

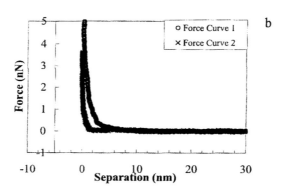

Figure 2. AFM image and force-versus-distance curve on OTS modified silicon wafer with 20° contact angle in M(D'E₁₂OH)M solution. (a) AFM image showing elongated rods and rod clusters. (b) Force-versus-distance curve.

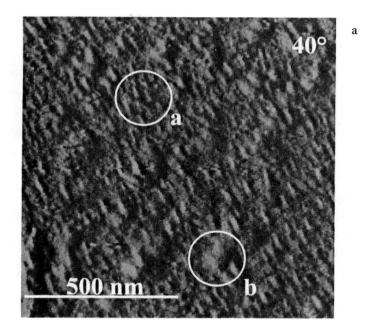

a

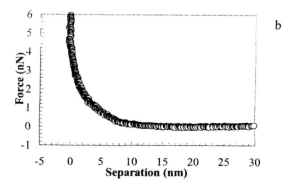

b

Figure 3. AFM image and force-versus-distance curve on OTS modified silicon wafer with 40° contact angle in $M(D'E_{12}OH)M$ solution. (a) AFM image showing elongated rods and rod clusters. Circle a and b highlight a region with individual rods and a region with clusters of rods respectively. (b) Force-versus distance-curve.

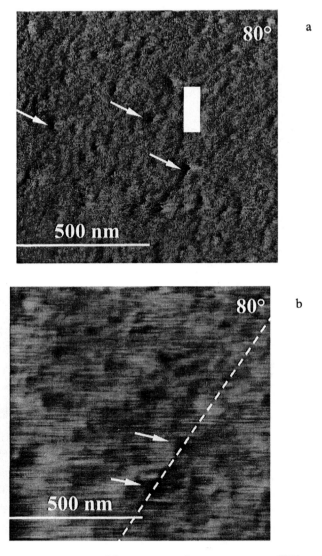

Figure 4. AFM images and force-versus-distance curve on OTS modified silicon wafer with 80° contact value in M(D'E$_{12}$OH)M solution. (a) AFM image showing holes in otherwise smooth surface. The holes are pointed out by the arrows. (b) The corresponding AFM height image. The arrows point out the holes. The dashed line in the image is where the sectional profile was taken. (c) The AFM sectional height profile along the dashed line in (b) showing the holes approximately 1.5 nm in depth. (d) Force-versus-distance curve. The jump-in is marked by the arrow.

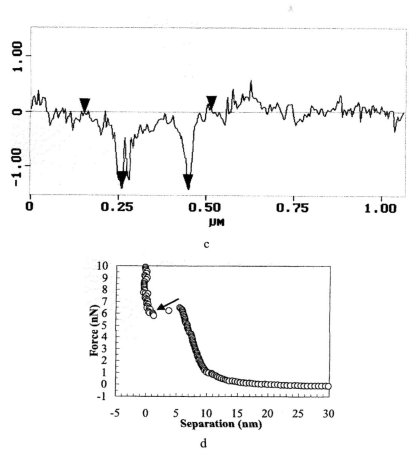

c

d

Figure 4. *Continued.*

different set point forces. This is in contrary to the case where holes were generated by the AFM tip *(7)*. The holes are an inherent feature of the surfactant film. Figure 4d shows a repulsive force with an onset at 14.9 nm and a force maximum of 6.5 nN at 5.4 nm. The jump-in event was clearly present. From the jump-in distance and the fact that only one jump-in event was observed, it was concluded that $M(D'E_{12}OH)M$ formed an imperfect monolayer on a substrate with 80° water contact angle. We would like to point out that almost identical force curves were obtained in the same surfactant solution on graphite surface with contact angle between 75° and 78° *(17)*. Figure 5a and 5b are the AFM image and force curve obtained on oxidized silicon with a fully covered OTS monolayer. Its water contact angle was 107°. The image remained smooth with no holes up to the highest allowable set point force. The repulsion started at 11.5 nm and reached a maximum of 20.2 nN at 3.8 nm. It should be pointed out that AFM-measured film thickness of nonionic polyoxyethylene surfactants often is larger than the thickness measured by other methods, probably due to surfactant adsorption on the AFM tip *(6,18)*. The jump-in event was also observed here indicating an instantaneous push-out of the monolayer segment from the interacting zone. Comparison between Figure 4 and Figure 5 suggests that $M(D'E_{12}OH)M$ monolayer packing density increased with increasing surface hydrophobicity.

Discussion

The AFM study revealed an evolution of $M(D'E_{12}OH)M$ aggregate structures on substrates with increasing substrate surface hydrophobicity. The AFM results of the trisiloxane surfactant agreed for the most part with previous studies of C_iE_j surfactants *(12)*: 1) $M(D'E_{12}OH)M$ formed surface micelles on silica surface with the oxyethylene head groups anchored at the hydroxyl surface and also exposed toward the aqueous solution; 2) $M(D'E_{12}OH)M$ formed a monolayer on fully covered OTS surface with the trisiloxane hydrophobe anchored at the methylated surface and the oxyethylene head group exposed toward the aqueous solution; and 3) monolayer packing density increased with increasing surface hydrophobicity. $M(D'E_{12}OH)M$ also displayed intriguing intermediate structures that are worth further discussion: 1) cylindrical-shaped aggregates at relatively hydrophilic surfaces between 20° and 40° contact angle; and 2) imperfect monolayer structure with holes at 80° contact angle. Though the various intermediate microstructures in C_iE_j surfactant bulk solution were never previously identified on surface, it was known that surface coverage is influenced by the surfactant chemical structure, i.e. the geometric packing

a

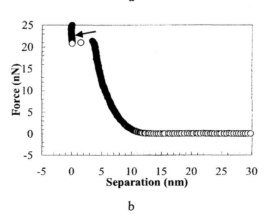

b

Figure 5. AFM image and force-versus-distance curve on OTS modified silicon wafer with 107° contact angle in M(D'E₁₂OH)M solution. (a) AFM image showing smooth surface. (b) Force-versus-distance curve. The jump-in is marked by the arrow.

parameter. Surfactants with geometric packing parameter that favors flat bilayers tend to have the highest adsorption plateau.

The molecular packing parameter S is defined as $S = V/(AL)$, where V and L are the volume and length of the hydrophobic chain respectively, and A is the head group area. S less than 0.33 favors spherical micelles, S between 0.33 and 0.5 favors rod-like micelles, and S larger than 0.5 favors disk-like micelles and vesicles (3). Increase in concentration or temperature of C_iE_j surfactant solutions causes an increase in S value (3).

At room temperature, $M(D'E_{12}OH)M$ forms a single isotropic phase at all concentrations in water, but its microstructure evolves with increasing concentration as determined by small angle neutron scattering, small angle x-ray scattering, cryo-transmission electron microscopy, pulse-gradient spin-echo NMR, and rheology measurement (13): 1) spherical micelles below 10 %w/w, 2) elongated micelles between 10 and 35 %w/w by speculation, 3) entangled long worm-like micelles between 35 and 45 %w/w, 4) multi-connected tubular network between 50 and 70 %w/w, 5) bilayer network between 70 and 80 %w/w, and 6) disordered and fluctuating bilayers above 80 %w/w. It was calculated that if the head group area is below 153 Å^2 (S = 0.33), $M(D'E_{12}OH)M$ forms sphere-shaped micelles; if the head group area is below 102 Å^2 (S = 0.5), $M(D'E_{12}OH)M$ forms rod-shaped micelles; and if the head group area is below 51 Å^2 (S = 1), $M(D'E_{12}OH)M$ forms disk-shaped micelles or infinite bilayer structure (14). One cannot help but notice the strong resemblance between the sequence of $M(D'E_{12}OH)M$ bulk microstructure and that observed on solid surface by soft-contact AFM. We speculate that surface interactions give rise to S value changes in $M(D'E_{12}OH)M$ surface aggregates in a similar fashion as those caused by concentration increase in solution. The intermediate surface aggregate structures must correspond to the microstructures at intermediate concentrations with comparable S values. In other words, the effective head group area of $M(D'E_{12}OH)M$ reduces from no smaller than 153 Å^2 on a silica substrate to less than 102 Å^2 when surface is covered by OTS. The decrease in head group area can be induced by the hydrophobic attraction between surfactant and solid surface, which dehydrates the head group and reduces the repulsion between head groups. It was argued that the hydrophobic attraction is the main driving force for nonionic C_iE_j surfactant adsorption on both hydrophobic and hydrophilic surfaces (10). The hydrophobic attraction becomes stronger with increasing OTS coverage, which causes gradual increase in $M(D'E_{12}OH)M$ S values from those favoring spherical micelles to those favoring rod-like micelles and planar structures.

Conclusions

The self-associating aggregate structure of a nonionic trisiloxane surfactant $M(D'E_{12}OH)M$ was studied by AFM soft-contact imaging and force measurements. The hydrophobicity of the oxidized silicon wafer substrate was gradually increased by increasing the amount of OTS monolayer coverage from vapor phase deposition. In the absence of OTS, spherical micelles were observed. On relatively hydrophilic surfaces with 20° and 40° water contact angles, rod-like micelles and their clusters were observed. On hydrophobic surfaces with 80° and 107° water contact angles, defected and perfect monolayers were observed. These surface aggregate structures are speculated to be two-dimensional analogues of the surfactant microstructures in bulk solution. The surface aggregate structural evolution bears close resemblance to that in bulk from spherical micelles, to worm-like micelles, tubular network, planar layer network, and ordinary planar layer with increasing concentration at room temperature. The surface structural evolution is attributed to the continuous increase in hydrophobic attraction between surfactant and solid surface with increasing surface hydrophobicity. The hydrophobic attraction dehydrates the head group, reduces the head group area, and increases the geometric packing parameter S. It will be interesting to find out if surfactants with equally rich bulk phase microstructures as $M(D'E_{12}OH)M$ will exhibit similar trend. The surfactant aggregate structures are not only interesting for fundamental study but also may serve as templates for novel nanostructures and applications.

References

1. *Surfactant Science Series: Silicone Surfactants;* Hill, R. M., Ed; Marcel Dekker: New York, NY, 1999.
2. Hill, R. M. *Curr. Opinion Colloids Interface Sci.* **1998**, *3*, 247-254.
3. Mitchell, D. J.; Tiddy, G. J. T.; Waring. L.; Bostock, T.; McDonald, M. P. *J. Chem. Soc., Faraday Trans. 1* **1983**, *79*, 975-1000.
4. Davis, H. T. *Statistical Mechanics of Phases, Interfaces, and Thin Films*; VCH: New York, NY, 1996; Chap. 6.
5. Manne, S.; Cleveland J. P.; Gaub, H. E.; Stucky G. D.; Hansma, P. K. *Langmuir* **1994**, *10*, 4409-4413.
6. Patrick, H. N.; Warr, G. G.; Manne, S.; Aksay, I. A. *Langmuir* **1997**, *13*, 4349-4356.
7. Dong, J.; Mao, G. *Langmuir* **2000**, *16*, 6641-6647.
8. Grant, L. M.; Ducker W. A. *J. Phys. Chem. B* **1997**, *101*, 5337-5345.
9. Grant, L. M.; Tiberg, F.; Ducker, W. A. *J. Phys. Chem, B.* **1998**, *102*, 4288-4294.
10. Grant, L. M.; Ederth, T.; Tiberg, F. *Langmuir* **2000**, *16*, 2285-2291.
11. Wolgemuth, J. L.; Workman, R. K.; Manne, S. *Langmuir* **2000**, *16*, 3077-3081.
12. Tiberg, F.; Brinck, J.; Grant, L. *Curr. Opinion Colloid Interface Sci.* **2000**, *4*, 411-419.
13. He, M.; Hill, R. M.; Doumaux, H.A.; Bates, F. S.; Davis, H. T.; Evans, D. F.; Scriven, L. E. In *Structure and Flow in Surfactant Solutions*; ACS

Symposium Series 578; American Chemical Society: Washington, 1994; pp 192-216.

14. He, M.; Hill, R. M.; Lin, Z.; Scriven, L. E.; Davis, H. T. *J. Phys. Chem.* **1993**, *97*, 8820-8834.

15. Kern, W., *J. Electrochem. Soc.* **1990**, *137*, 1887-1892.

16. Tortonese, M.; Kirk, M. SPIE **1997**, *3009*, 53-60.

17. Unpublished results.

18. Wang, A.; Jiang, L.; Mao, G.; Liu, Y. *J. Colloid Interface Sci.* **2002**, *256*, 331-340.

Chapter 2

Cryogenic Temperature Transmission Electron Microscopy Study of Self-Aggregation of Sodium Lithocholate Single-Molecular Walled Nanotubes

Y. Talmon[1], J. Schmidt[1], and P. Terech[2]

[1]Department of Chemical Engineering, Technion-Isreal Institute
of Technology, Haifa 32000, Israel
[2]UMR 5819 CEA-CNRS-Université J. Fourier, Département de Recherche
Fondamentale sur la Matière Condensée, 17, rue des Martyrs, 38054
Grenoble Cédex 09, France

We describe an ongoing study of nanoscopic self-aggregation in aqueous solutions of sodium lithocholate, leading to the formation of nanotubules. Using modern cryogenic temperature transmission electron microscopy (cryo-TEM) augmented by digital imaging (the state-of-the-science is described) and small-angle x-ray scattering (SAXS) we have shown that micrometer-long nanotubules form spontaneously with monodisperse cross-sections (D_o=52 nm, D_i=49 nm) in alkaline aqueous solutions of sodium lithocholate (SLC). The shell of these tubules, 1.5 nm thick, is made of a monomolecular sheet of the bile salt. Such SLC assemblies could be used to develop functional materials based on 1-D structures, and as supramolecular templates for synthesis of inorganic materials in nanotechnology. Time-resolved cryo-TEM has started to elucidate the mechanism of formation of these nanotubules. Intermediate nanostructures are multi-walled tubules with a wide range of diameters and lengths, which mature into uniform micron-long single-walled nanotubules.

We describe here the formation of long, uniform, single-molecular walled nanotubules with monodisperse cross-sections, which form spontaneously by a very simple bile acid (a steroid derivative) in alkaline aqueous solutions. Such formation of organic tubular aggregates is important in nanoscience because it may find applications in catalysis, selective separations, sensors, conducting devices in nanoelectronics, and in opto- or iono-electronics. While many molecules may produce tubular aggregates in the micron or atomic range *(1)*, tubular particle formation in the nanoscopic range is much less common. Certain lipids *(2-4)* and complex mixtures of steroids *(5,6)* are the best known examples. Self-assembly has a number of advantages over other means of nanoparticle production as it is performed at mild conditions, is quite often simple, and in many cases reversible. Besides its possible applications nano-tubular self-assembly is also interesting from the basic scientific point of view. The mechanism of formation is not well understood. Recent theoretical models, taking into account the coupling of chirality and topology, have been able to describe metastable species, helical ribbons, which are intermediates in the formation of tubules *(7)*.

Our work combines several characterization techniques. Here we emphasize the application of cryogenic temperature transmission electron microscopy (cryo-TEM). We describe the basics of the techniques, emphasize its advantages in elucidation the structure of complex aggregates such as the bile acid nanotubules and their intermediates, and discuss its synergistic combination with other techniques such as small-angle x-ray scattering (SAXS) and rheological measurements.

Cryo-TEM Combined with Other Techniques

Complete structural characterization of a fluid system requires accurate direct information on the 'building blocks' of which the system is composed, and quantitative information on the sizes in the system. The most effective experimental approach combines cryo-TEM (and possibly light microscopy) with indirect techniques such as SAXS. While imaging provides the microstructural building blocks of the system, the indirect techniques provide the quantitative measures of the system and their statistical investigation on a millimetric length scale of bulk specimens *(8)*.

To examine fluid specimens by TEM they must be made compatible with the instrument. While staining-and-drying techniques (unfortunately, still used) cannot preserve the original microstructure of these labile systems *(9)*, ultra-rapid cooling, which leads to specimen vitrification, does preserve it. The vitrified specimens are then examined at cryogenic temperature. Alternatively, the specimen is fractured, and the fracture surface is replicated by a metal/carbon replica which is examined in the TEM at room temperature. To assure preservation of their microstructure we prepare specimens for direct-imaging cryo-TEM under controlled humidity and temperature in the 'controlled environment vitrification system' (CEVS), as first described by Bellare et al. *(10)*. More recent developments of the technique are described by Talmon *(11)*. Ultra-rapid cooling rates (about 10^5 K/s) lead to vitrification, not freezing, of the specimen. This means that the liquid becomes a glass without undergoing a phase transition. While most of cryo-TEM work to date has been performed on aqueous systems, we have shown recently that, with some restrictions, the technique can be extended to nonaqueous systems as well *(12)*. The precautions taken prior to quenching by ultra-rapid cooling of the specimen, during transfer into the TEM, and during observation, to minimize electron-beam radiation-damage, assure that the structures observed in these cryo-specimens (typically kept at 100 K during examination) are indeed the original ones.

All our cryo-TEM data is recorded now digitally, using a cooled slow-scan digital camera. Digital imaging as an extension of cryo-TEM is explained and demonstrated elsewhere *(13)*. The advantages of digital imaging are:

- the ease of finding suitable areas;
- the ability to record the image with minimum electron dose;
- ease of real-time and post-microscope image processing;
- ease of archiving, retrieval and sharing of data;
- a cleaner microscope vacuum and thus clean cryo-specimens, as sheet film, a major source of vacuum contamination, is eliminated from the microscope column.

One advantage of direct-imaging cryo-TEM is the ability to combine imaging with electron diffraction. This allows crystalline structure (or the lack of crystallinity) of very small objects, down to tens of nanometers, to be determined and related to crystal morphology. An example is given in Figure 1. Figure 1A is a relatively low magnification image of a cholesterol monohydrate crystal (taken from a study of nucleation and growth of cholesterol crystal in human and in model biles; see for example earlier results in *(14)*). The diffraction pattern taken from the center of the image of Figure 1A is shown in Figure 1B. The black "arrow" in both images is the "beam stop", used to block the intense undiffracted beam. In Figure 1A it indicates the center of the area from which the diffraction data was collected.

Small-angle scattering of x-rays (SAXS) or neutrons (SANS) evaluates the intensity of radiation scattered at angles typically in the range of 0.05-5°. This yields microstructural information on the length scale of 1-100 nm, comparable to that obtained by TEM. Contrary to TEM, scattering techniques are almost nondestructive and impose no stringent requirements of sample preparation and environment. Thus samples can be studied in a variety of states (solutions, suspensions, gels, or solids) and if needed under shear, tension, electric or magnetic fields. High-flux sources (synchrotron x-ray facilities or spallation neutron facilities) allow dynamic measurements of structural transformations. The observed scattering pattern is a Fourier transform of the auto-correlation function of the density differences in the system under study. The relevant densities are electron density for x-rays and scattering-length density for neutrons. However, extraction of structural information from scattering measurements is not unique. Very reliable quantitative information is best achieved using a mathematical model of a structure supported by microscopic observations.

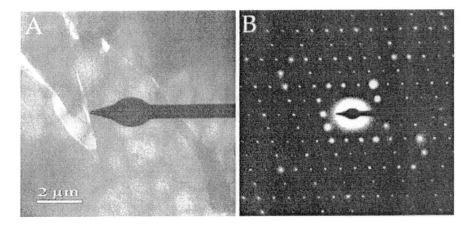

Figure 1: Cryo-TEM image of a cholesterol monohydrate crystal (A), and the corresponding diffraction pattern (B).

Rheological experiments give extensive information on viscous and elastic properties of complex fluids. These measurements are quite sensitive to the arrangement of clusters within the system, as for example random distribution of slender bodies versus their alignment in the direction of the flow.

In this chapter we concentrate on cryo-TEM results pertaining to SLC nanotubules. Earlier cryo-TEM, SAXS and rheology results on 'mature" system are described elsewhere *(15,16)*.

Materials and Methods

Materials

Lithocholic acid (Aldrich, 98% pure) was dispersed directly into aqueous solutions of sodium hydroxide at appropriate concentrations (in the range of 0.025 to 0.100 wt.%), resulting in alkaline (pH ~ 12.3) solutions. The solutions of the sodium salt of lithocholic acid thus formed were either examined at very short times after mixing to capture intermediate structures, or stabilized for at least one week to be examined as a 'mature' dispersions.

Cryo-TEM

The technical details of cryo-TEM are described in detail elsewhere *(11)*. In brief, we apply a 3 microliter drop of aqueous dispersion on a TEM copper grid coated with a perforated carbon film. Excess liquid is blotted with a filter paper to produce a thin (about 300 nm) liquid film supported on holes (several micrometer in diameter) in the carbon film. The specimen is plunged into liquid ethane at its freezing point, and vitrified. We prepare specimen in a Controlled Environment Vitrification System (CEVS) at 25 °C and 100% relative humidity. The vitrified specimens are stored under liquid nitrogen, transferred to an Oxford CT3500 cooling-holder of a Philips CM120 transmission electron microscope, where they are examined at a temperature of about -180 °C. To minimize specimen damage by the electron beam, we always use 'low-dose imaging' protocols, exposing our specimens to less than 20 electrons per $Å^2$. Images are recorded at 120 kV with a Gatan MultiScan 791 cooled CCD camera, using the Gatan DigitalMicrograph 3.1 software package.

Results and Discussion

As we have shown before *(15,16)*, mature nanotubes of sodium lithocholate (SLC) are uniform, straight, open, several micrometer-long cylinders, 52 nm in outer diameter and 49 nm in inner diameter. Thus the wall of the tubules is only one molecule thick. It is fascinating that the ordering on the monomolecular level dictates the tubule diameter, over thirty times larger. Electron diffraction from those nanotubes showed no indication of crystallinity, and no crystalline contrast effects were observed (i.e., moiré patterns expected from the superposition of the two walls of the tubule in projection, or when images of two tubules are superposed). In some micrographs narrow twisted ribbon were seen coexisting with the tubules. In some rather rare micrographs we saw ribbons of widths close to the diameter of the tubules opening up from (or closing down) to form the SLC nanotubules. That was an indication that the tubules may form by the coiling of wide ribbons, in a manner similar to making cardboard tubes to support paper or fabric rolls.

To try understand how SLC nanotubules self-assemble spontaneously, we followed their microstructure shortly after dispersion of the lithocholic acid in dilute NaOH aqueous solutions. We chose rather dilute SLC solutions of 0.025 to 0.100% to allows us image single tubules in the act of forming with minimal overlap. No difference was noticed between the nanostructures observed in the more and less dilute solutions; they all let to the same mature nanotubules. This type of series of experiments constitute what we call 'time-resolved cryo-TEM'.

Figure 2A shows a vitrified specimen of 0.050% SLC. The background is vitreous ice. The nanostructure of the aggregates is quite different from that of the mature tubules previously imaged: they are not uniform in diameter, and many of them are multiwalled. For example, the image of the tubule indicated with an asterisk (lower left corner) shows three lines (tubule walls in projection) above the astersik, but only two below it. This suggests the aggregate is made of a monomolecular sheet rolled on itself. Several other similar examples are seen in the field of view. One is in a particular "disintegrated state" (white arrow). Edges of rolled SLC sheets are also visible (black arrows). In Figure 2B we observe side by side a multiwalled tubule and a single walled one. The latter bears the sign of a wide ribbon rolled helically to form the tubule. Black arrows point to where the edges of the putative ribbon are seen. Inside the multiwalled aggregate there are also indication to the original presence of wide ribbons in the formation of the sheets (white arrow).

An entirely different nanostructure is seen in Figure 2C. This particular object was seen in 0.025% SLC, 8 minutes after dispersion. It seems to be disintegrating into narrow ribbon or threads. It is juxtaposed to 'normal' nanotubules. As mentioned above individual threads had been seen in preparation of mature SLC nanotubules.

Four hours after mixing lithocholic acid in NaOH aqueous solution the nanotubules are almost fully developed. Multiwalled tubules are not observed at that point in time, but the diameter of the single-walled tubules is still not uniform, as can be seen in Figure 3A. Superposition of tubules has occurred in this image, so arrowheads serve to delineate the contour of one tubule of diameter considerable wider than that of the average mature ones. Another example of SLC nanotubules four hours after mixing is shown in Figure 3B. In this case we observe two other features of immature tubules: fusion between adjacent tubules (black arrows), and a fine thread emerging from the end of a monomolecular wall (white arrow). The walls in this image seem thicker because

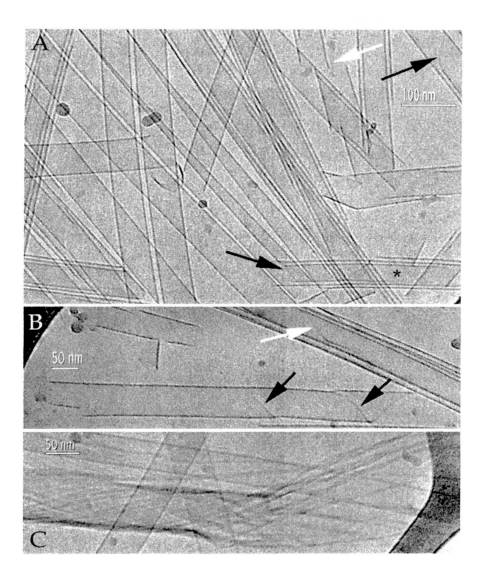

Figure 2: Vitrified SLC dispersions showing intermediate nanostructures. (A) and (B) 0.050% SLC, five minutes after mixing of lithocholic acid in dilute NaOH aqueous solution (pH 12.4); (C) 0.025% SLC eight minutes after dispersion

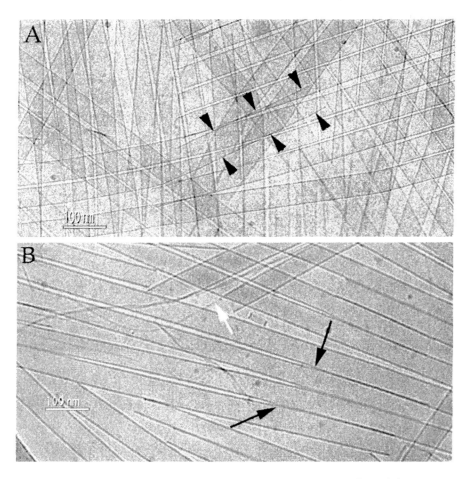

Figure 3: Vitrified 0.10% SLC dispersions four hours after mixing.

of a greater underfocus of the TEM objective lens. Two days later the aggregates observed are indistinguishable for those that are several weeks old or months, and described in our earlier publications.

These results are insufficient to explain the mechanism of assembly of SLC into long uniform cylinders of monomolecular layer walls. It seems that several mechanisms are at play, and are related to the chirality of the bile acid molecule. While intermediate microstructures such as rolled-up sheets, helically rolled wide ribbons, and narrow threads have been identified (narrow twisted ribbons are observed coexisting with mature tubules), it has not been clear yet what is the dominant mechanism, and how do the intermediates transform/fuse into the final mature nanotubes. The key questions regarding the sequence of micellar equilibrium reactions that eventually determine the diameter and length of the mature nanotubes also remain unanswered. Work is under way now to elucidate those remaining issues. One approach is to disperse crystallized SLC (rather than lithocholic acid) in the appropriate pH water. We plan also to use freeze-fracture-replication as a complimentary technique to allow collecting more information on the structure of the tubule wall, details that probably elude us because of low contrast they exhibit in direct-imaging cryo-TEM.

Acknowledgements

The cryo-TEM work was performed at the "Cryo-TEM Hannah and George Krumholz Laboratory for Advanced Microscopy" at the Technion, part of the "Technion Project on Complex Fluids". We thank Daphne Weihs for the images of Figure 1, and Berta Shdemati for excellent technical assistance.

References

1. Bong, D. T.; Clark, T.D.; Granja,J. R.; Ghadiri, M.R. *Angew. Chem. Int. Ed.* **2001**, *40*, 988-1011.
2. Schnur , J.M.; Shashidhar, R. *Adv. Mater.* **1994**, *6*, 971-974.
3. Schnur, J.M. *Science* **1993**, *262*, 1669-1676.
4. Spector, M.S.; Singh, A.; Messersmith, P.B.; Schnur, J.M. *Nano. Lett.* **2001**, *1*, 375-378.
5. Chung, D. S.; Benedek, G. B.;. Konikoff, F. M.; Donovan, J. M. *Proc. Natl. Acad. Sci. USA* **1993**, *90*, 11341-11345.

6. Zastavker, Y. V.; Asherie, N.; Lomakin, A; Paned, J.; Donovan, J. M.; Schnur, J. M.; Benedek, G.B. *Proc. Natl. Acad. Sci. USA* **1999**, *96*, 7883-7887.

7. Selinger, J. V.; MacKintosh, F.C.; Schnur, J.M. *Phys. Rev. E* **1996**, *53*, 3804-3818.

8. Burns, J.L.; Cohen, Y.; Talmon, Y. *J. Phys. Chem.* **1990**, *94*, 5308- 5312.

9. Talmon, Y. *J. Colloid Interface Sci.* **1983**, *93*, 366-382.

10. Bellare, J.R.; Davis, H.T.; Scriven, L.E.; Talmon, Y. *J. Electron Microsc. Tech.* **1988**, *10*, 87-111.

11. Y. Talmon, *Modern Characterization Methods of Surfactant Systems,*, Binks, B. P., Ed.; Marcel Dekker, New York, 1999; pp. 147-178

12. Danino, D.; Gupta, R.; Satyavolu, J.; Talmon, Y. *J. Colloid Interface Sci.* **2002**, *249*, 180-186.

13. Danino, D.; Bernheim-Groswasser, A.; Talmon, Y. *Colloids and Surfaces A* **2001**, *183-185*, 113-122.

14. Konikoff, F.M.; Danino, D.; Weihs, D.; Rubin, M.; Talmon, Y. *Hepatology* **2000**, *31*, 261-268.

15. Terech, P.; de Geyer, A.; Struth, B.; Talmon, Y. *Adv. Mat.* **2002**, *14*, 495-498.

16. Terech, P.; Talmon, Y. *Langmuir* **2002**, *18*, 7240-7244.

Chapter 3

Characterization Mesoscale Structure and Phenomena in Fluids Using NMR

Peter Stilbs

Physical Chemistry, Royal Institute of Technology, S–100 44 Stockholm, Sweden

NMR techniques can provide valuable insight into the structure and dynamics of mesoscale structures. Line broadening and spin relaxation effects, which researchers interested in small molecule NMR often work hard to eliminate, can yield detailed information about molecular motion, particularly reorientation processes. They have been used to characterize molecular dynamics and order in alkyl chains, water, or counter-ions participating in mesoscale structures, and to study protein dynamics. Multi-component self-diffusion techniques (such as FT-PGSE) provide information on molecular displacements over longer time scales (picoseconds to microseconds). The self-diffusion rate of a molecule can reveal its aggregation state (is it part of a larger structure?). Under carefully controlled conditions it can also reveal information about the medium through which it is diffusing (viscosity, concentration and structure of materials obstructing its path). Multi-component FT-PGSE has been used to characterize mixed micelles and to study polymer/ surfactant aggregates. Recent developments have extended the utility of these methods. Continued and expanded application to the study of mesoscale phenomena can be foreseen.

Introduction

As a concept, nuclear magnetic resonance (NMR) recently celebrated its 50-year anniversary. In the continued development of NMR during the last decades the majority of new methods were in the field of multidimensional NMR, for the purpose of structural determination of organic compounds or biological macromolecules. We have also witnessed extensive new use of NMR in biomedicine, both as a spectroscopic method and as an imaging tool (MRI) that complements X-ray and positron emission tomograph (PET) methods. To a large extent, these new NMR-based procedures rely on the general developments in the field of electronics and, in particular, on that of digital computing.

With regard to studies of mesoscopic structure and dynamics in solution, the main NMR 'tools' are quite different from those used in the determination of molecular structure. NMR offers many other experimental dimensions. The multinuclear detectability (protons (^{1}H), deuterium(^{2}H), ^{13}C, ^{17}O, ^{19}F, ^{23}Na etc.) is of course one unique characteristic of 'Nuclear Magnetic Resonance'. Furthermore detailed and quantitative dynamic information that is often not accessible from other sources is available from either NMR spin relaxation data or from NMR-based, multi-component studies of molecular self-diffusion. These are the most valuable NMR-tools in the present context. Here the basic spin relaxation approach(1) may provide detailed information on local molecular dynamics (picosecond to microsecond) and order of alkyl chains, water or counter-ions in the system. In principle, it can be based on data for any NMR-accessible type of nuclei in the system, at any location in a molecule.

The multi-component self-diffusion technique(2) (Fourier transform pulsed-gradient spin-echo: FT-PGSE or just PGSE) provides information on overall molecular displacements on a much longer timescale (typically the order of 5 to 500 milliseconds, depending on experimental conditions and the instrumental set-up. The two families of methods are both complementary and highly selective and applicable to very complex systems. Isotopic labeling can sometimes be applied to further enhance the selectivity and sensitivity of either of the methods in question. Particularly common in the present context is selective deuterium labeling for the purpose of studying deuterium spin relaxation at a particular location, at a high sensitivity and spectral selectivity. This strategy provides quantitative information on the reorientation and order of a particular C-H bond vector, in a very unambiguous and quantitative way.

Application examples in this paper are from studies of polymer or surfactant systems in aqueous solution.

Intrinsic Problems in NMR Investigations of Complex Fluids

Polymers or surfactants in solution do not really have well-resolved NMR-spectra to begin with - a problem that becomes amplified in mixed systems. This is not only because of added signal overlap, but also because of line-broadening and faster spin a relaxation rate that can occur as a direct result of enhanced aggregation and specific interactions. In general, line-broadening effects reduce sensitivity quite extensively in all types of NMR experiments, and they are normally the limiting experimental parameter.

Going to higher magnetic field might seem like an obvious cure for the basic 'spectral resolution of peaks'-problem, since one would assume that NMR-spectral dispersion should be linear with the applied magnetic field. However, for macromolecular systems this is only partly true, since the overall reorientation dynamics is normally at such a spectral range that bandwidths increase with magnetic field, and thereby do change 'unfavorably' from a mere resolution point of view. On the other hand, the line broadening and spin relaxation effects described actually do provide a rich source of detailed quantitative information about characteristics of the aggregates. Data on size, shape and local chain dynamics is accessible this way - provided a proper interpretation of the magnetic field-dependence and other experimental parameters is made. This was generally overlooked in early studies.

Studies of local and global dynamics in fluids through nuclear spin relaxation rates

Due to the low energy difference between nuclear spin states, the equilibrium Boltzmann distribution between them only differs by an order of 1 in 10^5. At the low spectral frequencies of NMR (1-1000 MHz, depending on nucleus and magnetic field) spontaneous emission is a totally ineffective mechanism for establishing an equilibrium Boltzmann distribution. Instead the pathway is via fluctuations of spin interactions, originating from molecular reorientation processes. There are several mechanisms for achieving spin state equilibrium this way, and the establishment and loss of phase coherence between individual nuclei is also a component here. The detailed theory is complex, and is far outside the scope of this overview. Its fundamentals can be found in classical textbooks on NMR(3,4). An excellent educational review on spin relaxation in liquids recently appeared.(5)

The useful consequence of spin relaxation rate information in the present context is that it contains detailed information about molecular motion, in particular reorientation processes. A specific overview with reference to complex surfactant systems was given some years ago(1). It should be consulted for a fuller treatment of the subject, together with some key references given in that review (pp 447-451of (1)).

Random molecular reorientation processes (overall reorientation of molecules, aggregates and macromolecules, and local chain motion) are modeled in terms of a correlation function for motion. This quantity is not determined directly. Instead one determines the Fourier transform at a limited number of frequencies. As a consequence one has to invoke models to proceed from the spectral densities (the intensity of reorientational fluctuations at a given frequency) to the correlation functions. Needless to say, such models should be chosen with care. Fortunately, the dynamics of polymer or surfactant systems in solution occur on timescales that are well matched to the frequencies of commonly available NMR spectrometers.

Two such models have become popular and are extensively used today i) the 'two-step model' ((6,7), mainly for surfactant systems) and ii) the 'model-free model' or Lipari-Szabo model ((8,9) for motional dynamics of proteins in solution). Both imply a global 'slow' motion of an aggregate or macromolecule (for example overall tumbling and diffusion on a surface, characterized by a correlation time τ_{slow}), order parameters (S) that quantify how rigidly a particular moiety (e.g. a $-CH_2$-group) is coupled to that 'slow' motional process, and a second correlation time τ_{fast}, characterizing the rate of local motions within that moiety. Mathematically, both approaches lead to the same type of expression. The somewhat subtle differences between the two, their parameters as well as their applicability to particular chemical systems are discussed in ref. (1).

Experimentally, one needs access to spectrometers operating at several magnetic fields to gather the required spin relaxation information, and a computer program to carry out a global fit of τ_{slow} and the individual S and τ_{fast} modeling parameters. Meaningful experiments that can actually be evaluated must be designed so that only one intramolecular spin relaxation mechanism operates on the studied nuclei. Feasible approaches include studies through ^{13}C spin relaxation (normally feasible without isotopic enrichment above the natural abundance of 1%) and/or deuterium NMR studies on selectively labeled alkyl groups. Protein backbone dynamics is normally quantified through spin relaxation studies on uniformly ^{15}N labeled molecules. Such isotope enrichment is made through modern biotechnology techniques. Typical examples of surfactant studies (through the two-step model) are in references (7,10-14) and of protein studies (through the 'model-free' approach) are in references (15-17).

PGSE-based studies of organization and dynamics in fluids through multi-component self-diffusion rates

A proven method for quantifying 'free/bound' situations and 'binding isotherms' is NMR-based multi-component self-diffusion monitoring by the FT-PGSE method (Fourier Transform Pulsed-Gradient Spin-Echo) (2,18,19). In a chemical sense this approach relies on the often vastly different self-diffusion rates for the same molecule in 'free' or 'bound/aggregated' state. The basic (1a) PGSE experiment(20) uses pulsed linear magnetic field gradients (of amplitude 'g', duration 'δ' and separation 'Δ') that are applied during a so-called spin-echo experiment(21), involving two or more radio frequency pulses (in the simplest case separated in time by 'τ' – see Figure 1).

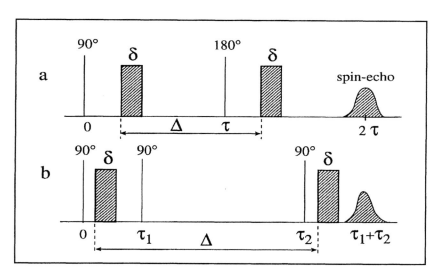

Figure 1: The basic two-pulse PGSE experiment (a) and the stimulated-echo variant (b). Radio frequency (rf) pulse phase cycling is required (to suppress other echoes in (b)), the details of which also determine the resulting sign of the echo. Subsequent Fourier transformation of the 2nd half of the echo produces a frequency-resolved NMR spin-echo spectrum (the individual traces in Figure 2)

Under these conditions the amplitude of the spin-echo (which occurs after a time 2τ after the initial radio frequency pulse) attenuates from its full value ($A(0)$) according to the so-called Stejskal-Tanner relation(20):

$$A(2\tau) = A(0) \exp(-2\tau/T_2) \exp(-D(\gamma g \delta)^2(\Delta-\delta/3)) \tag{1a}$$

T_2 represents the (disturbing and experimentally limiting) transverse spin-spin relaxation rate of the nuclei in question, and γ their magnetogyric ratio. As evident from this equation, protons are the easiest nuclei to study, owing to the high magnetogyric ratio. Applications to other nuclei are quite rare, due to a combination of instrumental factors, short T_2 values and low sensitivity (another effect of a lower magnetogyric ratio). 'Possible' nuclei include ^{19}F, ^{2}H, ^{13}C and ^{23}Na. PGSE experiments are always run at constant rf timing parameters (τ), varying δ, Δ or g only. This separates self-diffusion attenuation from spin relaxation attenuation, but does not eliminate the disturbing echo attenuation effect of transverse spin relaxation. For small molecules in solution at room temperature T_1 normally equals T_2, but in the presence of slower motional processes T_1 can become much longer than T_2 (which decreases monotonically with slower motion). Then, the three-pulse stimulated echo PGSE variant (1b) becomes favorable from a detection point of view, despite the intrinsic 50% reduction in the effective echo amplitude(*21*):

$$A(\tau_1+\tau_2) = \tfrac{1}{2}\, A(0)\, \exp(-(\tau_2-\tau_1)/T_1)\, \exp(-2\tau_1/T_2)\, \exp(-D\,(\gamma g\delta)^2\,(\Delta-\delta/3)) \qquad (1b)$$

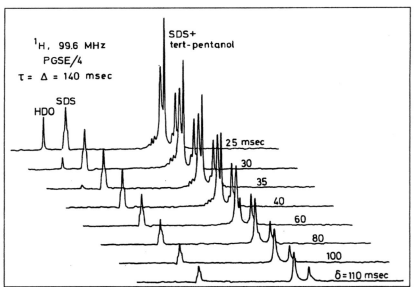

Figure 2: *A sequence of proton FT-PGSE 90-180 spin-echo spectra at sequentially increasing magnetic field gradient duration ('δ' in Eqn. 1a) of partially solubilized neopentanol in sodium dodecyl sulfate (SDS) micelles in D_2O solution. The individual molecular 'decay rates' translate into individual self-diffusion coefficients through Eqn. 1a.*

Although more complex rf phase cycling schemes (typically 8 cycles) have to be applied in the stimulated-echo variant, it is almost always the preferred of the above two. Subsequent Fourier transformation of the composite echo in the time-domain separates the contributions at each frequency, just as in the normal basic pulsed FT-NMR experiment(2). Figure 2 illustrates a typical application of FT-PGSE NMR to a micellar system, in the context of solubilization.

FT-PGSE NMR self-diffusion studies for quantifying mesoscopic structure.

Self-diffusion data in itself is a very direct source of information about aggregation processes. For unrestricted diffusion during a selected time span the displacement probability in space of a given component 'k' (ρ_k ($\mathbf{r}$,t)) is Gaussian and given by the equation:

$$\rho_k (\mathbf{r},t) = const\ (1/8(\pi D_k t)^{3/2})\ exp(-r^2/4D_k t) \tag{2}$$

The characteristic self-diffusion coefficient (D) simply translates into a mean square displacement in space ($<\Delta r^2>$) during the observation time (Δt), through the Einstein relation:

$$<\Delta r^2> = 6\ D\ \Delta t \tag{3}$$

The mean square displacement of even large macromolecules during the selected (method-dependent) time span of the experiment (Δt) is generally much larger than the average macromolecular diameter. Therefore the quantity ($<\Delta r^2>$) requires no further interpretation and becomes easy to visualize. In the case of restricted diffusion (for example in confined geometries) the situation will be different, but parameters like pore sizes and diffusion coefficients within porous structures are still accessible and quantifiable; see e.g. (22-25). One should note that self-diffusion coefficients are easily extracted from computer molecular dynamics simulation trajectories through the velocity autocorrelation function. This provides an easy path between theory and simulations and experimental self-diffusion data. A relatively recent study of this kind is (26). This is much more difficult with nuclear spin relaxation data.

Simple binding studies

The self-diffusion approach to binding studies relies on a relative comparison of time-averaged self-diffusion rates between a 'bound' and 'free' state. In the simplest two-state situation the effective self-diffusion coefficient D(obs) will be given by:

D(obs) = p D(bound) + (1-p) D(free)   (4)

where the degree of binding (p) may assume values in the range 0<p<1. This equation can be rewritten into the form:

p=(D(free)-D(obs))/(D(free)-D(bound))   (5)

Such experiments provide the information needed for obtaining binding isotherms by a simple comparison of the experimental self-diffusion coefficients. It is evident that the method is most accurate around p-values of the order of 0.5, and becomes uncertain near p-values of 0 or 1. The experimental detail and precision needed for more complex situations than a simple 'free/bound' may be difficult to achieve. However, this would be true for most other approaches as well.

Before the development of the FT variant (2,18,19) of the PGSE NMR technique(20), the acquisition of multi-component self-diffusion data was a very cumbersome task, requiring synthetic work to provide radioactive isotope labeling of the compounds to be studied, as well as tedious and time-consuming measurement procedures through *e.g.* radioactive tracer methods. Today, a simple NMR measurement of multi-component self-diffusion can be completed in a matter of minutes (c.f. Figure 2 above)

Some Selected Application Families

Micellization of surfactants

By applying Eqn (5) on FT-PGSE-based data, it is a straightforward matter to fully quantify self-aggregation processes such as micellization of surfactants. Early studies of this kind were a series of investigations on surfactants with organic (proton-bearing) counter-ions, for example dodecylammonium dichloroacetate(27). PGSE based diffusion data on this model surfactant very nicely confirmed current theories of micellezation and polyelectrolyte counter ion binding. The constancy of the percent of bound counter ions above the cmc is one predicted feature, and here it is high (about 80%), due to the hydrophobicity of the counter-ion. Later studies on mixed micelles of the same model surfactant type, but with counter-ions of the series acetate, monochloroacetate and dichloroacetate confirmed the effects of hydrophobicity on competitive counter-ion affinity to the micellar assembly(28). Mixed micellization of fluorocarbon and hydrocarbon surfactants was also investigated. These studies(29) were followed-up with several complementary techniques(30).

Applications of electrophoretic NMR to polymer and surfactant systems

Through controlled application of high-voltage electric DC fields over the sample in a modified PGSE pulse sequence (ENMR) one can separate out and quantify the coherent electrophoretic transport of ions and charged assemblies.

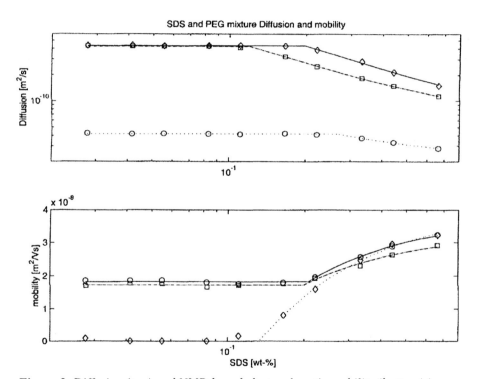

Figure 3. *Diffusion (top) and NMR-based electrophoretic mobility (bottom) in a SDS/poly(ethylene oxide (PEO)) system in aqueous solution. Diffusion: circles = PEO, squares = SDS in mixed system, diamonds = SDS in water. Electrophoresis: diamonds = PEO, squares = SDS in mixed system, circles = SDS in water.*

Despite pioneering methodological developments several years ago,(*31-34*) the occurrence of real applications in the literature is still scarce, probably due to the many practical and instrumental difficulties involved. The potentially available chemical information is exciting and is not accessible any other way. Figure 3 illustrates some mobility data on the sodium dodecyl sulphate-poly(ethylene oxide) system.(35).

When combined with NMR diffusion data on the same systems, one can quantify ion binding and other details on aggregation processes in a much more complete and quantitative fashion than possible from standalone measurements.

Aggregation of hydrophobically modified polymers – 'associative thickeners'.

Studies of the diffusion behavior of associative thickeners in aqueous solution have proven to be highly rewarding. With increasing concentration, unimers form 3-dimensional networks that manifest themselves in the rheological behavior – *i.e.* in 'thickening', that may be desirable for some practical application or formulation. The aggregation can be further modified through the addition of surfactants. General ideas on association patterns had been established(36,37) before the application of PGSE NMR self-diffusion techniques, but a lot of new information on the molecular details of aggregation have emerged from PGSE or spin relaxation investigations. The most notable finding is that the unimers exchange very slowly between aggregates, since their diffusion rates are not time-averaged. Rather, a wide distribution of diffusion rates results from NMR diffusion studies.(*38*) Such effects very clearly show up in ^{19}F chemical shifts of fluorine-containing associative polymers and also in parallel ^{19}F-based PGSE measurements on the same systems.(*39*)

Alkyl chain order and dynamics in aggregates from nuclear spin relaxation

The overall reorientational dynamics and surface diffusion of aggregates of micellar size matches the typical inverse NMR frequency range. For related reasons the spin relaxation behavior becomes NMR frequency dependent. Through a global analysis of spin relaxation data (^2H, ^{13}C) from micellized alkyl chain moieties of surfactants or in solution one can conveniently quantify the local alkyl chain dynamics along the chain, together with the order profile of the chain relative to the micellar surface. A large number of studies of this kind were summarized, together with the pertinent theory in a review some 10 years ago.(*1*)

Aggregate 'micro-viscosity' from spin relaxation data.

Macroscopic concepts like solution viscosity have been extrapolated to a 'micro-viscosity' of molecular assemblies like the micellar interior, membranes and liquid crystals. A feasible NMR approach for such modeling and quantification is to study the spin relaxation (^{2}H or ^{13}C) of a rigid solubilized probe molecule, like *trans*-decalin (Figure 4), whose rotational diffusion tensor can be evaluated from such data.

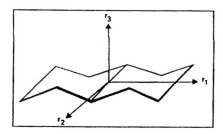

Figure 4: trans-*Decalin, and the orientation of its principal axes of rotation.(Motion around r_3 is fastest and around r_2 is slowest).*

Viscosity calibration reference is provided through parallel spin relaxation measurements on the same probe molecule, dissolved in suitable hydrocarbon liquids of known macroscopic viscosity (*40*). Basically the same approach has been used for fluorescence depolarization studies of suitable molecules, but here it suffers from fundamental interpretation difficulties. Suitable fluorescent molecules are larger than e.g. micelles and also are likely to orient in respect to the micellar entity, locate the surface region and also to protrude into the aqueous phase. NMR data on trans-decalin in micelles made up by various surfactants nicely demonstrate that the micellar interior has closely the same 'micro-viscosity' as a bulk hydrocarbon liquid made up by alkanes of the same chain length as that of the surfactant part.(40)

Issues in the NMR Characterization of Mesoscale Structure

Viscosity corrections

The macroscopic viscosity of e.g. polymer systems in solution varies greatly with solution composition, which is a quite central problem in the present type of experimental approach. There is still controversy and confusion in the literature on whether one should account for the macroscopic solution viscosity to 'correct' experimental self-diffusion (or spin relaxation) data. While such a 'viscosity correction' is highly justified for molecules in normal solution (c.f. Stokes law for a molecule with radius r in a continuous medium with a viscosity η; $D=6\pi r\eta$), it is clearly irrelevant for a small molecule in a solution of macromolecules.

Experimental observations show unambiguously that the self-diffusion rate of the solvent or any small molecule is only marginally correlated with the typically rapid and huge increase of solution viscosity with polymer concentration. Studies on gel systems point to the same conclusion - small molecules (like water) in a rigid gel framework diffuse almost as rapidly as they do in a normal aqueous solution (see e.g. (*41,42*)).

Obstruction effects

The presence of macromolecules or other barriers in a solution or gel framework function as an obstacle for small molecules in the system, forcing them to use a longer path in the normal process of diffusing between two points in space. The time-scale of the diffusion experiment is normally very much longer (milliseconds to days, depending on the technique used) than it takes to diffuse a macromolecular diameter. For this reason the apparent experimental self-diffusion coefficient from the FT-PGSE NMR experiment (with a time-scales around 5-500ms), and almost all other techniques, is simply reduced in proportion to this obstruction effect. An exception is the neutron spin-echo experiment, which has very much shorter characteristic time-scale in the nanosecond range (see other papers in this publication for applications and more information).

The factor to account for in describing the resulting apparent lowering of the effective time-averaged self-diffusion coefficient here is therefore 'the obstruction effect', rather than 'the solution viscosity'. Several attempts to quantify such obstruction effects on self-diffusion have been made to date. To first approximation it is linearly related to the volume fraction of obstructing objects in solution, and to their general shape. The most recent and elaborate treatment of obstruction effects on self-diffusion, together with a literature update on the subject can be found in a recent paper by Johannesson and Halle(*43*).

It would appear that the "obstruction-effect approach" could be an interesting pathway for the investigation of aggregate shapes in solution. However in practice this rarely works unless the system chemistry allows self-diffusion monitoring of relatively high concentration spans without any change in aggregate shape. Also a high experimental precision is required to allow a distinction between various aggregate shapes.

Clearly, the underlying effect behind the typically very strong increase in solution viscosity with concentration of a high polymer solution is a manifestation of entanglement effects. In a case of gradual self-association of polymers or polymer-surfactant systems into different families of 3-dimensional networks and similar, the situation becomes much more complex with regard to self-diffusion behavior. Attempts to quantify of experimental data would really be futile, were it not for quite dramatic self-diffusion coefficient changes (of three orders of magnitude or so) which often result from changes in solution concentration and composition. Under such conditions the experimental data will again become interpretable in terms of general association patterns or solution structure at a reasonable level of confidence, because of the very large effects seen.

Recent Instrumental and Methodological Progress

Modern magnetic field gradient NMR equipment has vastly better experimental performance than only a few years ago. The main improvements are the introduction of i) self-shielded magnetic field gradient coils and ii) much more powerful and stable current-regulated magnetic field gradient drivers (see e.g. references cited in (1) and (44)). For narrow-bore magnet systems or less ideal measurement conditions it has been suggested that one should use more complex NMR pulse sequences like LED(45) or bipolar magnetic field gradient sequences(46) to lessen e.g. the destructive interference of extraneous eddy current-related magnetic field gradients on the PGSE experiment. These procedures necessarily add complexity and time (at least a factor of 4) to the experiment and are not normally of any value with properly self-shielded wide-bore magnetic field gradient equipment. One should rather be aware that there are several additional sources of artifacts in PGSE experiments, for example effects of radiation damping and finite sample size. Some have recently been discussed, together with some partial remedies(47).

A continued and extended application of FT-PGSE NMR-based studies of multi-component self-diffusion is foreseeable. Instrumental developments make measurements much easier, and widen the experimental range quite extensively. This is essential to provide the necessary experimental information to allow conclusions on more complex interaction conditions than previously considered. Efficient solvent (water) signal suppression techniques are often desirable(48-50) and recent advances with regard to data processing of FT-PGSE NMR data sets will definitely also be needed in case of heavy signal overlap in multi-component systems, and in the case of poor signal/noise ratios. Antalek recently published a thorough review on PGSE studies of multi-component systems and mixtures(51).

The so-called COmponent-REsolved (CORE) data analysis approach(52) in particular, allows a straightforward and confident application of FT-PGSE to very complex systems. Utilizing prior knowledge it also generates an effective Signal/Noise increase of a factor of 10 or so in comparison with previous methods of data evaluation, at no extra cost with regard to experimental time. The data processing also results in a complete unraveling of highly overlapping component bandshapes, without any assumptions with regard to their actual form. Such strategies are essential in future studies of e.g. polymer-surfactant and protein-surfactant systems by the FT-PGSE technique.

Another new development, named GRAM (the Generalized Rank Annihilation Method)(53) can be used for similar purposes. It is based an elegant simplification of a previously suggested multivariate FT-PGSE data processing approach(54). GRAM-processing is much less computer-intensive than CORE-processing, but is only applicable for a separation of purely exponential decay patterns. In the present form it is therefore inapplicable for FT-PGSE studies of polymer-surfactant aggregate diffusion under the prevailing conditions of slow polymer inter-aggregate exchange, and a great effective polydispersity in diffusion behavior at the typical time-scale of the experiment (5-500 ms).

It has recently become popular to display FT-PGSE results in a 2D manner with a normal chemical shift scale on the x-axis, the self-diffusion data on the y-axis and the corresponding spectral intensity on the z-axis (or as a contour x-y plot). This approach (34,55,56) has been named 'DOSY'(for diffusion-ordered spectroscopy). Its introduction has unfortunately led to considerable confusion in the literature, and newcomers are often led to believe that diffusion techniques in NMR are sub-sets of DOSY, rather than the opposite. As such, DOSY is not a method; it is in essence just a nice *display mode* for FT-PGSE results – which has helped to popularize the use of NMR diffusion techniques for chemical applications.

The key concept behind 'DOSY' is the intrinsic property of FT-PGSE that the whole NMR bandshape of a particular molecule 'decays' (according to Eqn. 1) at exactly the same 'rate' in a FT-PGSE experiment.(57) This is simply because each part of the molecule has the same self-diffusion coefficient during the time span in question (typically 5-500 milliseconds). Component spectra thus can be separated and ordered in a diffusion dimension, as visualized in 'DOSY spectra'. One should note that the information for DOSY displays could be generated in many ways (*e.g.* via CORE- or GRAM-processing of FT-PGSE data sets). In its original and most widespread form 'DOSY' does not use the sensitivity enhancing and 'data stabilizing', prior knowledge-approach of such bandshape-extracting techniques. Rather it relies on raw inverse Laplace transform-evaluated diffusion coefficient distributions (Contin(58) or variants thereof)(56) at individual frequencies in the x-y grid, followed by numerical 2D interpolation and contourization in the chemical shift and diffusion coefficient dimensions.

References

1. Söderman, O.; Stilbs, P. *Prog.Nucl.Magn.Reson.Spectrosc.* **1994,** *26*(5), 445-482.
2. Stilbs, P. Prog.Nucl.Magn.Reson.Spectrosc. **1987,** 19 1-45.
3. Abragam, A. *The Principles of Nuclear Magnetism;* Clarendon Press: Oxford, 1960.
4. Slichter, C. P. *Principles of Magnetic Resonance;* Harper and Row Press: New York, 1963.
5. Murali, N.; Krishnan, V. V. *Concepts Magn.Reson., Part A* **2003,** *17A*(1), 86-116.
6. Halle, B.; Wennerström, H. *J.Chem.Phys.* **1981,** *75* 1928.
7. Wennerström, H.; Lindman, B.; Söderman, O.; Drakenberg, T.; Rosenholm, J. B. *J.Am.Chem.Soc.* **1979,** *101* 6860-6864.
8. Lipari, G.; Szabo, A. *J.Am.Chem.Soc.* **1982,** *104* 4546-4559.
9. Lipari, G.; Szabo, A. *J.Am.Chem.Soc.* **1982,** *104* 4559-4570.
10. Heatley, F. *J.Chem.Soc., Faraday Trans.1* **1987,** *83* 2593-2603.
11. Söderman, O.; Canet, D.; Carnali, J.; Henriksson, U.; Nery, H.; Walderhaug, H.; Wärnheim, T. *Surfactant Sci.Ser.* **1987,** *24* 145-161.
12. Söderman, O.; Carlström, G.; Olsson, U.; Wong, T. C. *J.Chem.Soc., Faraday Trans.1* **1988,** *84* 4475-4486.
13. Söderman, O.; Carlström, G.; Monduzzi, M.; Olsson, U. *Langmuir* **1988,** *4* 1039-1044.
14. Monduzzi, M.; Ceglie, A.; Lindman, B.; Söderman, O. *J. Colloid Interface Sci.* **1990,** *136* 113-123.
15. Palmer, A. G.; Williams, J.; Mcdermott, A. *J.Phys.Chem.* **1996,** *100*(31), 13293-13310.
16. Gardner, K. H.; Kay, L. E. *Annu.Rev.Biophys.Biomol.Struct.* **1998,** *27* 357-406.
17. Case, D. A. *Acc.Chem.Res.* **2002,** *35*(6), 325-331.
18. Vold, R. L.; Waugh, J. S.; Klein, M. P.; Phelps, D. E. *J.Chem.Phys.* **1968,** *48* 3831-3832.
19. James, T. L.; McDonald, G. G. *J.Magn.Reson.* **1973,** *11* 58-61.
20. Stejskal, E. O.; Tanner, J. E. *J.Chem.Phys.* **1965,** *42* 288-292.
21. Hahn, E. L. *Physical Review* **1950,** *80* 580-594.

42

22. Callaghan, P. T.; Coy, A. PGSE and Molecular Translational Motion in Porous Media, in *NMR Probes of Molecular Dynamics*, Tycko, P., editor; Dordrecht: Kluwer Acad. Publishers, 1994; Chapter 11, pp. 489-523.
23. Kärger, J.; Pfeifer, H. *NMR.Techniques.in Catalysis.55.* **1994,** -137.
24. Gladden, L. F. *Chem.Eng.Sci.* **1994,** *49* 3339-3408.
25. Kärger, J. Diffusion in Porous Media, in *Encyclopedia of Nuclear Magnetic Resonance*, Grant, D. M.; Harris, R. K., editors; Wiley: New York, 1996; Vol. 3, pp. 1656-1663.
26. Laaksonen, A.; Stilbs, P.; Wasylishen, R. E. *J.Chem.Phys.* **1998,** *108*(2), 455-468.
27. Stilbs, P.; Lindman, B. *J.Phys.Chem.* **1981,** *85* 2587-2589.
28. Jansson, M.; Stilbs, P. *J.Phys.Chem.* **1985,** *89* 4868-4873.
29. Carlfors, J.; Stilbs, P. *J.Phys.Chem.* **1984,** *88* 4410-4414.
30. Kadi, M.; Hansson, P.; Almgren, M.; Furó, I. *Langmuir* **2002,** *18*(24), 9243-9249.
31. Holz, M. *Chemical Society Reviews* **1994,** *23* 165-174.
32. Saarinen, T. R.; Johnson, C. S., Jr. *J.Am.Chem.Soc.* **1988,** *110* 3332-3333.
33. He, Q.; Johnson, C. S., Jr. *J.Magn.Reson.* **1989,** *81* 435-439.
34. Johnson, C. S., Jr. Transport Ordered 2D-NMR Spectroscopy, in *NMR Probes of Molecular Dynamics*, Tycko, R., editor; Dordrecht: Kluwer Acad. Publ., 1994; Chapter 10, pp. 455-488.
35. Pettersson, E.; Topgaard, D.; Stilbs, P.; Söderman, O. *to be published* **2003**.
36. Goddard, E. D.; Ananthapadmanabhan, K. P. *Surfactant Sci.Ser.* **1998,** *77* 21-64.
37. Glass, J. E. *Polym.Mater.Sci.Eng.* **1987,** *57* 618-621.
38. Walderhaug, H.; Hansen, F. K.; Abrahmsén, S.; Persson, K.; Stilbs, P. *J.Phys.Chem.* **1993,** *97* 8336-8342.
39. Furó, I.; Iliopoulos, I.; Stilbs, P. *J.Phys.Chem.B* **2000,** *104*(3), 485-494.
40. Stilbs, P.; Walderhaug, H.; Lindman, B. *J.Phys.Chem.* **1983,** *87* 4762-4769.
41. Brown, W.; Stilbs, P. *Chem.Scr.* **1982,** *19* 161-163.
42. Brown, W.; Stilbs, P.; Lindström, T. *J.Appl.Polym.Sci.* **1984,** *29* 823-827.
43. Johannesson, H.; Halle, B. *J.Chem.Phys.* **1996,** *104*(17), 6807-6817.
44. Stilbs, P. Diffusion studied using NMR Spectroscopy, in *Encyclopedia of Spectroscopy and Spectrometry*, Lindon, J. C.; Tranter, G. E.; Holmes, J. L., Eds; Academic Press: London, 1999; pp. 369-375.
45. Gibbs, S. J.; Johnson, C. S., Jr. *J.Magn.Reson.* **1991,** *93* 391-402.
46. Wu, D.; Chen, A.; Johnson, C. S., Jr. *J.Magn.Reson.Ser.A* **1995,** *115*(2), 260-264.
47. Price, W. S.; Stilbs, P.; Jönsson, B.; Söderman, O. *J.Magn.Reson.* **2001,** *150*(1), 49-56.
48. Price, W. S.; Elwinger, F.; Vigouroux, C.; Stilbs, P. *Magn.Reson.Chem.* **2002,** *40*(6), 391-395.

49. Price, W. S.; Walchli, M. *Magn.Reson.Chem.* **2002,** *40* S128-S132.
50. Price, W. S. Water Signal Suppression in NMR Spectroscopy, in *Annual Reports on NMR Spectroscopy*, Webb, G. A., editor; Academic Press: London, 1999; Vol. 38, pp. 289-354.
51. Antalek, B. *Concepts Magn.Reson.* **2002,** *14*(4), 225-258.
52. Stilbs, P.; Paulsen, K.; Griffiths, P. C. *J.Phys.Chem.* **1996,** *100*(20), 8180-8189.
53. Antalek, B.; Windig, W. *J.Am.Chem.Soc.* **1996,** *118*(42), 10331-10332.
54. Schulze, D.; Stilbs, P. *J.Magn.Reson.* **1993,** *105* 54-58.
55. Morris, K. F.; Johnson, C. S., Jr. *J.Am.Chem.Soc.* **1992,** *114* 3139-3141.

56. Johnson, C. S. *Prog.Nucl.Magn.Reson.Spectrosc.* **1999,** *34*(3-4), 203-256.
57. Stilbs, P. *Anal.Chem.* **1981,** *53* 2135-2137.
58. Provencher, S. W. *Comput.Phys.Commun.* **1982,** *27* 229-242.

Chapter 4

Characterization down to Nanometers: Light Scattering from Proteins and Micelles

Ulf Nobbmann

Proterion Corporation, One Possumtown Road, Piscataway, NJ 08854 (email: nobbmann@proterion.com)

The visible light wavelength ranges from ~400nm to ~700nm, just slightly larger than the main topic of this book: Mesoscale phenomena, roughly defined as 1 – 100nm. Nevertheless, light is an ideal probe to determine the size, shape, structure, or aggregation state of mesoscale phenomena in complex fluids. A brief overview of the theory behind light scattering is followed by 'real life' application examples from various proteins and surfactants.

Light Scattering Theory

The scattering of light has been a fascinating topic spanning centuries *(1)*. Just looking up into the sky on a sunny day often reveals two scattering phenomena at once: The blue color of the sky is caused by scattering from molecules in the atmosphere (blue is scattered more than red) in bright contrast to the white light eminating from clouds. Here multiple scattering happens at all wavelengths, and no preferential wavelength reaches the eye. Static light scattering investigates the intensity of the scattered light signal, for example as a function of wavelength or of the scattering angle *(2)*. Dynamic light scattering on the other hand is concerned with the intensity fluctuations of the scattered light *(3),(4)*. Brownian motion leads to faster or slower signal variations, and these can be used to characterise the scattering from a different perspective.

Part of a typical light scattering setup is shown schematically in Figure 1. Laser light illuminates a sample, and scattered light is emitted in all directions. A typical sample is a solution of a fluid with a solute or solvent-like component and a dissolved second component, e.g. polymer dissolved in an organic solvent, or protein dissolved in an aqueous buffer. The scattered light originates from scattering objects inside the sample, ideally mostly from the dissolved component. A screen placed behind the sample would show a speckle pattern of bright and dark dots. The transmitted light consists of photons which have (almost) not interacted with the sample molecules. As light interacts with matter dipoles are induced due to the polarizability of all materials. A part of the energy is absorbed, another part re-emitted. We are interested here in the scattered light, we are not concerned with the absorbed radiation. The scattered light is detected with a single photon counter, for example an avalanche photo diode, or a photo multiplier tube *(5)*.

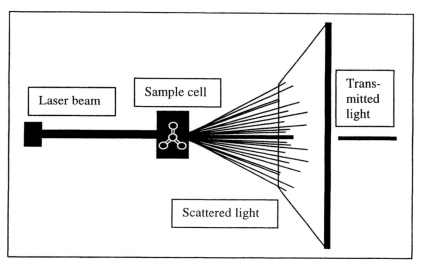

Figure 1: Scattering of light. Scattered light appears on a screen placed after the sample, incident light passing straight through the sample is transmitted without scattering.

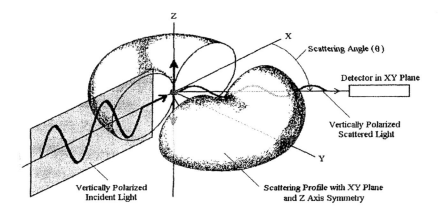

Figure 2: Dipole interaction. The incident light is re-radiated in the scattering plane, in a characteristic pattern. The magnitude and the shape of this scattering pattern depends on the polarizability, size, and concentration of the scattering objects.

The ideal case of linearly polarized light is shown in Figure 2. The oscillating electromagnetic waves from the incident light induce an oscillating dipole in the electron cloud surrounding the scattering object. In turn, the oscillating electrons of this charge cloud re-radiate (provided there is no absorption) in all directions. If the charge cloud is very small compared to the wavelength of the incident light then the scatterer can be considered a point source, and the radiation is emitted in a doughnut-shaped Hertz-dipole profile, as depicted in Figure 2.

The study of this scattering profile is the regime of static light scattering. With a few simplifying assumptions complete solutions are known and used to characterize scattering molecules and particles in solution *(1)*.

Static light scattering

For simplification let us assume that only single scattering occurs in the experiment. The opposite regime is called multiple scattering, requires usually rather larger sample concentrations of mesoscale particles, and shall be ignored in this chapter (it is covered in other chapters in this book). In a dilute sample the strength of the scattered signal is proportional to the concentration of the sample. Clearly, a more concentrated sample should scatter more light. A further simplification is the assumption that the particles are much smaller than the wavelength of the incident light. The scattering profile shown in Figure 2 is isotropic, the scattering in the scattering plane is independent of the scattering angle. This no longer holds true when the particle is larger. In that case, particles can not be considered point-like and intraparticle interferences of the dipole radiation lead to a characteristic anisotropic profile. This anisotropy (also called form factor) can be described by the normalised intensity as a function of scattering angle. In the isotropic case the form factor is 1, independent of scattering angle.

Scattering intensity as a function of sample concentration has been studied for many years. Only mention of the popular Debye plot is made here. The key equation is:

$$K \cdot c / R_{90} = 1/MW + A_2 \cdot c$$

where K is an instrument constant, c is the concentration of the solute, R_{90} is the Rayleigh ratio and describes the absolute intensity scattered at 90° in excess of the pure solvent, MW is the molecular weight, and A_2 is the second virial coefficient (a thermodynamic parameter). By measuring the scattered intensity of the sample as a function of concentration (and appropriate normalization to a known standard to obtain the Rayleigh ratio) it is possible to obtain the average molecular weight of the sample. This is just the inverse of the intercept of concentration divided by intensity versus concentration. The slope of this plot is related to the solubility of the sample in the solution. For ideal solvents, A_2 is 0.

If the sample is not very soluble, higher concentrations produce aggregation and lead to a stronger scattering signal than "expected", which corresponds to a negative second virial coefficient A_2. In order to quantitate this concentration dependence static light scattering measurements are often performed in concentration series. This requires preparation of several samples at different concentrations, and can be a hindrance when only minute sample amounts are available. Fortunately, there is another property of the scattered light which can be analyzed for just a single sample.

Dynamic light scattering

The picture of the scattered light on the screen shown in Figure 1 is not really "static". When looking at very small times, fluctuations in the speckles can be observed. The speckles are 'dancing' due to Brownian motion, the interference pattern moves because the particle arrangement causing the pattern is moving. We can analyze the time scale of these pattern fluctuations to determine the speed of the molecules. This technique is called dynamic light scattering since it relies on the dynamics, the movements inside the sample. We restrict further discussion to an ideal fluid sample in which Brownian motion is constantly reshuffling the scattering objects. This would not be the case in, for example, a solidly frozen or a gelled sample.

Dynamic light scattering (DLS) measures the diffusion of particles by illuminating the particles with coherent laser light, and then analysing the fluctuations in the scattered light using the autocorrelation technique *(6)*. The expressions photon correlation spectroscopy (PCS) or quasi-elastic light scattering (QELS) have been used synonymously with DLS.

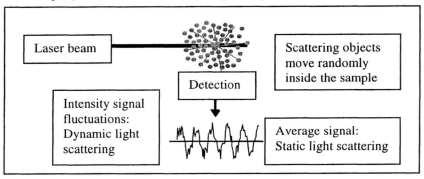

Figure 3: Different light scattering techniques. Dynamic light scattering analyses the statistical fluctuations in the detected signal, whereas static light scattering is concerned with the average, total signal strength.

The autocorrelation technique compares a signal with itself at a shifted time, slightly later. An analogy to this technique is to think of instantaneous photos of the particles being overlayed: When particles are essentially still the overlap of the second photo taken a very short time after the first photo is very good. However, if the second photo is taken at a later time when many particles have diffused away from the original location, then the photo overlay will no longer show any clear features, the correlation has vanished.

Experimentally, the signal from the photon detector is sent to a corrolator, usually a computer plug-in board or external device, which performs all the required calculations.

In mathematical terms we compare the intensity at time t with the intensity at a slightly later time $t+\tau$ to obtain the intensity autocorrelation function $g(\tau)$:

$$g(\tau) = <I(t) \cdot I(t+\tau)> / <I(t)>^2$$

where the <> brackets denote the average, and are in effect the value of the quantity integrated over the duration of the measurement. The average intensity $<I(t)>$ is used to normalize the correlation function. An example of such a function is shown in Figure 4.

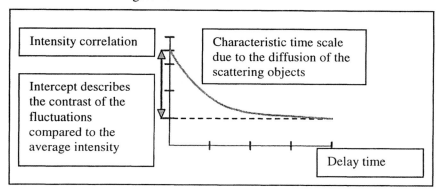

Figure 4: Typical autocorrelation function. Time scales are on the order of • s due to the fast diffusion of the molecules in solution.

A perfect sample of identical scattering particles will yield a single exponential decay function given by

$$g(\tau) = A + B \cdot exp\{-2 \cdot q^2 \cdot D_T \cdot \tau\}$$

with the fitting constants A (base line), B (amplitude), and D_T (translational diffusion coefficient). The wave vector q is known and given by

$$q = 2 \cdot \pi \cdot n \cdot sin(\theta/2) / \lambda$$

with the refractive index n, the scattering angle θ, and the wave length λ of the incident laser radiation. The desired fitting quantity is the diffusion coefficient D_T. This parameter characterizes the diffusion of the molecules and allows comparing different samples easily. For an ideal sphere the diffusion coefficient is directly related to the hydrodynamic radius of the diffusing particle. The Stokes-Einstein equation can be used to determine the hydrodynamic radius R_H

50

when the viscosity η of the solvent surrounding the particle and its diffusion coefficient are known

$$R_H = k_B \cdot T / (6 \cdot \pi \cdot \eta \cdot D_T)$$

k_B is the Boltzmann constant and T the absolute temperature in Kelvin. The hydrodynamic radius is the radius of a hypothetical sphere that diffuses with the same coefficient as the particle under investigation. This does not mean that the particles have to be round! The hydrodynamic radius describes the average size of the complete diffusing object, and includes the particles with adjacent solvent molecules, as shown in Figure 5.

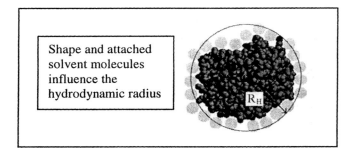

Figure 5: Visualization of the hydrodynamic radius. The average equivalent radius of the complete scattering system.

So far, we have considered a single species of scattering particles in the sample. The analysis can be broadened to look at complete size distributions inside the sample. In that case, it is still useful (and mathematically simplest) to obtain an overall mean radius of the whole sample. However, one has to keep in mind that each species is weighed according to its intensity, and larger species therefore contribute significantly more signal to the overall measurement, even when only present in small amounts. Thus an overall "cumulant" radius will often be larger than expected when samples are not clean, contaminated or dusty. With the help of essentially an inverse Laplace transformation with 'regularization' one can obtain the size distribution directly from the autocorrelation function (7). The distribution might consist of only one peak (monomodal) or several peaks (multimodal), and an individual peak can be very narrow (monodisperse) or rather broad (polydisperse). This technique is well-established but has limitations:

a) The resolution is limited to peaks separated by a factor two. Peaks which are closer together will appear as one, albeit broader distribution in the regularization analysis.

b) For particularly broad distributions or samples consisting of many different species regularization should be used with caution. In general, the more fitting parameters are used the less trust one should place in the result.

In practice, regularization analysis works well for system consisting of up to four separate peaks.

Established applications

The direct information obtained from the static light scattering are molecular weight and second virial coefficient. Dynamic light scattering directly measures the diffusion coefficient – or the hydrodynamic radius – and a polydispersity of the sample. From these directly obtained quantities information about shape, conformation, aggregation, melting point, quarternary structure, or in general terms the solutions composition can be deduced. Established mesoscale applications of light scattering can be found in protein characterization, structural biology, pharmaceutical research & development, and the studies of biological polymers, surfactants and micelles.

Proteins, shape, structure

Proteins are complex biological molecules with multiple levels of structure beyond their amino acid sequence and can be studied by light scattering *(8)*. The hydrodynamic radius of a few globular proteins is shown in Figure 6. An empirical line through the data points allows the prediction of the molecular weight for any unknown protein – under the assumption of globular shape *(9)*. In other words, this empirical fit can be used to estimate the molecular mass from the measured hydrodynamic radius of an unknown protein sample.

The scattering intensity for smaller molecules is much less than for larger molecules. In order to detect lower molecular weight proteins, a certain minimum concentration of the sample has to be present. A typical light scattering setup would show a sensitivity curve similar to Figure 7

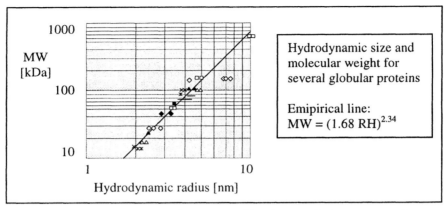

Figure 6: Hydrodynamic radius obtained by dynamic light scattering from several globular proteins of different molecular mass. Molecular weight expected from gel chromatography versus measured radius. In order of increasing expected molecular weight the symbols describe Ribonuclease, Lysozyme, Myoglobin, Chymotrypsinogen, Carbonic Anhydrase, Ovalbumin, Hb subunit of Hemoglobin, BSA, Transferrin, Amyloglucosidase, Horse ADH, Hexokinase, DAA, GDPase, Yeast ADH, IgG, Thyroglobulin.

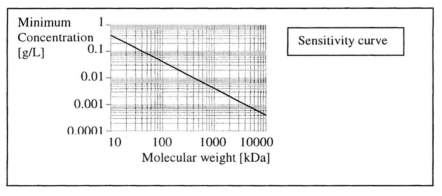

Figure 7: Sensitivity curve for a typical dynamic light scattering setup. For this system a protein of 20 kDa would have to be present at a concentration of 0.2 g/L to detect its hydrodynamic radius reliably.

The lower detection limit for dynamic light scattering is slightly less than 1 nm. Such small radii correspond to molecules of a few kDa. And at this end of the mesoscale we start to get interferences from solvated buffer molecules, e.g. hydrated sugars, salt ions, or other additives. It is best to ignore radii significantly below 1nm in the size distribution. This advice is especially important when operating at the detection limit of a light scattering system, i.e. at or below the minimum concentration of the sensitivity curve.

After obtaining the hydrodynamic radius it is often of interest to get an idea of the approximate molecular weight of the molecule. The standard globular molecular weight model can be used to determine if a protein is present in an oligomeric state *(10)*. As an example, insulin is a stable protein and can be prepared at buffers covering a wide range of pH values. Performing dynamic light scattering of insulin in acidic conditions at pH 3 shows a hydrodynamic radius of 1.7nm which suggests an estimated molecular weight of 12 kDa. It is known that the monomer of insulin is 5.7 kDa, which suggests that the molecule is present as a dimer $(2 \cdot 5.7 = 11.4)$. The same sample prepared in a neutral buffer of pH 7 (thus closer to physiological conditions) exhibits a hydrodynamic radius of 2.6nm, a significant increase from the radius at pH 3. Again using the empirical estimate for globular proteins we predict a molecular weight of 32 kDa which suggests the formation of a hexameric state $(6 \cdot 5.7 = 34)$. Insulin has been crystallized both as a dimer at pH 3 and a hexamer at pH 7 confirming the pH dependent dissociation observed with light scattering.

The standard model is not ideal when the scattering objects are "fluffy", have long arms sticking out or are simply not very globular. Transducin is a fibril-forming protein, and is known from strucutral data to be brick-shaped and not globular. For such cases the hydration and shape of the molecule can be taken into account in a volume shape hydration model *(11)*. In this model the density of the hydrated protein is used to determine an equivalent radius for a known molecular weight, and then this radius is enlarged by a shape factor due to the non-sphericity of the molecule. The shape (or also called Perrin) factor can be calculated for oblate or prolate ellipsoids. Conversely, it is possible to use the known molecular weight of a molecule and its known specific volume to determine the shape factor. We can again look at transducin, this time in a dimeric state. The measured hydrodynamic radius of transducin dimer with a molecular weight of 80 kDa and a specific volume of 0.73 mL/g would support an axial ratio of 5.7:1 for an oblate ellipsoid. This measurement in turn then suggests that the transducin bricks connect end to end rather than side by side in this dimeric stage. More generally, light scattering can be used for the study of self-assembly and aggregation of proteins *(12)*.

Most molecular structures from proteins have been obtained through crystallography, but finding the right buffer conditions under which proteins form crystals is still one of the time consuming steps on the way to the molecular structure. Here light scattering has been of used with success to investigate the crystallizability *(13)*. This approach is based on checking the solubility of proteins, ideally the absence of any aggregates, in native buffer conditions prior to crystallization *(14)*. Aggregation and polydispersity can easily be observed in the size distribution und various solvent conditions. The main advantage of this technique is the small sample requirement (~10μL) and compared to the actual crystallization conditions, the lower concentration *(15)*. The chance of obtaining crystals is under 10% when the protein is polydisperse under dilute buffer conditions yet increases to about 70% when dynamic light scattering shows a monomodal and monodisperse size distribution.

A similar crystallizabilty assessment can be performed with static light scattering. Here, the main idea is to measure the second virial coefficient, and obtaining crystals corresponds to a relatively narrow window for the value of this parameter *(16)*. If the second virial coefficient is large, then the molecules will attract each other so much that aggregates are formed. On the other hand, if the coefficient is low the molecules have no incentive to fall into their lattice spaces and prefer to stay in solution. As an example, lysozyme in a no-salt buffer will show strong interaction, but the same molecule shows much less aggregation when salt is added to the buffer. The only disadvantage of this technique is the preparation of a concentration series, and therefore larger sample and time requirement.

Light scattering has also been used in conjunction with other techniques like analytical ultracentrifugation to investigate purification *(17)* and protein interactions *(18)*.

Micelles, cmc, and other applications

A particular class of proteins called membrane proteins proves to be quite difficult to crystallize. One reason for this difficulty lies in their amphiphilic nature: the molecule is made up of both a hydrophilic and a hydrophobic part. This is more complex but similar to surfactant molecules. Above a critical concentration surfactants self-assemble into micelles because these often spherical structures are energetically favored *(19)*.

Figure 8: Micelle formation. Surfactants self-assemble into micelles under certain conditions.

This onset of micelle formation, i.e. the critical micelle concentration (cmc), can be detected by light scattering *(20)*. The formation of the larger micelle structures is directly visible from the hydrodynamic radius and corresponding increase in the scattering signal, or can be studied with other optical techniques, e.g. fluorescence *(21)*. The data shown in Figure 9 agree well with cmc determinations by fluorescence, and the aggregate size is in alignment with analytical ultracentrifugation results.

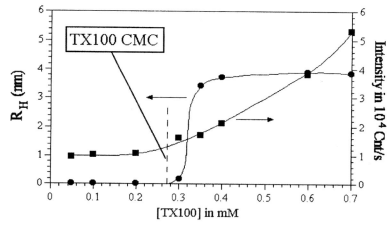

Figure 9: Dynamic and static light scattering from Triton X-100. Hydrodynamic radius and light scattering signal are plotted versus the concentration of the surfactant.

Detergents are complex systems sensitive to concentration, additives, temperature and other parameters, and therefore the use of static light scattering is problematic (for example when the micelle structure changes with concentration). Of particular practical concern is filtration: Even though in general the quality of light scattering data benefits from the elimination of dust by filtering the sample prior to a measurement, this can have an influence on the size distribution, or even lead to the formation of larger structures (for example worm-like micelles, vesicles). This effect can sometimes be reduced by the choice of larger pore size filters, but it is best to avoid filtration where possible. Below is a listing of some commercially available detergents which have been used in solubilizing membrane proteins.

Table 1: Hydrodynamic radius and other properties of Hampton detergent screen of nonionic (N) or zwitterionic (Z) type. Results for the radius are listed at 10 times the cmc, at the cmc, and after filtration through 10nm pore size. The notes field indicates whether the detergent is suitable for filtration. Complete lack of or immeasurable detergent is indicated by '0', highly questionable results due to strong aggregation and/or widely fluctuating numbers were entered as ' – '.

Detergent	MW [Da]	CMC [mM]	type	10x / @cmc / filtered	notes, filter
n-Undecyl-ß-D-maltoside	496.6	0.59	N	2.8/3.0/3.3	suitable
n-Nonyl-ß-D-maltoside	468.4	6.0	N	2.3/2.2/3.2	suitable
n-Decyl-ß-D-thiomaltoside	498.6	0.9	N	2.7/2.8/3.2	suitable
n-Nonyl-ß-D-thiomaltoside	484.6	3.2	N	2.3/3.0/7.4	no filter
n-Octyl-ß-D-thiomaltoside	470.6	9.0	N	2.3/2.1/2.2	suitable
Cymal®-4	480.5	7.6	N	2.3/2.3/3.3	suitable
Cymal®-2	452.5	120.0	N	2.7/1.3/1.6	suitable
1-s-Nonyl-ß-D-thioglucoside	322.4	2.9	N	2.0/10.5/0	no filter
1-s-Heptyl-ß-D-thioglucoside	294.4	29.0	N	5.0/ – / –	no filter
FOS-Choline®-12	351.4	1.5	Z	2.2/3.2/2.9	suitable
FOS-Choline®-10	323.4	13.0	Z	1.7/2.3/2.4	suitable
FOS-Choline®-9	309.4	19.0	Z	1.4/ – /3.2	no filter
FOS-Choline®-8	295.4	102.0	Z	156/ – /29	big mess

The table shows the hydrodynamic radius for some Hampton screen kit detergents at tenfold concentration, at cmc, and after filtration through a 20nm pore size Whatman Anodisk filter membrane. The samples were obtained from dilution of the tenfold stock solution with distilled water. The samples generally follow the expected trends *(22), (23), (24),* for example the nonionic maltosides and the phosphocholines show smaller micelle sizes with decreasing tail length. Nonionic detergents (except thioglucosides) and longer chain lengths seem more suitable to filtration than the ones with zwitterionic head groups and short chain lengths. Ionic strength of the dilution buffer can have a strong effect on the micellar size, and care has to be taken when trying to use the results for membrane protein / detergent mixes. The lack of additional ions in the distilled water could be the reason why some of the zwitterionic compounds were difficult to measure at their expected critical micelle concentration. Further discussion of the relationship between surfactant molecular structure and mesoscale (aggregate) structure is found in later chapters of this book.

All previous examples and results came from batch measurements where a sample was filled into a cuvette, and then placed into the light scattering instrument. It is also possible to perform light scattering under flow conditions. A flow cell can be attached in-line to a chromatography column, and the eluting material is then characterized while (slowly) flowing through the cuvette. This approach is limited to such samples which can be separated by chromatography, however, it can produce very monodisperse samples, since the column separates the aggregates from the main peak of interest.

Although the majority of this chapter's discussion was for relatively small scattering objects (up to ~10nm), light scattering can also be used very well for larger phenomena. Viruses are assemblies of proteins, and can reach 50nm in size. Polymer samples come in many sizes, shapes, and distributions, even inks, toners, paints and other colloidal systems can be investigated by looking at their scattered light. Some additional uses of light scattering are shown in other chapters of this book.

Summary

Light scattering is a powerful tool to characterize mesoscale particles in solution. The non-perturbing technique is fast and easy to use, while requiring very little sample volume. Particularly valuable is the study of precious samples 'under native conditions' due to the high sensitivity at small sizes and low concentrations. Light scattering is the most sensitive technique of detecting large aggregates in solution.

Dynamic light scattering measures the hydrodynamic radius, the size distribution, and an estimated molecular weight. Static light scattering measures the molecular weight and predicts molecular interactions. On the mesoscale light scattering is a well established application for protein and surfactant characterization.

Acknowledgements

Thanks are due to Fiona Case for making this symposium possible.
Current and former coworkers at Protein Solutions provided data, samples, advice and support: Kevin Mattison, Benoît Aragon, Bob Collins, and Dan Snyder. Bob Cudney from Hampton Research provided the detergent samples.

References

1. Van de Hulst, H.C. *"Light scattering by small particles"*, John Wiley & Sons, New York, **1957**.
2. Johnson, C.S.; Gabriel, D.A. *"Laser light scattering"*, Dover, **1995**.
3. Pecora, R. *"Dynamic light scattering: Applications of photon correlation spectroscopy"*, Plenum Press, **1985**.
4. Brown, W. *"Dynamic light scattering: The method and some applications"*, Clarendon, **1995**.
5. Brown, R. G. W. "Miniature laser light scattering instrumentation for particle size analysis", *Applied Optics* **1990,** *29(28),* 1-4.
6. Watling, R. G. W. US patent 4,975,237, **1989**.
7. Braginskaya, T.G.; Dobitchin, P.D.; Ivanova, M.A.; Klyubin, V.V.; Lomakin, A.V.; Noskin, V.A; Shmelev, G.E.; Tolpina, S.P. *Phys. Scripta.* **1983,** *vol. 28,* 73-79.
8. Liu, T.; Chu, B. "Light scattering by proteins", *Encyclopedia of Surface and Colloid Science,* Dekker, **2002,** 3023-3043.
9. Claes, P.; Dunford, M.; Kenney A.; Vardy, P. "An on-line dynamic light scattering instrument for macromolecular characterization", pp. 66-76 in *"Laser Light Scattering in Biochemistry"*, ed. S.E.Harding, D.B.Satelle and V.A.Bloomfield, Royal Society of Chemistry publishers, Cambridge, UK, **1992**.
10. Mullen, C.A.; Jennings, P.A. "Glycinamide ribonucleotide transformylase undergoes pH-dependent dimerization", *J. Molecular Biology* **1992,** *262(5),* 746-755.

11. Cantor, C.R.; Schimmel, P.R. *"Biophysical Chemistry Part II: Techniques for the Study of Biological Structure and Function"*, W.H. Freeman and Co, New York; **1980**.
12. Moradian-Oldak, J.; Leung, W.; Fincham, A.G. "Temperature and pH-dependent supramolecular self-assembly of amelogenin molecules: a dynamic light-scattering analysis", *J. Structural Biology* **1998**, *122(3)*, 320-327.
13. D'Arcy, Allan "Crystallizing proteins – a rational approach", *Acta Cryst.* **1994**, *D50*, 467-471.
14. Ferre-D'Amare, A.R.; Burley, S.K. "Dynamic light scattering in evaluating crystallizability of macromolecules", *Methods in Enzymology* **1997**, *276*, 157-166.
15. Bergfors, T.M. *"Protein crystallization: Techniques, strategies, and tips"*, International University Line, La Jolla, CA; **1999**, 27-38.
16. George, A.; Chang, Y.; Guo, B.; Arabshahi, A.; Cai, Z.; Wilson, W.W. "Second virial coefficient as predictor in protein crystal growth", *Methods in Enzymology* **1997**, *276*, 100-109.
17. Schönfeld, H.J.; Pöschl, B.; Müller, F. "Quasi-elastic light scattering and ultracentrifugation are indispensable tools for the purification and characterization of recombinant proteins", *Biochem. Soc. Trans.* **1998**, *26*, 753-758.
18. Schönfeld, H. J.; Behlke, J. "Molecular chaperones and their interactions investigated by analytical ultracentrifugation and other methodologies", *Methods in Enzymology* **1998**, *290*, 269-296.
19. Zana, R. "Critical micellization concentration of surfactants in aqueous solution and free energy of micellization", *Langmuir* **1996**, *12*, 1208.
20. Schmitz, K.S. "Photon correlation spectroscopy: Multicomponent systems", *SPIE Proceedings vol.1430*, SPIE Publishers, Bellingham **1991**.
21. Guenoun, P.; Lipsky, S.; Mays, J. W.; Tirrell, M. "Fluorescence study of hydrophobically modified polyelectrolytes in aqueous solution: Effect of micellization", *Langmuir* **1996**, *12*, 1425.
22. Rosen, M.J. "Surfactants and Interfacial Phenomena", Wiley Interscience, 2nd edition, **1989**.
23. Israelchvili, J.N.; Mitchell, D.J.; Ninham, B.W. "Theory of self-assembly of hydrocarbon amphiphiles into micelles and bilayers", *J. Chem. Soc. Faraday Trans.* **1976**, *72*, 1525-1568.
24. Holmberg, K; Jönsson, B.; Lindman, B. *"Surfactants and Polymers in Aqueous Solution"*, John Wiley & Sons, New York, 2nd edition, **2002**.

Chapter 5

Small-Angle Scattering Characterization of Block Copolymer Micelles and Lyotropic Liquid Crystals

Paschalis Alexandridis

Department of Chemical Engineering, University at Buffalo, The State University of New York, Buffalo, NY 14260–4200

The self-assembly mode of amphiphiles and the corresponding structure-property relations can be readily modulated by the quality and quantity of selective solvents present in the system. Small-angle X-ray and neutron scattering are very useful tools for the structural characterization of micelles and lyotropic liquid crystals thus formed. Drawing on examples from our research on poly(ethylene oxide)-poly(propylene oxide) (PEO-PPO) block copolymers in the presence of water (selective solvent for PEO) and water-miscible polar organic solvents (e.g., ethanol, glycerol), we discuss the utilization of small-angle scattering data for the determination of the structure and length-scales of the block copolymer assemblies and for the assessment of the degree of block segregation and the solvent location in the block copolymer domains.

Amphiphilic molecules (surfactants, lipids, block or graft copolymers) are well known to form a variety of self-assembled structures in the presence of selective solvents, e.g., water for ionic and PEO-containing surfactants or block copolymers (*1-5*). Such structures are based on micelles (spherical, cylindrical or planar) which consist of amphiphilic molecules that are typically segregated to form a solvent-incompatible (hydrophobic, in the case of aqueous solvents) core, and a solvated (hydrated) corona (shell) dominated by solvent-compatible (hydrophilic) segments. The micelles have a diameter of 4-20 nm, depending on the molecular dimensions of the amphiphile. The micelles can be randomly dispersed in solution or can form ordered arrays (lyotropic liquid crystals) of, e.g., lamellar, hexagonal, or cubic symmetry. The hierarchical organization that leads from amphiphilic (macro)molecules to micelles to ordered arrays of micelles can further result into higher-order structures such as spherulites (*6*).

Several functional properties of self-assembled systems (also referred to as "complex fluids" or "soft condensed matter") depend on the local environment afforded by the micelles and also on the dimensions, geometry, connectivity and spatial arrangement of the micelles (*1,2,7-9*). Small-angle scattering of neutrons, electrons, or light has the ability to probe length-scales in the mesoscale (1-100 nanometers) and is ideally suited for the study of self-assembly (*10-13*). Small-angle scattering is widely being used to characterize both the internal structure and the spatial organization of micelles. Scattering of X-rays originates from differences in the electron density between different domains, whereas scattering of neutrons depends on the contrast between hydrogenous and deuterated species. Micellar solutions of hydrogenous surfactants (*14,15*) or block copolymers (*16,17*) in deuterated solvent are typical self-assembled systems studied by small-angle neutron scattering (SANS). Lyotropic mesophases formed by ionic surfactants or lipids in water (*18,19*), and solvent-free block copolymers that are segregated (*3,5*) are typical self-assembled systems studied by small-angle X-ray scattering (SAXS).

The quality and quantity of the solvent have been shown to be controlling factors in the self-assembly of block copolymers (*20-26*). In particular, we have established that the addition to block copolymers of two immiscible solvents, selective for different copolymer blocks, can give rise to a remarkable structural polymorphism, and thus provides extra degrees of freedom in tailoring various properties (*27-30*). Such observations have more recently been extended to blends of block copolymers with homopolymers (*31-34*). Variation of the solvent (water) quality in the case of water-soluble amphiphiles such as PEO-PPO block copolymers and nonionic surfactants is often achieved by a change in the temperature (*35*) or by the addition of organic solvents miscible (wholly or partially) with water (*36*). The self-assembly of amphiphiles in mixtures of water and organic solvents less polar than water is of fundamental and practical importance in the elucidation of the driving force for self-assembly (*1*) and in water-borne formulations (e.g., pharmaceutical, cosmetics, coatings, inks) (*37-39*), respectively. Moreover, solutions of polymers in mixed solvents are of interest as analogues of industrially relevant polymer-solvent systems and as systems in which complex multi-component behavior may arise due to the interplay of competing types of binary interactions (*39,40*).

Significant research efforts in our group have been directed toward the fundamental understanding and practical utilization of the self-assembly afforded by macromolecular amphiphiles such as block copolymers consisting of poly(ethylene oxide) and poly(propylene oxide) (PEO-PPO) (*4,17,21,22,27-30,35,41-56*). PEO-PPO block copolymers are commercially available as Poloxamers or Pluronics in a range of molecular weights (2000-20000) and PEO contents (10-80%), and find numerous applications on the basis of their ability to form micelles and gels in aqueous solutions and to modify interfacial properties (*4,41,42*). A recent focus has been on the self-assembly of PEO-PPO block copolymers in mixtures of water and polar organic solvents such as formamide, ethanol, butanol, propylene glycol, or glycerol (*51-56*). These solvents were selected on the basis of fundamental considerations and/or of their relevance in pharmaceutics (*37,56*), water-borne coatings (*39,57*) and inks (*38,58*), or materials synthesis (*59*).

We highlight here the utilization of small-angle scattering techniques for the elucidation of block copolymer self-assembly. Our studies (i) probe the block copolymer organization in both micellar solutions (*51,52*) and lyotropic liquid crystals (*53-56*), (ii) are concerned with both the types of structure formed (e.g., spheres, cylinders) and their characteristic dimensions (nearest neighbor distances), and (iii) combine macroscopic observations (*51,53,55*) (e.g., composition-temperature phase boundaries) with microscopic information obtained from small-angle neutron (*51,52*) and X-ray (*54*) scattering experiments. The various findings are discussed in terms of preferential localization of the solvents in the self-assembled domains.

While the specific examples discussed here are obtained for a block copolymer having the structure $EO_{37}PO_{58}EO_{37}$ (Pluronic P105, nominal molecular weight = 6500, 50% PEO), the observations and conclusions have been shown valid for other PEO-PPO block copolymers (*56*) as well as for copolymers consisting of polydimethylsiloxane backbone and PEO grafts (*60,61*). Furthermore, we believe that the observations presented here are reflections of the general behavior of block copolymers in mixed solvents, and are not limited to specific classes of polymers and solvents.

Small-Angle Neutron Scattering

The formation and structure of PEO-PPO block copolymer micelles were probed by SANS measurements performed at the National Institute of Standards and Technology (NIST) Center for Neutron Research (NCNR), beam guide NG3. In the temperature - concentration region where micelles are formed, a monodisperse core-corona sphere form factor and a hard-sphere interaction structure factor were used to fit the SANS scattering patterns in order to extract information about the micelle size, intermicellar interaction distance and micelle association number i.e., the number of block copolymer molecules which (on the average) participate in one micelle. The polymer volume fractions in the micelle core and the corona were calculated on the basis of the above fitting parameters as discussed below (*51,52,60*).

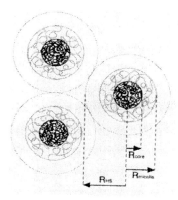

Figure 1. Schematic (cross-section) of block copolymer micelles. R_{core} and $R_{micelle}$ are the radii of the micelle core and the whole micelle, respectively. R_{HS} is the hard-sphere interaction distance. When the micelle concentration is relatively low, micelles are well separated and $R_{HS} > R_{micelle}$.

The absolute SANS intensity can be expressed as a product of the form factor $P(q)$, the structure factor $S(q)$, and the number density of the scattering particles (in our case micelles), N:

$$I(q)=NP(q)S(q) \tag{1}$$

The form factor $P(q)$ accounts for the intramicelle structure (*11-13*). Several models have been proposed for the form factor of block copolymer micelles; we summarize below the ones relevant to the system considered here. The simplest form factor that describes a micelle with a well-defined core devoid of solvent is that of a hard sphere (*61,63*). However, most block copolymer micelles are recognized as having a core-corona (shell) structure (*64*). A way to account for this is to have a form factor that represents a spherical core and a corona layer of a given thickness, with the polymer/solvent content across the corona thickness to be constant (but adjustable); in this model there is a step change in the solvent concentration at the core-corona interface and at the corona-solution interface (*65,66*). A variation of this theme is a form factor that uses a hard sphere to describe the micelle core and Gaussian chains to describe the micelle corona (67). A recent improvement is the so-called cap-gown form factor where the micelle core is again represented as a hard sphere but the corona consists of a Gaussian distribution of polymer segments with no abrupt boundary between the micelle core and the solvent (68). Perhaps the best approach would be to use no model at all for the micelle structure, but to extract the scattering density correlation function using the generalized indirect Fourier transformation (GIFT) method (10,69). However, the GIFT method is computationally involved compared to form factors for which analytical expressions are available. Note that when the GIFT method was applied to PEO-PPO block copolymer micelles, the results were indicative of spherical micelles, relatively monodisperse in size (69).

In this work we have used the core-corona form factor model to describe the structure of the micelles formed by the $EO_{37}PO_{58}EO_{37}$ block copolymer in mixed solvents (51,52). A schematic of the core-corona micelle structure is shown in Figure 1. In particular, the core-corona form factor is based on the contrast (i) between the polymer-rich micelle core and the micelle corona of intermediate polymer concentration, and (ii) between the micelle corona and the solvent-rich surroundings, thus allowing for different solvent contents in the micelle core and the micelle corona:

$$P(q) = \{(4\pi R_{core}^{3}/3)(\rho_{core} - \rho_{corona})[3J_{1}(qR_{core})/(qR_{core})] +$$

$$(4\pi R_{micelle}^{3}/3)(\rho_{corona} - \rho_{solvent})[3J_{1}(qR_{micelle})/(qR_{micelle})]\} \quad (2)$$

Here R_{core} and $R_{micelle}$ are the radii of the micelle core and whole micelle, respectively; ρ_{core}, ρ_{corona}, and $\rho_{solvent}$ are the scattering length densities (SLD) of the core, corona, and solvent (assuming a homogeneous solvent distribution in each of the domains); $J_1(y)$ is the first-order spherical Bessel function. In fitting the core-corona model to the scattering data, we view the micelles as consisting of a core composed of PPO segments (with little or no solvent present) and a relatively solvated corona where the PEO segments reside. The SLD of the core, ρ_{core}, and of the corona, ρ_{corona}, are then a function of the average (over the core radius and over the corona thickness) volume fraction of PPO segments in the core (α_{core}) and of PEO segments in the corona, respectively:

$$\rho_{core} = \alpha_{core}\rho_{PPO} + (1 - \alpha_{core})\rho_{solvent}$$

$$\rho_{corona} = \alpha_{corona}\rho_{PEO} + (1 - \alpha_{corona})\rho_{solvent} \quad (3)$$

ρ_{PPO} (=0.325 x 10^{10} cm^{-2}) is the SLD of PPO, ρ_{PEO} (=0.547 x 10^{10} cm^{-2}) is the SLD of PEO, and $\rho_{solvent}$ is the SLD of the water – polar organic solvent mixture (ρ_{D2O} = 6.33 x 10^{10} cm^{-2}, $\rho_{Dethanol}$ = 5.95 x 10^{10} cm^{-2}, and $\rho_{Dglycerol}$ = 7.48 x 10^{10} cm^{-2}) (52). The use of the PPO SLDs to determine the SLD of the micelle core and of the PEO SLD to determine the SLD of the micelle corona does not preclude the presence of small amount of PEO in the core and/or of small amount of PPO in the corona region; this is because the SLDs of the hydrogenous PPO and PEO are very different than those of the deuterated solvents. Also, the use of the initial (fixed) water – polar organic solvent composition in determining the SLD of the solvent in the micelle core, micelle corona, and micelle-free solvent regions, does not preclude different ratios of organic solvent/water to exist in each of the three regions (again, because the SLD values of the deuterated solvents are similar to each other relatively to the SLD of the polymer); in fact, as we discuss below, we may have partitioning of the organic solvent in the micelle core or corona domains.

The volume fraction of polymer in the core (α_{core}) and the volume fraction of polymer in the corona (α_{corona}) can be expressed in terms of the core and micelle radii (R_{core} and $R_{micelle}$) and the micelle association number, N_{assoc}. (52). The R_{core}, $R_{micelle}$, N_{assoc}, α_{core}, and α_{corona} parameters extracted from the model fits to our data, while accurate for the model used here (a variation of more than ±1 in the values of R_{core}, $R_{micelle}$, and N_{assoc} would result in a discernible worsening of the fit), are dependent on the choice of model used to fit the data, as different models use different physical descriptions for the fitting parameters. However, the physical picture that emerges for the micelle (when one considers the parameters obtained from the fitting in the context of the specific model) ought to be independent of the model used.

In addition to the form factor that describes the intramicelle structure as discussed above, the structure factor must be accounted for in order to describe intermicellar interactions. Interactions between the micelles are manifested in a correlation peak that appears in the neutron scattering patterns generated from 8 % EO$_{37}$PO$_{58}$EO$_{37}$ block copolymer solutions, as seen in Figure 2. Note that, while common for ionic surfactant micelles with long-range electrostatic interactions, such correlation peaks are usually not encountered in nonionic surfactant micellar solutions at such relatively low concentrations. PEO-PPO block copolymers, however, are different, because of their relatively high micelle size and volume fraction that the hydrated micelles occupy. To describe such intermicellar interactions, we utilized the structure factor, $S(q)$, for hard spheres, which has been described in detail in several publications (52,65,69). Representative fits of the monodisperse core-corona sphere form factor and hard sphere structure factor model to SANS patterns in the q range 0.01-0.1 Å^{-1} are shown in Figure 2 for EO$_{37}$PO$_{58}$EO$_{37}$ solutions in water and ethanol or glycerol.

66

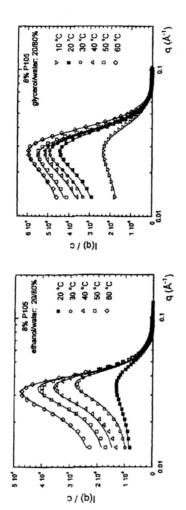

Figure 2. Representative examples of fits of the core-corona model (discussed in the text) to the SANS scattering intensities, used to extract information on the micelle structure. Data are shown for 8 wt % $EO_{37}PO_{58}EO_{37}$, in ethanol-water mixed solvent with 20 vol % ethanol, (left) and in glycerol-water mixed solvent with 20 vol % glycerol (right), in the temperature range 10-60 °C. (Adapted from reference 52. Copyright 2000 American Chemical Society.)

lyotropic liquid crystalline microstructures

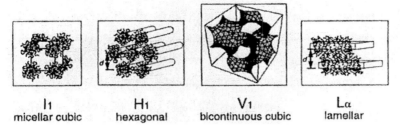

| I₁ | H₁ | V₁ | Lα |
| micellar cubic | hexagonal | bicontinuous cubic | lamellar |

*Figure 3. Schematic of various lyotropic liquid crystalline structures attained
by the block copolymer self-assembly in the presence of selective solvents. The
Gyroid minimal surface is used as a representation of the bicontinuous cubic
structure. The lattice parameter d in the lamellar structure and the distance, d,
between the planes of the centers of two adjacent rows of cylinders in the
hexagonal structure are indicated in the schematics of the respective structures.*

Small-Angle X-Ray Scattering

The effects of solvents on the stability range, type of structure (see Figure 3 for schematics of various structures) and characteristic dimensions of the lyotropic liquid crystalline assemblies formed by PEO-PPO block copolymers were probed by systematic SAXS measurements performed on series of samples at various block copolymer-water-solvent systems, keeping the polymer weight fraction constant and varying the water-solvent ratio (*45,54,56*). The Bragg diffraction peaks obtained from the ordered phases were relatively sharp (see Figures 4 and 5), and the difference between slit-smeared and desmeared data small (a few Å) (*54*). The structure of the liquid crystalline phases can be determined from the relative positions of the Bragg diffraction peaks (*21,30,53,55*). For the lamellar and hexagonal structures the positions of the peaks relative to the primary peak should obey the relationship $1 : 2 : 3 : 4...$ and $1 : 3^{1/2} : 2 : 7^{1/2} : 3...$, respectively. Typical SAXS patterns obtained from samples of hexagonal structure in the $EO_{37}PO_{58}EO_{37}$ - water - ethanol and $EO_{37}PO_{58}EO_{37}$ - water – glycerol systems are shown in Figure 4.

Two parameters are characteristic of the structure: the lattice parameter (the repeat distance, d, in the lamellar structure and the distance, a, between the centers of adjacent cylinders in the hexagonal structure) and the interfacial area per block copolymer molecule, a_p (the area that a PEO block of the PEO-PPO block copolymer occupies at the interface between PEO-rich and PPO-rich domains). When we compare the lattice parameters in the lamellar and the hexagonal structures, instead of a in the hexagonal structure, we prefer to use the distance between the planes of the centers of two adjacent rows of cylinders, d, which is related to a through a simple geometrical relation. In the following, as lattice parameter in the hexagonal structure we will discuss only the distance d.

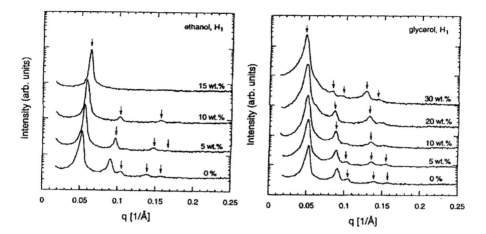

Figure 4. SAXS diffraction patterns obtained from hexagonal (H₁) samples at 50 wt % EO₃₇PO₅₈EO₃₇ and at various polar organic solvent contents (indicated just above the corresponding curve). The diffraction patterns have been shifted by multiplying them by an appropriate number to avoid overlap. (Adapted from reference 53. Copyright 2000 American Chemical Society.)

In the hexagonal and lamellar structures, the lattice parameter is given directly by the position, q^*, of the first and the most intense diffraction peak:

$$d=2\pi/q^* \qquad (4)$$

The determination of the interfacial area per block copolymer molecule involves some assumptions as to the definition of the PEO-rich and PPO-rich domains. For details on the calculation of the interfacial area in the lamellar and hexagonal structures and the assumptions made, see reference (54).

The establishment of the crystallographic space group of the cubic liquid crystalline phases (of both micellar or bicontinuous structure, shown schematically in Figure 3) is more complicated than that of the lamellar and hexagonal structures. The proper indexing to a crystallographic space group is often ambiguous due to the small number of reflections observed in SAXS and their relatively lower intensities. It is usually necessary to consider not only the relative positions of the diffraction peaks but also their relative intensities as well as other criteria, such as the missing peaks and the values for the lattice parameter and the interfacial area per block copolymer molecule (21,30,53).

The positions of the Bragg diffraction peaks obey different relationships for the different space groups. However, all space groups can be classified in three main families of cubic lattices (70): primitive (P...), body-centered (I...), and face-centered (F...). The indexing of the SAXS diffraction peaks to a given crystallographic group was assessed by plotting the reciprocal spacing, $1/d_{hkl}$, of the reflections versus the sum of the Miller indices, $(h^2 + k^2 + l^2)^{1/2}$. For a correct assignment this plot is linear and passes through the origin (21,30,53). The lattice parameter, a, is obtained from the slope of such plot. The interfacial area per block copolymer molecule, a_p, in the water-continuous ("normal") micellar cubic structure is a function of the lattice parameter, the number of the spherical micelles (structural elements) in the cubic cell, the volume fraction of the PPO-rich domains (micelle cores), the block copolymer volume fraction in the ternary system, and the block copolymer molecular volume (53).

SAXS patterns obtained from micellar cubic samples in the $EO_{37}PO_{58}EO_{37}$ - water - ethanol and $EO_{37}PO_{58}EO_{37}$ - water – glycerol systems are shown in Figure 5. The face-centered cubic lattice can be easily ruled out as a possible structure leading to these diffraction patterns, however, both I... and P.. assignments were possible (53). In simple primitive or body-centered cubic lattices, the first peak is the most intense one, while in more complex structures this rule does not hold. Indeed, in the SAXS data for the glycerol system the first diffraction peak is the most intense; therefore, we have simple primitive or body-centered structures in this case. The ethanol system apparently has a more complex structure. Let us first review the case of glycerol. In the assignment to the body-centered structure, the Bragg diffraction peak with Miller indices hkl = 200 is systematically missing in all SAXS patterns except for those at the highest glycerol concentration. In the assignment to the primitive structure, the Bragg peak with hkl = 110 is missing. Dealing with simple structures, however, allowed us to compare both the lattice parameter, a, and the interfacial area per block copolymer, a_p, with the corresponding values obtained for the neighboring hexagonal structures. Such a comparison led us to consider the primitive cubic structure as more plausible for the glycerol system (53). In the ethanol system we considered the crystallographic space-group $Pm3n$ as more likely. The $Pm3n$ space group is common in surfactants and lipids (70); it has also been observed in some PEO-PPO block copolymer systems (22,71). The cubic lattice of the crystallographic space group $Pm3n$ consists of eight short rodlike (with axial ratio of 1.2-1.3) micelles per unit cell, two of them having complete rotational freedom and the other six having only lateral rotational freedom (71).

The bicontinuous cubic structure can be represented by a multiply connected bilayer, the midplane of which can be modeled as a minimal surface (see Figure 3), and the polar (PEO-rich) / apolar (PPO-rich) interface can be described as two parallel surfaces displayed on the opposite sides of the minimal surface (27-30). Such a structure is observed in the ethanol system, but not in the glycerol system at 25 °C (53,54). The most common crystallographic space group for the bicontinuous cubic phases formed by surfactants and lipids (70) as well as by PEO-PPO block copolymers (27-30) is $Ia3d$. It features two characteristic Bragg peaks: the first and most intense at Miller indices hkl = 211, and a "shoulder" at hkl = 220. The cubic lattice of the $Ia3d$ space group is consistent with that afforded by the Gyroid minimal surface (27-30).

Results and Discussion

We highlight here features of the self-assembly exhibited by a PEO-PPO block copolymer (Pluronic P105: $EO_{37}PO_{58}EO_{37}$) in mixed solvents consisting of water and a polar organic solvent, e.g., ethanol or glycerol, with an emphasis on

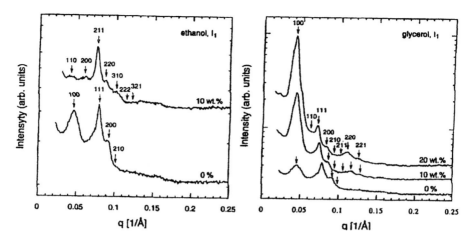

Figure 5. SAXS diffraction patterns obtained from micellar cubic (I₁) samples at 50 wt % EO₃₇PO₅₈EO₃₇ and at various polar organic solvent contents (indicated just above the corresponding curve). The diffraction patterns have been shifted by multiplying them by an appropriate number to avoid overlap. The identified diffraction peaks used in the assignment of the cubic lattice are indicated with arrows and denoted with the corresponding Miller indices. (Adapted from reference 53. Copyright 2000 American Chemical Society.)

the effect that the organic solvent has on the self-assembled structure. We first consider the micellar solutions and then the ordered (lyotropic liquid crystalline) phases. We utilize both macroscopic observations (51,53,55) (e.g., composition-temperature phase boundaries) and microscopic (mesoscale) information obtained from SANS (51,52) and SAXS (54) as described above.

The formation of micelles is a spontaneous process driven by the interaction between the solvophobic segments of the amphiphiles and the selective solvent (44). The onset of the micellization is often reflected in a dramatic change of a number of properties, e.g., scattering intensity, and can be detected by monitoring such properties (35). In the case of PEO-PPO block copolymers, temperature is an important variable because it affects the water solvent quality for both blocks: the PPO block is water-soluble at low (4 °C) temperatures but becomes insoluble at room temperature, while PEO remains water-soluble at

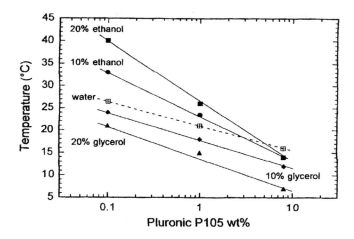

Figure 6. Micellization boundary for $EO_{37}PO_{58}EO_{37}$ (Pluronic P105) dissolved in water and in ethanol-water or glycerol- water- mixtures. At temperatures and concentrations below the micellization boundary, the block copolymers do not associate (unimers). Above the micellization boundary, micelles are formed which coexist in equilibrium with unimers.

temperatures as high as 100 °C but eventually phase separates. At low temperatures, when both PPO and PEO blocks are more soluble in water, the PEO-PPO block copolymers are present as unassociated unimers. The unimer solution has very low scattering length density contrast due to the relatively even distribution of the deuterated solvent, and thus gives rise to low neutron scattering intensity. The onset of micelle formation leads to strong scattering originating from the contrast between the solvent-poor micelle core and the solvent phase. Thus, we can readily obtain CMT values by following the temperature dependence of the relative scattering intensity, normalized with respect to the block copolymer concentration (*51,52*). As shown in Figure 6, the addition of ethanol to water shifted the micellization boundary to higher block copolymer concentrations and temperatures, while the addition of glycerol reduced the critical micellization concentration (CMC) and temperature (CMT). The ethanol-water mixture appears a better solvent than water for the PEO-PPO block copolymer, whereas the glycerol-water mixture a worse solvent.

The better solvent quality afforded by the ethanol-water mixtures ought to be reflected in the PEO-PPO block copolymer micelle structure. Indeed, the micelle core radii and corona thicknesses, the micelle association numbers, and the polymer volume fractions in the micelle core and corona decreased with increasing ethanol-water ratio, as shown in Figure 7 (*52*). On the contrary, the addition to water of glycerol led to higher micelle association numbers and

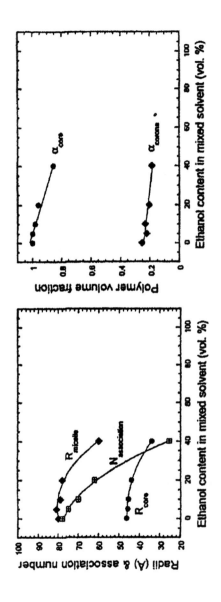

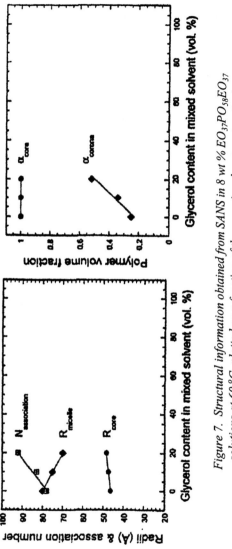

Figure 7. Structural information obtained from SANS in 8 wt % $EO_{37}PO_{58}EO_{37}$ solutions at 60 °C, plotted as a function of the organic solvent content in the solvent mixture. Left panels: radii of core and micelle, and micelle association number; right panels: polymer volume fraction of core and corona. (Adapted from reference 52. Copyright 2000 American Chemical Society.)

74

higher polymer volume fractions in the micelle corona. Judging from its octanol/water partition coefficient ($\log P = -0.32$), ethanol, although fully miscible with water, is hydrophobic compared to the PEO blocks, and has a higher affinity to the PPO blocks than water. Thus the presence of ethanol leads to an increase of the solvent content in the micelle interior. Glycerol has a very negative $\log P$ ($= -2.55$) and a strong affinity to water. In the presence of glycerol, the amount of water hydrating the PEO-rich micelle corona decreases due to the strong tendency of glycerol to mix with water (52).

These changes in the micelle structure upon the addition of ethanol or glycerol, are in excellent agreement with our results on the effects of these solvents on the lyotropic liquid crystals formed by PEO-PPO block copolymers (53,56), as discussed below in the context of data presented in Figures 8 and 9.

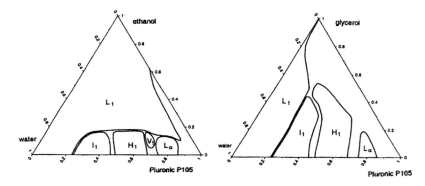

Figure 8. Effect of polar organic solvents on the phase behavior of $EO_{37}PO_{58}EO_{37}$ (Pluronic P105) in water at 25 °C: (left) Pluronic P105 – water - ethanol system, (right) Pluronic P105 – water - glycerol system. The boundaries of the one-phase regions are drawn with solid lines. L_1 denotes the ternary composition region of liquid clear isotropic solution. I_1 denotes the region where the samples are clear, optically isotropic gels and the structure is that of spherical micelles arranged in a cubic lattice. H_1 denotes the region where the samples are clear birefringent gels and the structure is that of cylindrical micelles arranged in a hexagonal lattice. V_1 denotes the region where the samples are clear isotropic (hard) gels and the structure is that of interconnected (bicontinuous) cylindrical micelles arranged in a cubic lattice. L_α denotes the region where the samples are clear birefringent (soft) gels and the structure is that of planar micelles (lamellae). Samples with compositions that fall outside the one-phase regions indicated above are two- or three-phase. The composition is given in weight fractions. (Adapted from reference 53. Copyright 2000 American Chemical Society.)

A pronounced difference between the effects of ethanol and glycerol on the concentration range of stability of the different lyotropic liquid crystalline structures formed by $EO_{37}PO_{58}EO_{37}$ in water is evident from Figure 8. Glycerol causes an expansion ("swelling") of the range of stability of the hexagonal structure toward lower copolymer contents to the expense of the micellar cubic structure, whereas the L_1-I_1 boundary remains unaltered at a constant block copolymer content (parallel to the water-glycerol axis of the triangle) up to 50% glycerol. Ethanol, on the other hand, causes a "deswelling" of the micellar cubic structure, as indicated by a bend of the L_1-I_1 boundary away from the water-ethanol axis. In fact, about 20% ethanol is sufficient to destabilize ("melt") all the ordered lyotropic liquid crystalline structures formed by $EO_{37}PO_{58}EO_{37}$. These opposing effects are reinforced by the appearance of a bicontinuous cubic phase in the ethanol system but not in the glycerol system.

Such solvent effects on the phase boundaries can be considered as indications for the location of the solvents in the self-assembled structures and the degree of swelling with solvent of the copolymer blocks, since they reflect changes of the interfacial curvature in the system. The interfacial curvature is defined as positive when the interface bends toward the apolar domains and as negative when the interface bends toward the polar domains (1). The interfacial curvature is zero in the lamellar structure. The transition from lamellar to hexagonal to micellar cubic structure, observed along the $EO_{37}PO_{58}EO_{37}$-water axis with increasing the water content is a result of an increase of the interfacial curvature: it changes from zero (planar micelles) to positive (cylindrical micelles) to highly positive (spherical micelles). This reflects an increase in the solvent swelling of the PEO block relative to the PPO block (54).

Therefore, even from the general appearance of the ternary phase diagrams ("macroscopic" information) a conclusion about the changes in the interfacial curvature, the degree of swelling of the different copolymer blocks, and, hence, the location of the solvents in the self-assembled structure can be drawn. The bend of the hexagonal structure toward the water-glycerol axis at constant block copolymer content (i.e., the micellar geometry changes from spheres to cylinders) is an indication of a decrease of the interfacial curvature due to a decrease of the solvation of the PEO block. The destabilization of the ordered structures caused by ethanol is an indication of an increased solvation of the PPO and also of the PEO blocks, solvation that decreases the block segregation and the resulting tendency for self-assembly.

A more sensitive and precise measure of the effects of solvents on the self-assembly can be obtained from systematic measurement of the characteristic length-scales of the liquid crystalline structures. As seen in Figure 9, at constant block copolymer content (i.e., lines in the phase triangle that are parallel to the water-solvent axis), glycerol increased the lattice parameter, whereas ethanol decreased the lattice parameter (54). These effects are manifested in all liquid

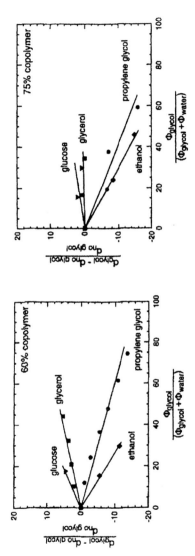

Figure 9. Percent variation of the lattice parameter at different polar organic solvent ("glycol") contents with respect to that at 0% solvent, plotted versus the percent solvent volume fraction relative to the total solvent volume fraction. Data for four different glycols (glucose, glycerol, propylene glycol, and ethanol) are shown on the same graph at constant block copolymer content: (left) 60% $EO_{37}PO_{58}EO_{37}$, H_1 phase, and (right) 75% $EO_{37}PO_{58}EO_{37}$, L_α phase. (Adapted from reference 54. Copyright 2000 American Chemical Society.)

crystalline phases and have related to changes in the interfacial area, the preference of the different solvents to locate in different domains of the block copolymer assemblies, and the solvent ability to modify the preferred curvature in the system by swelling different blocks of the PEO-PPO block copolymer (54). In particular, the increase in the spacing (i.e., the separation distance between self-assemblies) that is caused by glycerol reflects a decrease in the interfacial area in the system (i.e., fewer assemblies are spaced further apart in order to fill the same volume). Since the interface between polar/apolar domains is created by the segregation of the PEO-PPO block copolymer, a decrease in the interfacial area means that the solvation (and effective volume) of the copolymer blocks decreases. Correspondingly, the decrease in the spacing that is caused by ethanol reflects an increase in the interfacial area in the system and an increased solvation of the copolymer blocks.

Concluding Remarks

Small-angle scattering measurements have been instrumental in providing structural information on self-assembly of amphiphiles that supports and complements "macroscopic" phase-behavior observations. The solvent effects observed on the lyotropic liquid crystalline structures formed by polyether block copolymers agree well with the effects of solvents on the micelle structure: glycerol tends to increase the polymer volume fraction in the micelle corona by pulling water away from the PEO-rich domains, whereas ethanol decreases the polymer volume fraction in the micelle core and corona since it tends to preferentially partition there (Figure 7). More recently (Lin, Y.; Alexandridis, P., unpublished data) we have carried out SANS measurements using combinations of hydrogenous and deuterated solvents (contrast variation) that provide direct evidence of solvent partitioning/depletion in the self-assembled domains, in good agreement with the data presented here. These trends correlate well with the hydrogen-bonding component of the Hansen solubility parameter for various solvents (61).

Solvent-induced phase behavior changes have interesting repercussions on aqueous formulations containing PEO-PPO block copolymers: for example, high-viscosity low-polymer-content (lyotropic liquid crystalline) gels can be formed upon dilution of low-viscosity high-polymer-content solutions (72). This allows for effective targeting, delivery, and retention of desired substances to moistened surfaces and aqueous environments in a form (gel) that resists erosion or run-off.

Acknowledgements

Funding for our fundamental and applied studies on block copolymer self-assembly has been generously provided by the National Science Foundation (CTS-9875848/CAREER, CTS-0124848/TSE), ACS Petroleum Research Fund

(33408-G7), Dow Chemical Co., Xerox Foundation, Procter & Gamble Co. University Exploratory Research Program (UERP), and Swedish Natural Science Research Council (NFR). The National Institute of Standards and Technology (NIST) provided the neutron research facilities used in this work. P.A. is grateful to all his coworkers listed as coauthors in the references cited below for their valuable contributions to this research endeavor.

References

1. Evans, D. F.; Wennerstrom, H. *The Colloidal Domain: Where Physics, Chemistry, Biology, and Technology Meet*, 2nd ed.; Wiley-VCH, 1999.
2. Holmberg, K.; Jonsson, B.; Kronberg, B.; Lindman, B. *Surfactants and Polymers in Aqueous Solution*, 2nd ed.; Wiley, 2003.
3. Hamley, I. W. *The Physics of Block Copolymers*; Oxford University Press, 1998.
4. Alexandridis, P.; Lindman, B. (Eds.) *Amphiphilic Block Copolymers: Self-Assembly and Applications*; Elsevier Science B.V., 2000.
5. Hadjichristidis, N.; Pispas, S.; Floudas, G. *Block Copolymers: Synthetic Strategies, Physical Properties, and Applications*; Wiley, 2002.
6. Zipfel, J.; Lindner, P.; Tsianou, M.; Alexandridis, P.; Richtering, W. *Langmuir* **1999**, *15*, 2599-2602.
7. Larson, R. G. *The Structure and Rheology of Complex Fluids*; Oxford University Press, 1999.
8. Stokes, R. J.; Evans, D. F. *Fundamentals of Interfacial Engineering*; Wiley-VCH, 1997.
9. Rajagopalan, R., Chapter 8 in *The Expanding World of Chemical Engineering*, 2nd ed., Furusaki, S.; Garside, J.; Fan, L. S. (Eds.); Taylor & Francis, 2002.
10. Glatter, O.; Kratky, O. (Eds.) *Small Angle X-ray Scattering*; Academic Press, 1982.
11. Lindner, P.; Zemb, Th. (Eds.) *Neutron, X-Ray and Light Scattering: Introduction to an Investigative Tool for Colloidal and Polymeric Systems*; North Holland, 1991.
12. Higgins, J. S.; Benoit, H. C. *Polymers and Neutron Scattering*; Oxford University Press, 1997.
13. Roe, R.-L. *Methods of X-Ray and Neutron Scattering in Polymer Science*; Oxford University Press, 2000.
14. Chen, S. H. *Ann. Rev. Phys. Chem.* **1986**, *37*, 351-399.
15. Kline, S. R.; Kaler, E. W. *Langmuir* **1996**, *12*, 2402-2407.
16. Pedersen, J. S.; Svaneborg, C.; Almdal, K.; Hamley, I. W.; Young, R. N. *Macromolecules* **2003**, *36*, 416-433.
17. Alexandridis, P.; Yang, L. *Macromolecules* **2000**, *33*, 3382-3391.
18. Ricoul, F.; Dubois, M.; Zemb, T.; Heck, M. P.; Vandais, A.; Plusquellec, D.; Rico-Lattes, I.; Diat, O. *J. Phys. Chem. B* **1998**, *102*, 2769-2775.

19. Seddon, J. M.; Templer, R. H.; Warrender, N. A.; Huang, Z.; Cevc, G.; Marsh, D. *Biochim. Biophys. Acta* **1997**, *1327*, 131-147.
20. Alexandridis, P.; Spontak, R. J. *Curr. Opin. Colloid Interface Sci.* **1999**, *4*, 130-139.
21. Alexandridis, P.; Zhou, D.; Khan, A. *Langmuir* **1996**, *12*, 2690-2700.
22. Alexandridis, P. *Macromolecules* **1998**, *31*, 6935-6942.
23. Laurer, J. H.; Khan, S. A.; Spontak, R. J.; Satkowski, M. M.; Grothaus, J. T.; Smith, S. D.; Lin, J. S. *Langmuir* **1999**, *15*, 7947-7955.
24. Hanley, K. J.; Lodge, T. P.; Huang, C.-I. *Macromolecules* **2000**, *33*, 5918-5931.
25. Lodge, T. P.; Pudil, B.; Hanley, K. J. *Macromolecules* **2002**, *35*, 4707-4717.
26. Lai, C.; Russel, W. B.; Register, R. A. *Macromolecules* **2002**, *35*, 841-849.
27. Alexandridis, P.; Olsson, U.; Lindman, B. *Macromolecules* **1995**, *28*, 7700-7710.
28. Alexandridis, P.; Olsson, U.; Lindman, B. *J. Phys. Chem.* **1996**, *100*, 280-288.
29. Alexandridis, P.; Olsson, U.; Lindman, B. *Langmuir* **1997**, *13*, 23-34.
30. Alexandridis, P.; Olsson, U.; Lindman, B. *Langmuir* **1998**, *14*, 2627-2638.
31. Washburn, N. R.; Lodge, T. P.; Bates, F. S. *J. Phys. Chem. B.* **2000**, *104*, 6987-6997.
32. Corvazier, L.; Messe, L.; Salou, C. L. O.; Young, R. N.; Fairclough, J. P. A.; Ryan, A. J. *J. Mater. Chem.* **2001**, *11*, 2864-2874.
33. Lee, J. H.; Balsara, N. P.; Chakraborty, A. K.; Krishnamoorti, R.; Hammouda, B. *Macromolecules* **2002**, *35*, 7748-7757.
34. Huang, Y. Y.; Chen, H. L.; Hashimoto, T. *Macromolecules* **2003**, *36*, 764-770.
35. Alexandridis, P.; Nivaggioli, T.; Hatton, T. A. *Langmuir* **1995**, *11*, 1468-1476.
36. Armstrong, J.; Chowdhry, B.; Mitchell, J.; Beezer, A.; Leharne, S. *J. Phys. Chem. B* **1996**, *100*, 1738-1745.
37. Martin, A.; Swarbick, J.; Cammarata, A. *Physical Pharmacy*; Lea & Febiger Press Inc., 1983.
38. Kang, H. R. *J. Imaging Sci.* **1991**, *35*, 179-188.
39. Lambourne, P. (Ed.) *Paint and Surface Coatings: Theory and Practice*; Ellis Horwood Ltd., 1987.
40. Solomon, M. J.; Muller, S. J. *J. Polym. Sci. B: Polym. Phys.* **1996**, *34*, 181-192.
41. Alexandridis, P.; Hatton, T. A. *Coll. Surf. A* **1995**, *96*, 1-46.
42. Alexandridis, P. *Curr. Opin. Colloid Interface Sci.* **1996**, *1*, 490-501.
43. Alexandridis, P. *Curr. Opin. Colloid Interface Sci.* **1997**, *2*, 478-489.
44. Alexandridis P.; Holzwarth, J. F.; Hatton, T. A. *Macromolecules* **1994**, *27*, 2414-2424.
45. Holmqvist, P.; Alexandridis, P.; Lindman, B. *J. Phys. Chem. B* **1998**, *102*, 1149-1158.

46. Svensson, B.; Olsson, U.; Alexandridis, P.; Mortensen, K. *Macromolecules* **1999**, *32*, 6725-6733.
47. Kositza, M. J.; Bohne, C.; Alexandridis, P.; Hatton, T. A.; Holzwarth, J. F. *Macromolecules* **1999**, *32*, 5539-5551.
48. Yang, L.; Alexandridis, P.; Steytler, D. C.; Kositza, M. J.; Holzwarth, J. F. *Langmuir* **2000**, *16*, 8555-8561.
49. Schmidt, G.; Richtering, W.; Lindner, P.; Alexandridis, P. *Macromolecules* **1998**, *31*, 2293-2298.
50. Zipfel, J.; Berghausen, J.; Schmidt, G.; Lindner, P.; Alexandridis, P.; Richtering, W. *Macromolecules* **2002**, *35*, 4064-4074.
51. Yang, L.; Alexandridis, P. *Langmuir* **2000**, *16*, 4819-4829.
52. Alexandridis, P.; Yang, L. *Macromolecules* **2000**, *33*, 5574-5587.
53. Ivanova, R.; Lindman, B.; Alexandridis, P. *Langmuir* **2000**, *16*, 3660-3675.
54. Alexandridis, P.; Ivanova, R.; Lindman, B. *Langmuir* **2000**, *16*, 3676-3689.
55. Ivanova, R.; Lindman, B.; Alexandridis, P. *Langmuir* **2000**, *16*, 9058-9069.
56. Ivanova, R.; Lindman, B.; Alexandridis, P. *J. Colloid Interface Sci.* **2002**, *252*, 226-235.
57. Annable, T.; Brown, R. A.; Padget, J. C.; van den Elshout, A. *JOCCA-Surface Coatings International* **1998**, *81*, 321-329.
58. Lin, Y.; Smith, T. W.; Alexandridis, P. *J. Colloid Interface Sci.* **2002**, *255*, 1-9.
59. Feng, P.; Bu, X.; Pine, D. J. *Langmuir* **2000**, *16*, 5304-5310.
60. Lin, Y.; Alexandridis, P. *J. Phys. Chem. B* **2002**, *106*, 12124-12132.
61. Lin, Y.; Alexandridis, P. *Langmuir* **2002**, *18*, 4220-4231.
62. Mortensen, K.; Brown, W. *Macromolecules* **1993**, *26*, 4128-4135.
63. Jain, N. J.; Aswal, V. K.; Goyal, P. S.; Bahadur, P. *J. Phys. Chem. B* **1998**, *102*, 8452-8458.
64. Zheng, Y.; Won, Y.-Y.; Bates, F. S.; Davis, H. T.; Scriven, L. E.; Talmon, Y. *J. Phys. Chem. B* **1999**, *103*, 10331-10334. 65. Wu, G.; Chu B.; Schneider, D. K. *J. Phys. Chem.* **1995**, *99*, 5094-5101.
66. Goldmints, I.; von Gottberg, F. K.; Smith, K. A.; Hatton, T. A. *Langmuir* **1997**, *13*, 3659-3664.
67. Pedersen, J. S.; Gerstenberg, M. C. *Macromolecules* **1996**, *29*, 1363-1365.
68. Liu, Y.; Chen, S.-H.; Huang, J. S. *Macromolecules* **1998**, *31*, 2236-2244.
69. Mortensen, K.; Pedersen, J. S. *Macromolecules* **1993**, *26*, 805-812.
70. Mariani, P.; Luzzati, V.; Delacroix, H. *J. Mol. Biol.* **1988**, *204*, 165-189.
71. Alexandridis, P.; Olsson, U.; Lindman, B. *Langmuir* **1996**, *12*, 1419-1422.
72. Dobrozsi, D. J.; Hayes, II, J. W.; Lindman, B. O.; Ivanova, R. H.; Alexandridis, P. United States Patent 6,503,955.

Chapter 6

X-ray Scattering Studies of Long-Chain Alkanol Monolayers at the Water–Hexane Interface

Mark L. Schlossman[1] and Aleksey M. Tikhonov[2]

[1]Department of Physics and Chemistry, University of Illinois at Chicago, 845 W. Taylor Street, Chicago, IL 60607 (email: schloss@uic.edu)
[2]Center for Advanced Radiation Sources, University of Chicago, and Brookhaven National Laboratory, National Synchrotron Light Source, Beamline X19C, Upton, NY 11973 (email: tikhonov@bnl.gov)

X-ray reflectivity and interfacial tension measurements demonstrate that long-chain alkanol monolayers at the water-hexane interface exhibit a well defined chain disorder and partial hexane mixing into the monolayer, in contrast to alkanol monolayers at the water-vapor interface that consist of close-packed rigid rod molecules. At the water-hexane interface triacontanol molecules form a condensed phase with progressive disordering of the chain from the $-CH_2OH$ to the $-CH_3$ group. At this interface the density in the head-group region is 10 to 15% greater than bulk water, an effect not seen for the ordered monolayer at the water-vapor interface. Monolayers of shorter length alkanols (consisting of 20, 22, and 24 carbons) and variations with temperature are also discussed.

Introduction

Solutions of water, oil, and surfactants often contain mesoscopic regions that can be studied with x-ray, light, or neutron scattering. The stability, shape, and ordering of these regions often relies upon the orientational and conformational ordering of the surfactants, as well as the water and oil molecules, at the water-oil interface. However, due to the complicated orientation of the interfaces of these mesoscopic structures it can be hard to probe the microscopic details of the molecular organization at the interfaces.

There has been much success in using x-ray and neutron scattering to study surfactant organization at a single, large water-vapor interface either by spreading a single layer of insoluble surfactant molecules on the water surface (Langmuir monolayers) or by studying soluble surfactants that adsorb to the water-vapor interface (Gibbs monolayers, see the article by Penfold). This sample geometry provides an oriented and easily controllable interface of surfactants for study. However, studies of these monolayers cannot account for interactions with an oil phase that may alter the molecular conformation and condensed matter phases of the surfactants at a macroscopic or mesoscopic water-oil interface.

One solution to this problem is to study oriented monolayers of surfactants at the water-oil interface instead of at the water-vapor interface. Although thermodynamic studies of surfactant monolayers have been conducted in the past (1-4), it has proven difficult to probe liquid-liquid interfaces on the molecular length scale. However, recent experimental advances have allowed for the investigation of molecular conformations and ordering at liquid-liquid interfaces. For example, surface-sensitive, non-linear optical spectroscopies have been used to study molecular orientation at the water-organic liquid interface (reviewed in (5)). X-ray and neutron reflectivity have been used to study molecular ordering at the water-alkane interface between bulk liquid phases and at the aqueous-aqueous interface between a thin film of an aqueous solution on top of a different, immiscible aqueous solution (reviewed in (6)).

We present a study of n-alkanols at the water-hexane interface. These molecules contain a single alkyl chain and a polar headgroup. Although simple, these molecules exhibit a very interesting conformation at the water-hexane interface and are, arguably, a good starting point to model the interfacial behavior of many surfactant molecules of industrial and biological interest that contain alkyl chains and polar head-groups. We have studied four different chain lengths of n-alkanols, but focus our attention on the longest, triacontanol $(CH_3(CH_2)_{29}OH)$, because it provides the highest resolution x-ray data. We find that triacontanol at the water-hexane interface exhibits a distinctive disorder likely due to a distribution of gauche conformers along the chain with a progressively greater density of these conformers towards the $-CH_3$ group near the bulk hexane. Our analysis also indicates the presence of hexane molecules mixed into the region of the alkyl chain closer to the $-CH_2OH$ group. Both of these results are in contrast to the close-packed, rigid rod ordering of triacontanol molecules at the water-*vapor* interface. In addition, we find that at the water-hexane interface the density in the head-group region is 10 to 15% greater than either bulk water or the ordered head-group region found at the water-vapor interface. It is conjectured that this higher density is a result of water penetration into the head-group region of the disordered monolayer. This latter result may be of particular interest in understanding the ordering of water near biological macromolecules.

Experimental and Analysis Methods

Sample Preparation and Interfacial Tension

Samples were prepared by placing a bulk solution of an *n*-alkanol in hexane on top of bulk water, (the four alkanols used in this study are $CH_3(CH_2)_{m-1}OH$, where $m = 20, 22, 24,$ and 30). When the hexane solution is poured on top of the water, the surfactants spontaneously self-assembled into a monolayer at the water-hexane interface. The liquid samples were stirred and equilibrated in a temperature-controlled, vapor-tight stainless steel sample cell described in detail elsewhere (7). Hexane was purified by passing it through a column of activated alumina several times; alkanols were purified by double crystallization from the purified hexane; water was from a Barnstead Nanopure system. Interfacial tensions were measured with a Wilhelmy plate (8) in the sample cell used for x-ray scattering.

X-ray Methods

X-ray reflectivity from the water-hexane interface was conducted at beamline X19C at the National Synchrotron Light Source (Brookhaven National Laboratory, USA) with a liquid surface spectrometer and measurement techniques described in detail elsewhere (9,10). A similar measurement at the water-vapor interface was conducted at the ChemMatCARS sector of the Advanced Photon Source (Argonne National Laboratory). The kinematics of reflectivity is illustrated in Figure 1A. For specular reflection, the wave vector transfer, $Q = k_{scat} - k_{in}$, is only in the z-direction, normal to the interface; $Q_x =$

$Q_y = 0$ where x and y are in the plane of the interface, and $Q_z = (4\bullet /\lambda)\sin(\alpha)$. Therefore, specular reflection probes the electron density normal to the interface and averaged over the region of the x-ray footprint on the sample. The x-ray wavelength was $\lambda = 0.825\pm0.004$ Å for water-hexane studies, $\lambda = 1.5507 \pm 0.0002$ Å for the water-vapor studies. The reflectivity data consist of measurements of the x-ray intensity reflected from the sample normalized to the incident intensity measured just before the x-rays strike the surface. These data are further modified by subtracting a background measured slightly off the specular condition (9). An incident slit (gap ≥ 10 μm) determines the beam size on the sample and a slit (gap ≥ 200 μm) before the detector sets the resolution of the measurement. The footprint of the beam on the sample varies from 0.5 to 1.5 cm long by 0.2 cm wide.

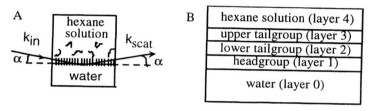

Figure 1. (A) Kinematics of X-ray reflectivity from the water-hexane interface. X-rays pass through the hexane solution of alkanols to probe the interface. (B) Nomenclature for interfacial layers as in Eq.(1).

The surfactant monolayer is described by three layers sandwiched between bulk water and bulk hexane (or vapor), see Figure 1B. Layer 1 is the head-group region ($-CH_2OH$), layers 2 and 3 are for the alkyl tail-group region ($-(CH_2)_{m-2}CH_3$); layers are ordered water–1–2–3–hexane (or vapor). A general formula for the electron density gradient normal to a surface with m layers is *(11)*

$$\frac{d\langle\rho(z)\rangle}{dz} = \sum_0^m (\rho_i - \rho_{i+1})\frac{1}{(2\pi\sigma_{i+1}^2)^{1/2}} e^{-(z-D_i)^2/2\sigma_{i+1}^2}, \tag{1}$$

where ρ_0 is the electron density of the water, ρ_{m+1} is the density of hexane, L_i is the thickness of the ith layer, $D_i = \sum_{j=1}^{i} L_j$ is the distance from the surface of the water to the interface between the ith and $(i+1)^{st}$ layers, and σ_{i+1} is the interfacial width between the ith and $(i+1)^{st}$ layers and is chosen to be the same value for all layers, $\sigma_{i+1} \equiv \sigma$. The interfacial width is primarily due to interfacial roughening by capillary waves. The values of the widths that result from the fitting are very similar to values calculated from capillary wave theory using the measured interfacial tension (12). Note that the quoted electron densities in this paper are normalized to the value for bulk water ($0.333e^-/\text{Å}^3$).

Given the electron densities of each layer and the subphase, as well as the widths for each interface, the specular reflectivity is calculated from the Born approximation for x-ray scattering. This approximation relates the reflectivity to the electron density gradient normal to the interface, $d<\rho(z)>/dz$ (averaged over the interfacial plane), and written as (13)

$$\frac{R(Q_z)}{R_F(Q_z)} \approx \left| \frac{1}{\Delta\rho_{bulk}} \int dz \frac{d\langle\rho(z)\rangle}{dz} \exp(iQ_z z) \right|^2 , \qquad (2)$$

where $\Delta\rho_{bulk}$ is the electron density difference between water and hexane, and $R_F(Q_z)$ is the Fresnel reflectivity predicted for an ideal, smooth and flat interface that has a step-function change in the electron density when going from one bulk phase to the other (14). As a guide to fitting the reflectivity data, the minimum number of layers is chosen that can reasonably account for the structure in the data.

Monolayers at the Water-Hexane Interface

Previous measurements of interfacial tension have shown that the interface between bulk water and an alkane solution of an alkanol undergoes a transition as a function of temperature (15-17). Consistent with these earlier measurements, Figure 2 illustrates measurements of interfacial tension γ as a function of temperature for the four alkanols we studied and for the pure water-hexane interface. The change ΔS in excess surface entropy across the transition is 2.0 (2.3, 2.4, or 4.3) mJ/m^{2}K, respectively for the 20 (22, 24, or 30) carbon alkanol (where $S = -d\gamma/dT$ and ΔS is given by the slope difference on either side of the kink). This is larger than $\Delta S = $ 1.16 (1.29, 1.39, or 1.72) mJ/m^{2}K measured previously for monolayer surface freezing at a pure C_{20} (C_{22}, C_{24}, or C_{30}) alkane-vapor interface (18). It is reasonable to expect a larger ΔS in our system since the transition occurs when alkanol molecules in a dilute bulk solution form a condensed monolayer at the interface, whereas surface freezing occurs when the topmost layer of molecules at a liquid alkane surface freezes into a solid monolayer. The thermodynamic measurement in Figure 2 indicates that the alkanol monolayers undergo a single transition from a low temperature condensed phase to a high temperature disordered phase.

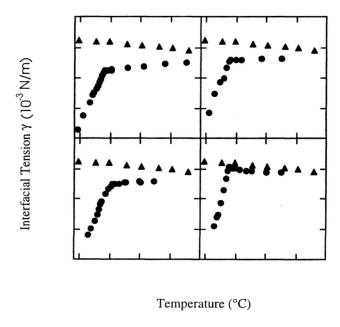

Temperature (°C)

Figure 2. Interfacial tension as a function of temperature for alkanols at the water-hexane interface (dots) and, for comparison, the pure water-hexane interface (triangles). Concentrations of the hexane solution of alkanols were chosen to place the transition temperature at a convenient value. The tensions were measured with a Wilhelmy plate.

Figure 3 illustrates x-ray reflectivity measurements (normalized to the Fresnel reflectivity) from the four alkanols at the water-hexane interface at nearly the lowest temperatures shown in Figure 2. These temperatures are slightly above the temperature at which the bulk hexane is saturated with the alkanols (as observed by the formation of crystallites one or two degrees lower in temperature). Under these conditions, the alkanol monolayers are close to their densest state for that temperature.

Oscillations in the reflectivity represent interference minima and maxima from x-rays scattered from different parts of the monolayer. The decrease in the oscillation period for longer chain length alkanols indicates that the monolayers get thicker as chain length increases. The appearance of three oscillations for the longest alkanol, triacontanol, provides the highest spatial resolution in

87

interpreting the x-ray reflectivity. We will focus on the interpretation of the triacontanol data (*19*).

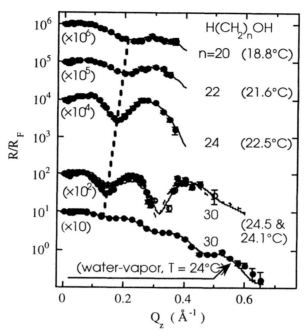

Figure 3. X-ray reflectivity (normalized to the Fresnel reflectivity) as a function of the wave vector transfer normal to the interface. The top four curves are for alkanols at the water-hexane interface, the bottom curve is for triacontanol at the water-vapor interface. Lines are fits described in the text. At the temperatures chosen the monolayers are in a condensed phase.

Figure 4A shows the electron density profile normal to the water-hexane interface for the triacontanol monolayer. Two layers are required to model the alkyl chain (plus one layer for the head-group) in order to fit the data (see Table 1 for fitting parameters). The normalized electron densities for the two layers modeling the chain are 0.95±0.01 and 0.79±0.01, both different from the close-packed value of 1.03 (or 1.00) for the alkyl chain density in close-packed *bulk* phases of long chain alkanols (*20*). However, the normalized electron density of 0.79 near the –CH$_3$ group corresponds closely to the value of 0.81 for bulk liquid alkyl chains near their freezing point (*20*). In the bulk it is believed that liquid *n*-alkanes near the freezing point are arranged in quasi-lamella regions

88

with gauche conformations to cause chain disorder, while still maintaining a good alignment of neighboring molecules (*21,22*).

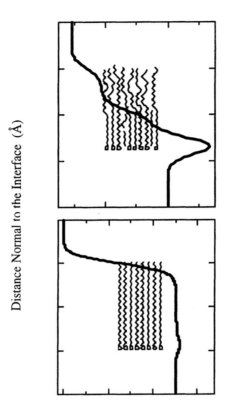

Figure 4. Electron density profile normal to the interface for a triacontanol monolayer at the (A) water-hexane and (B) water-vapor interfaces. At the water-hexane interface the molecules have a well defined disorder described in the text. At the water-vapor interface the molecules are nearly rigid rods.

Although not unambiguously proven, it seems that a similar arrangement occurs in the part of the triacontanol chain near the $-CH_3$. A disordered liquid-like part of the chain would occupy a surface area of at least 23.4 $Å^2$/chain, an increase of ~25% over the close-packed area/molecule of 18.5 $Å^2$/chain. In the region of the alkyl chain close to the $-CH_2OH$ group, the chain has to be more ordered to yield a normalized electron density of 0.95. In addition, either water or hexane must be mixed into this part of the monolayer to properly account for the available volume in the monolayer region. A compelling arrangement is a well ordered chain near the $-CH_2OH$ group that is mixed with hexane (20% to 25% by volume) and a progressively more disordered chain towards the $-CH_3$ end of the tailgroup (as illustrated in Figure 4A). This arrangement is quantitatively consistent with our measured densities.

The disorder in the chain will account, at least partially, for the overall monolayer thickness (36±2Å) being less than the length of an all-trans triacontanol molecule (40.7Å). For example, a single kink defect (gtg' or g'tg conformation) will maintain the overall chain orientation while reducing its length by 0.6Å to 0.7Å. In addition, the triacontanol molecules may be tilted slightly from the normal to the interface, but the reflectivity does not directly tilt if the monolayer was sufficiently ordered, the background scattering from the bulk precludes its measurement probe this tilt. Although grazing incidence diffraction could directly probe this

Table I Fitting Parameters for Triacontanol Monolayers

Interface	Layer 1		Layer 2		Layer 3				
	L (Å)	ρ_{max}	L_2 (Å)	ρ_2	L_3 (Å)	ρ_3	σ	σ_{cap}	L_{total} (Å)
Water-Hexane T=24.52°C	5 ±4	1.13 ±0.01	13 ±2	0.95 +0.01/-0.02	18 ±1	0.79 ±0.01	3.6 ±0.3	3.9	36 ±2
Water-Vapor T=24 °C	5 ±4	1.04 ±0.01	24 ±5	1.014 ±0.003	11 ±4	0.99 +0.01/-0.04	3.25 +0.1/-0.25	3.24	40 ±2

Table I shows fitting parameters for two of the fits in Figure. 3 (similar parameters are obtained for the water-hexane data at T = 24.07°C and T=24.52°C). Layer 1 is the head-group region ($-CH_2OH$), layers 2 and 3 are for the tail-group region; layers are ordered water–1–2–3–hexane (or vapor); L is the layer thickness; ρ is the electron density; σ is the interfacial roughness; σ_{cap} is the roughness calculated from the measured interfacial tension using the capillary wave theory. The electron densities are normalized to the value for bulk water ($0.333e^-/\text{Å}^3$). The normalized hexane density is 0.692. The length of an all-trans triacontanol molecule is 40.7Å. For the head-group (layer 1) the maximum electron density is quoted rather than the density of the layer because the density and layer thickness fitting parameters are strongly correlated for this thin layer, but the resultant profile is well determined. The error bar on the total thickness, L_{total}, is small because of correlations between fitting parameters.

Triacontanol Monolayer at the Water-Vapor Interface

The bottom curve of Figure 3 shows an x-ray reflectivity measurement of a triacontanol monolayer spread at the water-vapor interface. The monolayer was spread on a home built Langmuir trough (23) from a 2.1 mM chloroform solution at a low density (50Å^2/molecule), then compression cycled eight times between surface pressures of 0 mN/m and 25 mN/m (with addition of pure chloroform at high pressures) to create a stable, homogeneous monolayer.

Figure 4B illustrates the electron density profile for the triacontanol monolayer at the water-vapor interface. This indicates that the molecular ordering at the water-hexane interface is very different from that at the water-vapor interface (see Table 1 for fitting parameters). The overall thickness of the monolayer at the water-vapor interface is $40\pm2\text{Å}$, nearly identical to the length of an all-trans triacontanol molecule, calculated to be 40.7Å (20,24,25). Most of the region of the monolayer corresponding to the alkyl chain has a normalized electron density of 1.014 ± 0.003 (normalized to the value for water of 0.333 e$^-$ /Å^3). This is comparable to literature values of 1.03 and 1.00 measured for the alkyl chain density in close-packed *bulk* phases of long chain alkanols (20) and indicates that most of the chain is close-packed. The fit shown for the bottom curve in Figure 3 requires a slightly lower electron density (0.99) towards the $-$CH$_3$ group. This is consistent with molecular dynamics simulations that predict a small percentage of gauche conformations in these nearly rigid rod monolayers with the gauche defects concentrated near the $-$CH$_3$ end (26,27).

This analysis demonstrates that the triacontanol monolayer at the water-vapor interface is close packed with nearly all-trans and nearly upright molecules (normal to the interface). In contrast, the triacontanol monolayer at the water-hexane interface has a well defined disorder along the chain with hexane mixed into the region of the monolayer.

The electron density in the head-group region ($-$CH$_2$OH) is larger at the water-hexane interface ($\rho_{max} = 1.13\pm0.01$) than at the water-vapor interface ($\rho_{max} = 1.04\pm0.01$). Since the area per head-group is larger at the water-hexane interface due to disorder in the monolayer, the additional electron density cannot be attributed to closer packing of head-groups. In addition, the higher density is not likely due to the interaction of water with hexane since x-ray measurements of the pure water-hexane interface do not reveal an enhanced interfacial density of water (28). The lower density of head-groups at the water-hexane interface may allow for water penetration into the head-group region, which then induces a higher density in this region. It has been recently proposed from experiments and molecular dynamics simulations that water in the first hydration shell of lysozyme and other proteins has an average density approximately 10% to 20% greater than the bulk density, similar in magnitude to the change we observed, and that this higher density is due to orientational ordering of water molecules in depressions on the protein surface (29,30). Orientational ordering of water has been observed near charged surfactants at the water–CCl$_4$ interface (31). We suggest that the additional space between the triacontanol –CH$_2$OH groups at the water-hexane interface allows for orientational ordering of water by the polar – CH$_2$OH.

Temperature Variation of Monolayers at the Water-Hexane Interface

The reflectivity data for the triacontanol monolayer at the water-hexane interface at high temperature (T=45.02°C, data not shown) are featureless and indicate that most of the triacontanol molecules have desorbed from the interface into the bulk hexane. These high temperature data are well fit by a one parameter fit to a model of a simple interface between two bulk media (i.e., no monolayer), with the interfacial width (or roughness) as the only fitting parameter. Similar high temperature data are obtained for all four alkanols studied. The width varies between 5.0Å and 5.5Å for these four alkanols and is much larger than the expected value of approximately 3.6Å computed from capillary wave theory and the measured interfacial tensions. Since it is known that the measured interfacial width of the water-hexane interface is the same as the prediction of the capillary wave theory, it is likely that the larger interfacial width is due to a low density of triacontanol molecules that remain at the interface at high temperatures (28). It is interesting to note that the measured interfacial width is similar to the value of 5.7±0.2Å measured between bulk docosane (C$_{22}$ H$_{46}$) and water at T = 44.6°C (32). The capillary wave contribution to the interfacial width of the neat water-docosane interface is 3.5Å and the larger width at the water-docosane interface is believed to be due to an intrinsic width determined by the bulk correlation length of the flexible docosane (32). A similar effect may be present in the high temperature phase of our alkanol systems.

We have also measured x-ray reflectivity from the triacontanol system for many other temperatures between 24°C and 45°C. A subset of these data are shown in Figure 5 which illustrates the reflectivity at fixed Q_z (= 0.15 Å^{-1}) as a function of temperature . The sharp change at T ≈ 29°C corresponds to the transition (the offset of 1.5°C from the transition temperature given by the tension measurements is attributed to a small difference in the sample concentration for the different measurements).

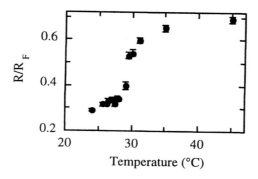

Figure 5. X-ray reflectivity normalized to the Fresnel reflectivity at a fixed Q_z ($=0.15 Å^{-1}$) as a function of temperature for a triacontanol monolayer at the water-hexane interface. The sharp change at $T \cdot 29°C$ is the phase transition

A full interpretation of these reflectivity measurements as a function of temperature (to be published elsewhere) indicate that below the transition temperature the interface is covered by a high density of domains of the condensed phase with a relatively sharp transition to an interface with a low density of these domains. Variation of the surface density occurs primarily near the transition temperature. Similar results for the temperature dependence were observed for the other three alkanol systems we studied and also for partially fluorinated alkanols as previously reported (33,34). The temperature variation of the reflectivity for the triacontanol system indicates that at low temperatures (T ≈ 24°C, as for the data in Figure 3) the nearly fully covered interface may contain a small fraction (~10%) of gaseous regions. If present, these gaseous regions will not affect our qualitative conclusions regarding the molecular ordering of triacontanol in the low temperature condensed phase, but will quantitatively modify them slightly.

Summary

Our conclusion regarding a progressive distribution of gauche conformations along the alkyl chain that increases away from the $-CH_2OH$ group is sensible considering the constraints of placing a headgroup at the water-hexane interface and orienting the alkyl chain towards the hexane. A similar effect was observed, though to a lesser extent, in molecular dynamics simulations of Langmuir monolayers in which there are a small number of gauche conformations that appear primarily at the chain ends (26,35). Also, in the liquid phase of bulk alkanols far from the freezing point, NMR experiments have shown that a 7 carbon long region of the alkyl chain near the $-CH_2OH$ group of 1-dodecanol ($CH_3(CH_2)_{11}OH$) has a constant degree of order with increasing chain disorder further out along the chain (20). In the bulk liquid, hydrogen bonding between nearest neighbor $-CH_2OH$ groups provides the constraint that establishes the pattern of chain ordering. This is a weaker constraint than that provided by a flat water-hexane interface.

These studies demonstrate that long-chain alkanols that form rigid-rod monolayers at the water-vapor interface instead form monolayers at the water-hexane interface with a well defined disorder. The molecular ordering is characterized by chain disorder that progressively varies from relatively well ordered chains near the $-CH_2OH$ group to liquid-like chain densities at the $-CH_3$ end of the molecule. Solvent mixing into the monolayer accompanies this chain disorder. For the monolayer at the water-hexane interface, but not at the water-vapor interface, the density in the head-group region is larger than bulk water by 10 -15%. These results have important implications for understanding water structuring near biological macromolecules and alkyl chain ordering in many systems.

Acknowledgments

We acknowledge the valuable assistance of Binhua Lin (UofChicago), Guangming Luo (UIC), Sai Venkatesh Pingali (UIC), Mati Meron (UofChicago), Tim Graber (UofChicago), David Schultz (UIC), and Jeff Gebhardt (UofChicago) at the APS. Financial support is gratefully acknowledged from the U.S. National Science Foundation DMR (for MLS), National Science Foundation Chemistry (for ChemMatCARS), and U.S. Department of Energy (for ChemMatCARS, Brookhaven, and Argonne National Laboratories).

Literature Cited

1. Yue, B. Y.; Jackson, C. M.; Taylor, J. A. G.; Mingins, J.; Pethica, B. A. *J. Chem. Soc. Faraday I* **1976**, *72*, 2685.
2. Mingins, J.; Taylor, J. A. G.; Pethica, B. A.; Jackson, C. M.; Yue, B. Y. T. *J. Chem. Soc. Faraday Trans. 1* **1982**, *78*, 323.

94

3. Hayami, Y.; Uemura, A.; Ikeda, N.; Aratono, M.; Motomura, K. *J. Coll. Int. Sci.* **1995**, *172*, 142.
4. Takiue, T.; Yanata, A.; Ikeda, N.; Hayami, Y.; Motomura, K.; Aratono, M. *J. Phys. Chem.* **1996**, *100*, 20122.
5. Richmond, G. L. *Chem. Rev.* **2002**, *12*, 2693.
6. Schlossman, M. L. *Curr. Opin. Coll. Int. Sci.* **2002**, *7*, 235.
7. Zhang, Z.; Mitrinovic, D. M.; Williams, S. M.; Huang, Z.; Schlossman, M. L. *J. Chem. Phys.* **1999**, *110*, 7421.
8. Adamson, A. W. *Physical Chemistry of Surfaces*; 5th ed.; John Wiley & Sons: New York, NY, 1990.
9. Schlossman, M. L.; Synal, D.; Guan, Y.; Meron, M.; Shea-McCarthy, G.; Huang, Z.; Acero, A.; Williams, S. M.; Rice, S. A.; Viccaro, P. J. *Rev. Sci. Instrum.* **1997**, *68*, 4372.
10. Schlossman, M. L.; Pershan, P. S. *Light Scattering by Liquid Surfaces and Complementary Techniques*; ed.; Langevin, D., Ed.; Marcel Dekker Inc.: New York, NY, 1992, pp 365-403.
11. Tidswell, I. M.; Ocko, B. M.; Pershan, P. S.; Wasserman, S. R.; Whitesides, G. M.; Axe, J. D. *Phys. Rev. B* **1990**, *41*, 1111.
12. Mitrinovic, D. M.; Tikhonov, A. M.; Li, M.; Huang, Z.; Schlossman, M. L. *Phys. Rev. Lett.* **2000**, *85*, 582.
13. Pershan, P. S. *Far. Disc. Chem. Soc.* **1990**, *89*, 231.
14. Born, M.; Wolf, E. *Principles of Optics*; 6 ed.; Pergamon Press: Oxford, England, 1980.
15. Jasper, J. J.; Houseman, B. L. *J. Phys. Chem.* **1963**, *67*, 1548.
16. Matubayasi, N.; Motomura, K.; Aratono, M.; Matuura, R. *Bull. Chem. Soc. Jpn.* **1978**, *51*, 2800.
17. Lin, M.; Firpo, J.-L.; Mansoura, P.; Baret, J. F. *J. Chem. Phys.* **1979**, *71*, 2202.
18. Ocko, B. M.; Wu, X. Z.; Sirota, E. B.; Sinha, S. K.; Gang, O.; Deutsch, M. *Phys. Rev. E* **1997**, *55*, 3164.
19. Tikhonov, A. M.; Schlossman, M. L. *J. Phys. Chem. B* **2003**, in press.
20. Small, D. M. *The Physical Chemistry of Lipids*; Plenum: New York, 1986.
21. Stewart, G. W.; Morrow, R. M. *Phys. Rev.* **1927**, *30*, 232.
22. Brady, G. W.; Fein, D. B. *J. Appl. Crystallogr.* **1975**, *8*, 261.
23. Acero, A. A.; Li, M.; Lin, B.; Rice, S. A.; Goldman, M.; Azouz, I. B.; Goudot, A.; Rondelez, F. *J. Chem. Phys.* **1993**, *99*, 7214.
24. 40.7Å = 29x1.27Å (C—C) + 1.5Å (—CH3) + 2.4Å (—CH2OH).
25. Israelachvili, J. N. *Intermolecular and Surface Forces*; Academic Press: London, England, 1992.
26. Harris, J.; Rice, S. A. *J. Chem. Phys.* **1988**, *89*, 5898.
27. Bareman, J. P.; Cardini, G.; Klein, M. L. *Phys. Rev. Lett.* **1988**, *60*, 2152.
28. Mitrinovic, D. M.; Zhang, Z.; Williams, S. M.; Huang, Z.; Schlossman, M. L. *J. Phys. Chem. B* **1999**, *103*, 1779.

29. Svergun, D. I.; Richard, S.; Koch, M. H. J.; Sayers, Z.; Kuprin, S.; Zaccai, G. *Proc. Natl. Acad. Sci. USA* **1998**, *95*, 2267.
30. Merzel, F.; Smith, J. C. *Proc. Natl. Acad. Sci. USA* **2002**, *99*, 5378.
31. Gragson, D. E.; Richmond, G. L. *J. Phys. Chem. B* **1998**, *102*, 569.
32. Tikhonov, A. M.; Mitrinovic, D. M.; Li, M.; Huang, Z.; Schlossman, M. L. *J. Phys. Chem. B* **2000**, *104*, 6336.
33. Tikhonov, A. M.; Li, M.; Mitrinovic, D. M.; Schlossman, M. L. *J. Phys. Chem. B* **2001**, *105*, 8065.
34. Tikhonov, A. M.; Li, M.; Schlossman, M. L., BNL National Synchrotron Light Source Activity Report 2001, http://nslsweb.nsls.bnl.gov/nsls/pubs/actrpt/2001/sec2_scihi_softmat_tikhonov.pdf, 2002.
35. Kaganer, V. M.; Mohwald, H.; Dutta, P. *Rev. Mod. Phys.* **1999**, *71*, 779.

Chapter 7

The Structure and Composition of Mixed Surfactants at Interfaces and in Micelles

J. Penfold[1], R. K. Thomas[2], E. Staples[3], and I. Tucker[3]

[1]CCLRC, Rutherford Appleton Laboratory, Chilton, Didcot, Oxon, United Kingdom
[2]Physical and Theoretical Chemistry, Oxford University, South Parks Road, Oxford, United Kingdom
[3]Unilever Research, Port Sunlight Laboratory, Quarry Road East, Bebington, Wirral, United Kingdom

The article describes the use of specular neutron reflectivity and small angle neutron scattering, SANS, in combination with H / D isotopic substitution, to investigate the composition and structure of mixed surfactants at interfaces and in micelles. The application of these techniques are illustrated in two broad areas of surfactant mixing; which in the anionic / nonionic surfactant mixture investigate the extent of non-ideal mixing over a broad range of conditions, and in the nonionic mixture quantify the role of structural changes at interfaces on mixing.

Results for the nonionic / anionic surfactant mixture of hexaethylene glycol monododecyl ether, $C_{12}E_6$ and sodium dodecyl sulphate, SDS, over a wide range of conditions, at the air / water, liquid /solid, and liquid / liquid interfaces, and in micelles, are presented and compared with theoretical predictions. The variations in the departure from ideal mixing that are observed are discussed, and used to highlight the limitations of some of the current theories of surfactant mixing.

Detailed structural measurements of the mixed nonionic surfactant monolayers of triethylene glycol monododecyl ether, $C_{12}E_3$ / octaethylene glycol monododecyl ether, $C_{12}E_8$ and hexaethylene glycol mono octadecyl ether, $C_{10}E_6$ / hexaethylene glycol monotetradecyl ether, $C_{14}E_6$ at the air / water interface are presented; and quantify the extent to which structural changes occur in a system that is close to ideal. The consequences of different ethylene oxide and alkyl chain lengths on the monolayer structure are discussed. Even for surfactant mixtures which behave closely to ideal, the results illustrate some of the additional factors that need to be taken into account in developments of the theories of surfactant mixing.

Introduction

The study of surfactant mixtures is of considerable current interest, for both practical and academic reasons (1,2). The many domestic, technological and industrial applications of surfactants invariably involve mixtures. This is because mixtures provide some synergistic improvement in properties or performance, or because commercial surfactants are inherently impure. That is, they often contain mixtures of different alkyl chain lengths, different isomers, and in the case of nonionic surfactants different ethylene oxide chain lengths. There is a rich background of information available on surfactant mixing, as a result of the classical measurements such as surface tension (3), and the extensive thermodynamic treatments based on the pseudo phase approximation (4). In recent years there has been a resurgence of interest in the nature of surfactant mixing. This is in part due to the continuing relevance of mixtures, an incomplete understanding of their behaviour, recent theoretical developments (5,6) and the application of new experimental techniques, such as neutron reflectometry (7), SANS (8), NMR (9), optical spectroscopy methods (10,11), and atomic force microscopy, AFM (12).

This paper focuses on the use of neutron scattering techniques, neutron reflectivity and SANS, in investigating surfactant mixing. Neutron reflectivity and SANS, in combination with hydrogen-deuterium isotopic substitution, provide the opportunity to obtain compositional and structural information over a wide range of solution compositions and concentrations; the former at the air-solution (13) and solid-solution (14) interfaces, and the latter at the liquid-liquid interface (15), and in micellar aggregates (8). The availability of the detailed information these methods can obtain provides the opportunity to confront theoretical treatments. For example, recent neutron reflectivity and surface tension measurements on the adsorption of SDS / dodecyl dimethylamino acetate, Betaine (16), and SDS / n-dodecyl-β-D-maltoside, Maltoside (17) mixtures at the air-solution interface are not well described by the pseudo phase approximation. In the Pseudo phase approximation, or Regular

Solution Theory, RST, approach, the departure from ideality is characterized by a single interaction parameter, β, and a central assumption is that the excess entropy of mixing is zero (18). The correlation between structure and composition in mixed surfactants systems suggests that structural changes and changes in hydration levels at the interface and in aggregates are important. Modifications to the RST approach have been developed (19) to address such issues. More recently Blankschtein et al (1,5,20) have developed a more fundamental approach, aimed at removing the phenomenological nature of RST, including specific molecular interactions, whilst retaining thermodynamic rigor. Their approach, and the related work of Hines (6) specifically for interfaces, is based on a two-dimensional lattice approach which also includes molecular detail.

In the first part of this paper we present data for the mixing of the anionic and nonionic surfactants of SDS and $C_{12}E_6$, at different interfaces and in micelles. This has provided an opportunity to explore the extent to which RST is applicable and highlighted the deficiencies of the thermodynamic based treatments. In the second part we focus on detailed structural measurements of the mixed monolayer of two different nonionic surfactant mixtures at the air-water interface. In a system close to ideal mixing it provides an opportunity to quantify the detailed structural changes that can arise on mixing, and provides important input for further theoretical refinements and developments.

Experimental Details

In recent years the neutron scattering techniques of neutron reflectivity and SANS have been extensively used in the study of surfactant adsorption at interfaces (7) and of surfactant aggregates (21). A central feature of the application of these techniques is the use of hydrogen / deuterium isotopic substitution. The deuterium labeling provides the ability to alter the visibility of selected components or parts of components, whilst leaving the chemistry essentially unaltered. It provides the sensitivity and selectivity for the study of mixtures, and enables both compositional information and structural details to be obtained.

Neutron Reflectivity

Specular neutron reflectivity provides information about inhomogeneities normal to a surface or interface (22). The basis of a neutron reflectivity measurement is that the variation of specular reflection with Q (the wave-vector transfer, defined as $Q = \dfrac{4\pi}{\lambda} \sin \theta$, where θ is the glancing angle of incidence, and λ the neutron wavelength) is simply related to the composition or concentration profile in the direction normal to the surface. In the kinematic approximation (23) the specular reflectivity is then given by,

$$R(Q) = \frac{16\pi^2}{Q^2} |\rho(Q)|^2 \qquad (1)$$

where $\rho(Q)$ is the one-dimensional Fourier transform of $\rho(z)$, the average scattering length density profile in the direction normal to the interface,

$$\rho(Q) = \int_{-\infty}^{+\infty} \rho(z) \exp(iQz) dz \qquad (2)$$

$$\rho(z) = \sum_i N_i(z) b_i \qquad (3)$$

where N_i is the number density of species i, and b_i its scattering length.

Hydrogen and deuterium have vastly different scattering lengths (and are of opposite sign), hence H / D isotopic substitution can be used to manipulate the scattering length density distribution, $\rho(z)$, in organic systems. This is particularly powerful in determining surface structure and in investigating mixtures, where by selective deuteration particular components of fragments can be highlighted or isolated. At the simplest level, neutron reflectivity can be used to determine adsorbed amounts at interfaces in single and multi-component mixtures, with good accuracy over a wide concentration range. For deuterium labeled surfactant in null reflecting water (nrw, water with a scattering length density or refractive index the same as air, 92 volume % H_2O / 8 volume % D_2O mixture) the reflectivity arises only from the layer of deuterated surfactant at the interface. The reflectivity can be analysed to sufficient accuracy as a single layer of uniform or homogeneous composition (24). Using the optical matrix method (25) this gives rise to a scattering length density and thickness such that,

$$\rho = \frac{b}{\tau A} \qquad (4)$$

where b is the scattering length density / adsorbed molecule, A is the area /molecule of the adsorbed surfactant monolayer, and ρ, τ are the scattering length density and thickness obtained from the model fit. For mixtures, where each component in the mixture is selectively deuterated in turn, the resulting reflectivity is then treated as arising from a uniform single layer; such that for a binary mixture,

$$\rho = \frac{b_1}{A_1 \tau} + \frac{b_2}{A_2 \tau} \qquad (5)$$

where b_i, A_i are the scattering length and area/molecule of each component.

Detailed structural information on the monolayer can be obtained by a direct method of analysis based on the kinematic approximation (26), which provides information about the volume fraction distributions of the individually labeled components. Writing the scattering length density profile, $\rho(z)$, for a simple binary surfactant mixture at the air-water interface we have,

$$\rho(z) = b_1 n_1(z) + b_2 n_2(z) + b_s n_s(z) \qquad (6)$$

where b_i, $n_i(z)$ are the scattering lengths and number density profiles of the two surfactant components, and s refers to the solvent. From equation 1 this gives,

$$R(Q) = \frac{16\pi^2}{Q^2} \left[\begin{array}{l} b_1^2 h_{11}(Q) + b_2^2 h_{22}(Q) + b_s^2 h_{ss}(Q) + \\ 2b_1 b_2 h_{12}(Q) + 2b_1 b_s h_{1s}(Q) + 2b_2 b_s h_{2s}(Q) \end{array} \right] \qquad (7)$$

The 3 h_{ii} factors are the self-partial structure factors, $h_{ii} = |\hat{n}_{ii}(Q)|^2$, the 3 cross-partial structure factors, h_{ij}, are given by $h_{ij}(Q) = \text{Re}\{\hat{n}_i(Q)\hat{n}_j(Q)\}$ and $n_i(Q)$ is the one-dimensional Fourier transform of $n_i(z)$. The self-partial structure factors relate directly to the distributions of the individual components at the interface, whereas the cross- partial structure factors relate to their relative positions at the interface. For a series of different reflectivity measurements, using differently labeled combinations, the different partial structure factors can be extracted. Simple analytical functions describe well those partial structure factors under most circumstances (27), and this approach has been applied successfully to a range of different systems (7).

The neutron reflectivity measurements have been made on the SURF reflectometer (28) at the ISIS pulsed neutron source, using the 'white beam time of flight' method. That is, the required Q range is obtained using a fixed angle of incidence (in the range 0.35 to 1.8°) and a neutron wavelength band of 1 to 6.7 Å. Details of the experimental procedures used are described in detail elsewhere (29).

The data in figure 1 shows typical data, measured for a binary surfactant mixture at the air-water interface in nrw, for the nonionic mixture $C_{12}E_3$ / $C_{12}E_8$. The combinations h,d and d,h (h,d refers to hydrogenous or deuterated alkyl chains) give information about the amount and extent of the $C_{12}E_3$ and $C_{12}E_8$ respectively at the interface, whereas the d,d combination provides information about the total adsorbed amount. The solid lines are calculated curves for fits to a model of a single layer with uniform composition (the details of the fits are included in the figure caption).

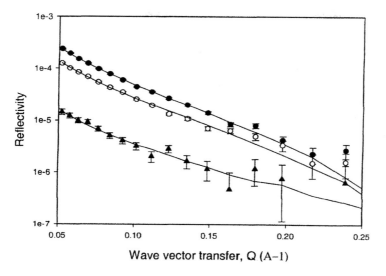

Figure 1: Specular reflectivity for 5 x 10^{-4}M 50 / 50 mole % C$_{12}$E$_3$ / C$_{12}$E$_8$ in nrw, ($\bullet$) dd, τ= 19.7 Å, ρ= 3.54 x 10^{-6}Å^{-2}, (o) dh, τ=18.7 Å, ρ=2.74 x 10^{-6} Å^{-2}, (Δ) hd, τ=14.1 Å, ρ=1.19 x 10^{-6} Å^{-2}

Small Angle Neutron Scattering, SANS

SANS is an established technique for the study of micellar structure and composition (21). We have determined micellar structure by the analysis of the scattering data using a well-established structural model for globular micelles (30). For globular poly-disperse interacting micelles the scattered intensity at small Q can be written in the 'decoupling approximation' as (30),

$$I(Q) = N\left[S(Q)\left|\langle F(Q)\rangle_Q\right|^2 + \left\langle |F(Q)|^2\right\rangle_Q - \left|\langle F(Q)\rangle_Q\right|^2 \right] \qquad (8)$$

where the averages $< >_Q$ denote averages over all particle sizes and orientations with respect to the scattering vector Q, N is micelle number density (where N=CNa/v. C=C$_0$-cmc, v is the aggregation number, and cmc the critical micellar concentration). S(Q) the inter-micellar structure factor (reflecting the strength of the intermicellar interactions), and F(Q) the micelle form-factor. The micelles are modeled using the standard core + shell model (30) where,

$$F(Q) = V_1(\rho_1 - \rho_2)F_0(QR_1) + V_2(\rho_2 - \rho_s)F_0(QR_2) \qquad (9)$$

where $V_i = 4\pi R_i^3$, $F_0(QR) = 3j_1(QR)/QR = 3[\sin(QR) - QR\cos(QR)]/(QR)^3$, ρ_1, ρ_2, and ρ_s are the core, shell and solvent scattering length densities, and R_1 and R_2 the core and outer radii (determined from the model constraints). Poly-dispersity, σ, is included using a Schultz size distribution (31), and $S(Q)$ is calculated within the RMSA approximation for a repulsive screened coulombic inter-micellar potential (32) (at low values of surface charge, $S(Q)$ approximates to the Percus-Yevick hard sphere structure factor (33)). The key parameters obtained from such an analysis are the micelle aggregation number, υ, and the surface charge, z; and the modeling approach is described in more detail elsewhere. The absolute value of the scattered intensity and its variation with H / D isotopic substitution are important constraints and enable the surfactant composition in mixed micelles to be determined to good accuracy (13). For a dilute micellar solution of mixed surfactants, in the limit of small Q (where $S(Q){\sim}P(Q){\sim}1.0$) the scattered intensity is (13),

$$I(Q) \approx \sum_i N_i V_i^2 \left(\rho_{ip} - \rho_s \right)^2 \qquad (10)$$

V is the micelle 'dry' volume, ρ_p and ρ_s are the micelle and solution scattering length densities, and the summation is over all micellar compositions. For a binary surfactant mixture two different measurements in D_2O, one with both surfactants hydrogenous, and one with one of the surfactants deuterium labeled, provides directly (by the application of equation 10) a ratio which is related to the micellar composition. In practice this approach works even for finite concentrations where $S(Q)$ is not ${\sim}1.0$. as $S(Q)$ does not vary with isotopic content of the micelle (13).

The SANS measurements reported here were made on the LOQ diffractometer (34) at the ISIS pulsed neutron source, and on the D22 diffractometer at the Institute Laue Langevin, Grenoble (35). The LOQ measurements were made using the 'white beam time of flight' method, which gave rise to a Q range of 0.02 to 0.2 Å^{-1}. The D22 measurements were made at a fixed wavelength of 4.6 Å and a flight path between 2.5 and 10.0 m, to give a Q range of 0.004 to 0.35 Å^{-1}. Samples were measured in 1 or 2 mm quartz cells, and the data corrected using standard procedures (36).

Results and Discussion

SDS / $C_{12}E_6$ surfactant mixtures

Neutron reflectivity and SANS have been used to study the mixing of SDS and $C_{12}E_6$ at different interfaces and in micelles, over a wide range of concentrations. This has provided the opportunity to explore the extent to which RST is applicable in such mixtures. Measurements have been made exclusively at concentrations > cmc, and in 0.1M NaCl (with the exception of the liquid / liquid interface where 0.01M NaCl was used). The electrolyte ensures that the cmc's of the SDS and $C_{12}E_6$ are more comparable, and that the departure from ideality is not dramatic.

Air-water interface

The composition of the adsorbed layer of SDS / $C_{12}E_6$ has been measured at the air-water interface over a wide range of concentrations, from ~ 2 x 10^{-4} M to ~ 100 times the cmc, and for solution compositions of 70 mole% SDS / 30 mole% $C_{12}E_6$ and an equi-molar mixture (*13*). The data in figure 2 shows the variation in surface composition as a function of concentration for a 70 mole % SDS / 30 mole % $C_{12}E_6$ mixture.

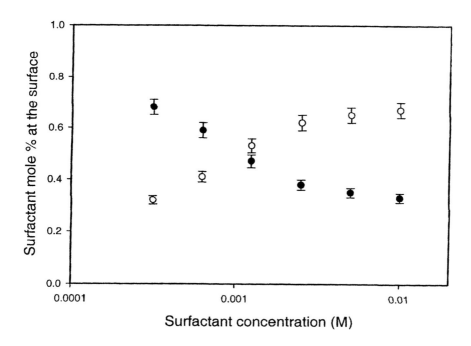

Figure 2 : Mole % SDS (o), and mole % $C_{12}E_6$ (•) as a function of surfactant concentration for a 70 mole % SDS / 30 mole % $C_{12}E_6$ solution mixture in 0.1M NaCl

The variation in surface composition with surfactant concentration follows the qualitatively expected trend. Just above the cmc the surface is rich in the most surface active component ($C_{12}E_6$ in this case), and as the surfactant concentration increases the surface evolves towards the solution composition. The relative amount of SDS at the surface increases whilst that of the $C_{12}E_6$ decreases. A direct comparison shows that the evolution of composition with concentration cannot be predicted by a single interaction parameter (*13*), and this is shown in figure 3 where

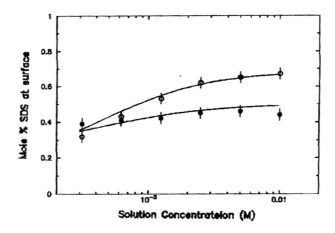

Figure 3: Variation of Mole % of SDS with surfactant concentration at the air-water interface for (o) 70 / 30 SDS / $C_{12}E_6$ and (•) 50 / 50 SDS / $C_{12}E_6$, the solid lines are RST calculations for β -2.5 and -1.5 respectively.

the interaction parameter is -2.5 for the 70 / 30 mole ratio solution and -1.5 for the 50 / 50 mole ratio solution.

The variation in the interaction parameter with composition indicates that the departure from ideality at the surface is more pronounced for solutions richer in SDS. This is consistent with a more pronounced interaction between SDS head groups in the monolayer, and has been implied in other studies on related systems (*19*). This is not strictly within the scope of RST, which assumes a single interaction parameter to describe the departure from ideality.

Liquid-solid interface

Neutron reflectivity has been used to study the adsorption of SDS / $C_{12}E_6$ surfactant mixtures at the hydrophilic silica-solution interface. The hydrophilic surface of silica is net negatively charged, and hence SDS does not adsorb in the absence of other surfactants due to its like charge (*14*). At a surfactant concentration of 10^{-3} M (>cmc) the adsorption of SDS / $C_{12}E_6$ mixtures at the hydrophilic surface has been investigated (*14*). Variations in the deuterium labeling of the surfactants and solvent enabled the structure of the mixed surface bilayer and its composition to be determined (see figure 4).

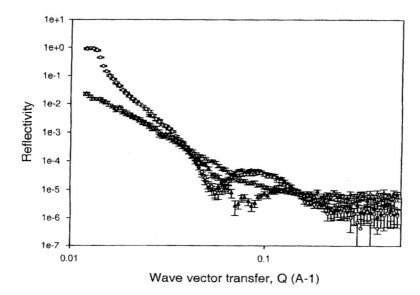

Figure 4 : Neutron reflectivity at the hydrophilic liquid-solid interface for 20 / 80 mole ratio SDS / $C_{12}E_6$ at a solution concentration of 10^{-3} M, ($\bullet$) hh, (o) dh, in D_2O, and ($\blacktriangle$) hd, and ($\triangle$) dh in H_2O.

For the isotopic combination h,h (h-SDS / h-$C_{12}E_6$) in D_2O both surfactants are visible, whereas for the combination d,h (d-SDS / h-$C_{12}E_6$) only the $C_{12}E_6$ is visible. The differénces in figure 4 indicate that qualitatively the surface layer is predominantly $C_{12}E_6$. The measurements in H_2O provide complementary information. Figure 5 shows the variation in the adsorbed amount and the surface composition with solution composition. The total amount adsorbed decreases markedly with increasing SDS mole % in solution, and for solutions richer than 50 mole % SDS there is essentially no measurable adsorption. Furthermore as a result of the specific interaction with the hydrophilic surface the surface composition is now markedly different to that expected on the basis of RST. The detailed structure of the adsorbed bilayer (obtained from an analysis of the different isotopic combinations measured) (*14*) also reflects the lack of affinity of the SDS for the surface; and there is systematically more SDS in the region of the bilayer adjacent to the solution phase.

Liquid-Liquid Interface

Measuring surfactant adsorption at the liquid-liquid interface by neutron reflection methods has proved difficult, and a different approach using SANS has been developed. We have produced stable and reproducible hexadecane in water

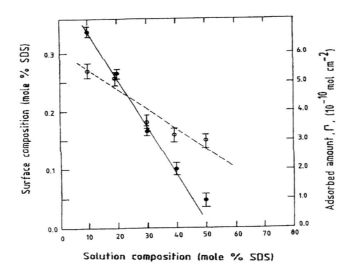

Figure 5 : Variation in surface composition (o) (mole % SDS) and adsorbed amount (•) as a function of solution composition for 10^{-3} M SDS / $C_{12}E_6$ in 0.1M NaCl and at a pH of 2.4, at the hydrophilic silica - solution interface.

emulsions with a diameter ~ 0.2 microns. For hexadecane index matched to D_2O the scattering arises only from the adsorbed layer of hydrogeneous surfactant at the interface. A wide range of mixed surfactant compositions and concentrations can be adsorbed at the emulsion droplet surface without destabilization. It is important to note that compared to the other measurements reported here for SDS / $C_{12}E_6$ mixtures, the measurements were made in 0.01M NaCl. The lower level of electrolyte was required to ensure stable emulsions.

As with the liquid-solid interface selectively deuterium labeling each surfactant component effectively matches it to the oil and solution phases, hence enabling the total adsorption and composition of the adsorbed layer to be determined. Measurements were made for a 70 / 30 mole ratio SDS / $C_{12}E_6$ mixture over a range of surfactant concentrations (15), see figure 6. The marked change in the shape of the scattering that arises for surfactant concentrations > 8.6 mM (6th curve from the bottom in figure 6) indicates the onset of micelle formation in solution, and the scattering curves for 11.4 and 14.3 mM arise from a combination of the adsorbed layer at the emulsion interface and from mixed SDS / $C_{12}E_6$ micelles in the aqueous phase. This provides the opportunity to characterize the adsorbed surfactant layer at the emulsion surface and the micellar phase that is in equilibrium with the emulsion.

Using an approach similar to that described for micellar solutions (equation 10), measurements for the isotropically labeled combinations, hh and hd, provide a

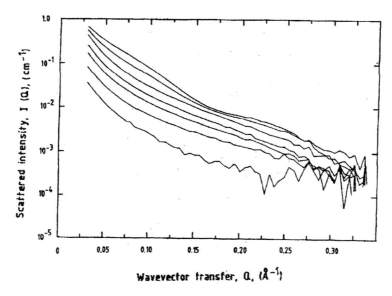

Figure 6 : Scattered Intensity, $I(Q)$ (cm^{-1}), for a 6.4 volume % d / h hexadecane in D_2O emulsion with SDS / $C_{12}E_6$ (70 / 30 mole ratio) in the concentration range 1.8 to 14.3 mM h-SDS / h-$C_{12}E_6$ in 0.01M NaCl.

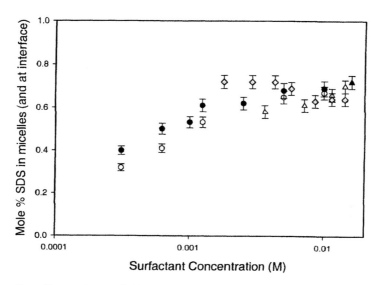

Figure 7 : Comparison of the composition of the oil-water ($\Diamond$), air-water (o) interfaces, mixed micelles ($\bullet$), mixed micelles in 0.01M NaCl ($\triangle$);, and mixed micelles in equilibrium with the emulsion ($\blacktriangle$), for 70 / 30 mole ratio SDS / $C_{12}E_6$ mixture in 0.1M NaCl (except that the emulsion measurements were made in 0.01M NaCl). -

measure of the composition of the adsorbed layer at the oil (hexadecane)–water interface (*15*). These are compared in figure 7 with those previously obtained at the air-water interface (*13*), in mixed micelles (*8,13,37*), and for the micelles in equilibrium with the emulsion.

Although a direct comparison is not entirely valid, because of the change in electrolyte concentration, the behaviour at the oil-water interface is markedly different to that obtained at the air-water interface and in micelles. Furthermore it is inconsistent with the predictions of RST, which have been calculated taking into account the changes in electrolyte concentration. It was attributed to the additional degrees of freedom associated with the emulsion, and is associated with the partitioning of the $C_{12}E_6$ into the oil phase, and solubilisation of hexadecane into the mixed micelles (*15*).

Micelles

Using the methodologies described by equations 8 and 10 the composition and structure of SDS / $C_{12}E_6$ micelles at a range of solution compositions and concentrations have been studied (*8,13,37*). Figure 8 shows some typical SANS data for a 7.2×10^{-3}M 70 / 30 mole ratio mixture of SDS / $C_{12}E_6$ in 0.01M NaCl.

The ratio of intensities, for the hh and hd isotopic combinations in D_2O, extrapolated to low Q values provide an accurate estimate of the micelle composition. The variation of the micelle composition with surfactant concentration for three different solution compositions for the SDS / $C_{12}E_6$ mixture in 0.1 M NaCl is shown in figure 9 (*13*).

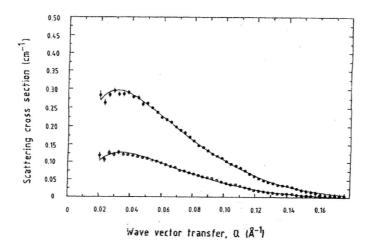

Figure 8 : Scattered Intensity, I(Q), for 7.2×10^{-3}M 70 / 30 mole ratio SDS / $C_{12}E_6$ in 0.1 M NaĊl for (•) hh, (o) hd SDS / $C_{12}E_6$. The solid lines are calculated curves for globular interacting micelles (see text).

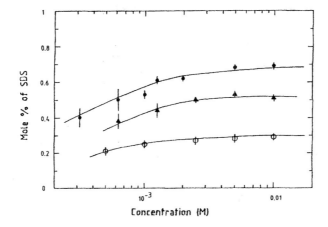

Figure 9 : Composition (mole % SDS) in mixed SDS / C₁₂E₆ micelles in 0.1M NaCl as a function of solution concentration, for solution compositions (•) 70 / 30, (▲) 50 / 50 and (o) 30 / 70. The solid lines are RST calculations for β -2.8 (70 / 30) and -2.0 (50 / 50 and 30 / 70)

The interaction parameter obtained from the mixed micelles is similar to that obtained for the air-water interface (8), and a similar variation with solution composition is observed. The departure from ideality is more pronounced for the solutions rich in SDS, and at a solution composition < 50 / 50 it is possible to describe the departure from ideality with a single interaction parameter.

A model analysis of the full scattering curves (using equations 8 and 9) provides details of the micelle structure and in particular the variation in aggregation number with solution concentration and composition (13,37). The variation with solution composition, for a 25 mM SDS / C₁₂E₆ solution in 0.1 M NaCl, is shown in figure 10.

For solution rich in SDS (>40 mole % SDS) there is little variation in aggregation number with composition, and changes in the interaction parameter shown in figure 9 are associated with subtle changes in packing. For solutions rich in C₁₂E₆ a marked change in aggregation number is observed (see figure 10), and its variation is highly dependent upon the electrolyte concentration. This was initially reported by Blankschtein et al (1), and rationalized in terms of the balance between the steric and electrostatic contributions to the free energy of micellisation. For micelles rich in C₁₂E₆ the free energy is dominated by the steric constraints of the nonionic head group. The addition of small amounts of SDS relaxes that constraint, reduces the mean area / molecule, and promotes micellar growth. As further SDS is incorporated the electrostatic interaction between the head groups will eventually dominate, and the micelle size will be reduced (see particularly the data for 0.1M NaCl). For the lower electrolyte concentrations the electrostatic contribution is larger and shifts the

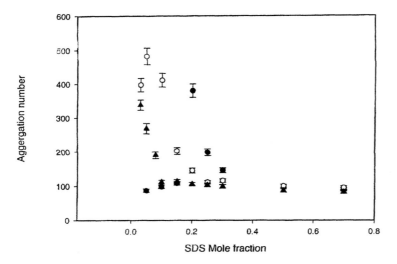

Figure 10 : Variation in micelle aggregation number with solution composition (mole % SDS) for 25 mM SDS / C₁₂E₆, (•) 0.1M NaCl, (o) 0.05M NaCl, (Δ) 0.01M NaCl

balance between the steric and electrostatic contributions to solutions richer in $C_{12}E_6$. The variation in aggregation number with concentration is consistent with this balance of steric and electrostatic contributions to the free energy of micellisation. For SDS / $C_{12}E_6$ in 0.1M NaCl it was observed (37) that for a solution composition close to the maximum in aggregation number with composition (20 mole % SDS) there is a pronounced increase in the aggregation number with surfactant concentration. Whereas for solution compositions away from that maximum (10, 30 mole % SDS) there is only a modest increase in the aggregation number with concentration.

Nonionic mixtures at the air-water interface

We have used neutron reflectivity, in combination with D / H isotopic substitution, to determine the detailed structure of the adsorbed monolayer of two different nonionic surfactant mixtures at the air-water interface, $C_{12}E_3$ / $C_{12}E_8$ and $C_{10}E_6$ / $C_{14}E_6$. In two different systems, which behave close to ideal mixing, we have explored the extent to which structural changes occur on mixing and have investigated the relative roles of changing the alkyl and ethylene oxide chain lengths. This provides an opportunity to quantify the extent to which structural changes and changes in hydration need to be incorporated into advanced theories of mixing.

$C_{12}E_3$ / $C_{12}E_8$ mixtures

Neutron reflectivity measurements were made for a 30 / 70 mole ratio mixture of $C_{12}E_3$ / $C_{12}E_8$ at a concentration of 5 x 10^{-5}M (*38*). This mixture results in an approximately equi-molar composition at the air-water interface (~70 Å^2 for $C_{12}E_6$ and ~ 120 Å^2 for $C_{12}E_8$). Deuterium labeling of the alkyl and ethylene oxide chains of each surfactant and of the solvent, and using the analysis method described by equations 1,6, and 7, the structure of the adsorbed layer was determined. That is, the extent and relative distributions of the different components at the interface were determined. Additional measurements, with more detailed labeling, of the outer C_6 of the alkyl chain, provides a coarse estimate of the conformation of the alkyl chain. These measurements were contrasted with similar measurements for pure $C_{12}E_3$ and $C_{12}E_8$ monolayers at an area / molecule equivalent to the mean area / molecule in the mixed monolayer (5 x 10^{-5}M for $C_{12}E_3$, and 2 x 10^{-4}M for $C_{12}E_8$). Figure 11 shows a comparison of the structure of the $C_{12}E_8$ in the mixed $C_{12}E_3$ / $C_{12}E_8$ monolayer with the pure $C_{12}E_8$ monolayer. Mixing with $C_{12}E_3$ results in a decrease in the E_8 / alkyl chain overlap. The E_8 group is less extended and more hydrated in the mixed monolayer compared to a pure $C_{12}E_8$ monolayer. In contrast the E_3 chain in the mixed monolayer has a similar extent, but is less hydrated compared to a pure $C_{12}E_3$ monolayer. It has been previously observed in the structure of a range of nonionic surfactant monolayers form $C_{12}E_2$ to $C_{12}E_{12}$ that the extent of the alkyl chain / ethylene oxide chain overlap increases with increasing ethylene oxide chain length. The changes in the E_8 group on mixing with the E_3 are consistent with that trend.

On mixing, the width of the alkyl chain distributions for both the $C_{12}E_3$ and $C_{12}E_8$ increase. Measurements with the C_{12} alkyl chain deuterium labeled and with the outer C_6 of the alkyl chain deuterium labeled show that although the C_{12} / solvent overlap remains constant, the C_6 / solvent separation increases. This implies that the alkyl chain conformation has changed as a result of mixing the E_3 and E_8 head groups. The result of the perturbation caused by packing two different ethylene oxide chain lengths in the mixed monolayer has an impact on the structure. The results show that the impact on the $C_{12}E_8$ is greater than on the $C_{12}E_3$; that is, the $C_{12}E_8$ structure changes more to accommodate the $C_{12}E_3$ than vice-versa.

$C_{10}E_6$ / $C_{14}E_6$ mixtures

The $C_{12}E_3$ / $C_{12}E_8$ measurements were contrasted with similar measurements on the $C_{10}E_6$ / $C_{14}E_6$ mixture (*39*). Using partial deuterium labeling of the alkyl and ethylene oxide chains of each surfactant and of the solvent, the distributions and relative positions of the alkyl chains at the interface were determined (*39*). A 97.3 mole % $C_{10}E_6$ / 2.7 mole % $C_{14}E_6$ mixture at a concentration of 4.2 x 10^{-4} M (the extreme solution composition was required to provide an approximately equi-molar composition at the interface, ~100 Å^2 for $C_{10}E_6$ and ~ 80 Å^2 for $C_{14}E_6$. due to the

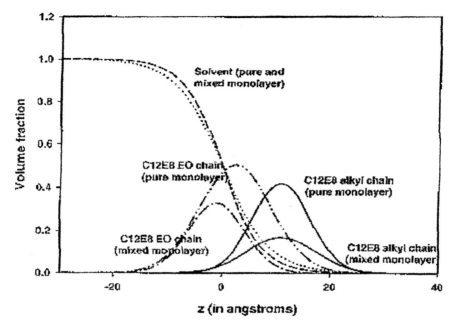

Figure 11 : Volume fraction distribution (derived from analysis of the partial structure factors) for $C_{12}E_8$ alkyl and ethylene oxide chains and solvent in $C_{12}E_3$ / $C_{12}E_8$ monolayer (30 / 70 mole ratio mixture at 5 x $10^{-5}M$) and $C_{12}E_8$ monolayer (2 x $10^{-4}M$) : (-,-..-s) ethylene oxide and alkyl chains in pure $C_{12}E_8$, (-.-, -) ethylene oxide and alkyl chains of $C_{12}E_8$ in $C_{12}E_3$ / $C_{12}E_8$ mixture, and (...,---) solvent distributions

large difference in cmc of the two surfactants) was compared with pure $C_{10}E_6$ and $C_{14}E_6$ (10^{-4} M) monolayers at an area/molecule (74, 50 $Å^2$ respectively) broadly similar to the mean area/molecule of the mixed monolayer (45 $Å^2$).

Figure 12 shows a comparison of the structure of the $C_{10}E_6$ monolayer and of the $C_{10}E_6$ / $C_{14}E_6$ mixed monolayer. The packing of the different alkyl chain lengths has a different impact on the structure of the mixed monolayer compared to the $C_{12}E_3$ / $C_{12}E_8$ mixture. In the $C_{10}E_6$ / $C_{14}E_6$ mixed monolayer the two surfactant distributions are essentially coincident. This is achieved by the $C_{10}E_6$ adjusting its position and conformation to a greater extent than the $C_{14}E_6$. The $C_{10}E_6$ is partially dehydrated on mixing, and the alkyl chain / ethylene oxide chain overlap decreases. This also results in a change in the $C_{10}E_6$ alkyl chain conformation, which is more extended in the mixed monolayer.

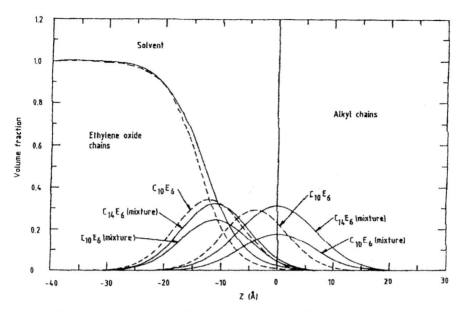

Figure 12 : Volume fraction distribution (derived from partial structure factor analysis) for 4.2 x 10⁻⁴ M (7.3 mole % $C_{10}E_6$ / 2.7 mole% $C_{14}E_6$ mixed monolayer and $10^{-4}M$ $C_{10}E_6$ monolayer, for $C_{10}E_6$ and $C_{14}E_6$ alkyl chain and ethylene oxide chains and solvent (as labeled in the figure), solid lines for $C_{10}E_6$ / $C_{14}E_6$ mixture and dashed lines for $C_{10}E_6$ monolayer.

Summary

In this article we have demonstrated how neutron scattering techniques (SANS and reflectivity) can be used to investigate surfactant mixtures at different interfaces (air-solution, solid-solution, and oil-solution) and in micelles.

For the anionic / nonionic surfactant mixture of SDS / $C_{12}E_6$ we have explored and quantified the extent to which RST is consistent with the mixing at the different interfaces, and in micelles, over a wide range of concentrations and compositions.

Detailed structural measurements on mixed nonionic surfactant monolayers of $C_{12}E_3$ / $C_{12}E_8$ and $C_{10}E_6$ / $C_{14}E_6$ at the air-water interface have demonstrated the extent to which structural changes and changes in hydration occur on mixing; and specifically on the effects of different alkyl and ethylene oxide chain lengths. This provides directly quantitative information that can be used to refine current theories of surfactant mixing at interfaces.

References

1. Shiloach, A.; Blankschtein, D. Langmuir **1998**, 14, 7166
2. Scamehorn, J, F. *Mixed Surfactant Systems*; Ogino, K; Abe, M., EDs.; Marcel Dekker, New York, 1992
3. Rosen, M. *Phenomena in Mixed Surfactant Systems*; Scamehorn, J,F.; Ed.; ACS Symposium Series, Washington DC, 1988
4. Rubingh, D, N. *Solution Chemistry of Surfactants*; Mittal, K.; Ed.; Plenum Press, New York, 1979, Vol 1
5. Shiloach, A.; Blankschtein, D. Langmuir **1997**, 14, 1618
6. Hines, J, D. Langmuir **2000**, 16, 7575
7. Lu, J, R.;Thomas, R.K.; Penfold, J. Adv Coll Int Sci **2000**, 84, 143
8. Penfold, J.; Staples, E.; Thompson, L, J.; Tucker, I.; Hines, J.; Thomas, R,K.; Lu, J, R.; Warren, N. J Phys Chem B **1999**, 103, 5204
9. Griffiths, P, C.; Stilbs, P.; Paulsen, K.; Howe, A, M.; Pitt, A, R. J Phys Chem B **1997**. 101, 915
10. Corn, R, M.; Higgins, D, A. Chem Rev **1994**, 94, 107
11. Bell, G, R.; Bain, C, D.; Ward, R, N. J Chem Soc, Faraday Trans **1996,** 92, 915
12. Mawie, S.; Cleveland, J, P.; Gaub, H, E.; Stucky, G, D.; Hausma, D, K. Langmuir **1994**, 10, 4409
13. Penfold, J.; Staples, E.; Thompson, L, J.; Tucker, I.; Hines, J.; Thomas, R, K.; Lu, J, R. Langmuir **1995**, 11, 2496
14. Penfold, J.; Staples, E.;Tucker, I.; Thomas, R, K. Langmuir **2002**, 18, 5755
15. Staples, E.; Penfold, J.; Tucker, I. J Phys Chem B **2000**, 104 , 606
16. Hines, J,D.; Garrett, P, R.; Rennie, G, K.; Thomas, R, K. J Phys Chem B **2000**, 101, 7121
17. Hines, J, D.; Thomas, R, K.; Garrett, P, R.; Rennie, G, K.; Penfold, J. J Phys ChemB **1998**, 102, 8834
18. Holland, P, M. Coll Surf A **1986**, 19, 171
19. Osbourn-Lee, L, W.; Schecter, R, S. *Phenomena in Mixed Surfactant Systems* Scamehorn, J, F. Ed. ACS Symposium Series 311 1986 30
20. Nikas, Y, F.; Pruuvada, S.; Blankschtein, D. Langmuir **1992**, 8, 2680
21. Penfold, J. *Encylopedia of Surface and Colloid Science* Marcel Dekker, 2002
22. Penfold, J.; Thomas, R, K. J Phys : Condens Matt **1996**, 2, 13698
23. Crowley, T, L.; Lee, E, M.; Simister, E, A.; Thomas, R, K. Physica B **1991**, 173, 143
24. Simister, E, A.; Thomas, R, K.; Penfold, J.; Aveyard, R.; Binks, B, P.; Fletcher, P, D, I.; Lu, J, R.; Sokolowski, A. J Phys Chem **1992**, 96, 1383
25. Heavens, O, S. *Optical Properties of Thin Films*, Butterworths, London, 1975
26. Simister, E, A.; Lee, E, M.; Thomas, R, K.; Penfold, J. J Phys Chem **1992**, 96, 1373
27. Lu, R, R.; Hromadova, M.; Simister, E, A.; Thomas, R, K.; Penfold, J. J Phys Chem **1994**, 98, 11519
28. Penfold, J. et al, J Chem Soc, Faraday Trans **1999**, 93, 3800
29. Lee, E, M.; Thomas, R, K.; Penfold, J.; Ward, R, C. J Phys Chem **1989**, 93, 381
30. Hayter, J, B.; Penfold, J. Coll Polym Sci **1983**, 261, 1022

31. Schultz, G, V, Z. Z Phys Chem **1935**, 43, 25
32. Hayter, J, B.; Penfold, J. Mol Phys **1981**, 42, 109, Hayter, J, B.; Hansen, J, P. Mol Phys **1982**, 42, 651
33. Ashcroft, N, W.; Lekner, J. J Phys Chem **1986,** 145, 83
34. Heenan, R, K.; King, S, M.; Penfold, J. J Appl Cryst **1997**, 30, 1140
35. *Neutron Beam Facilities at the high flux reactor available to users*, Institute Laue Langevin, Grenoble, France, 1994
36. Ghosh, R, E.; Egelhaaf, S, U.; Rennie, R, R. Institute Laue Langevin Report, 1998
37. Penfold, J.; Staples, E.; Tucker, I. J Phys Chem B **2002**, 106, 8891
38. Penfold, J.; Staples, E.; Tucker, I.; Thomas, R, K. J Coll Int Sci **1998**, 201, 223
39. Penfold, J.; Staples, E.; Tucker, I.; Thomas, R, K.; Woodling, R.; Dong, C, C. J Coll Int Sci **2003** in press

Chapter 8

Particle Characterization by Scattering Methods in Systems Containing Different Types of Aggregates

Aggregation of an Amphiphilic Poly(paraphenylene) in Micellar Surfactant Solutions

Tobias Fütterer, Thomas Hellweg, and Gerhard H. Findenegg

Stranski-Laboratorium für Physikalische und Theoretische Chemie, Technische Universität at Berlin, Strasse des 17, Juni 112, D–10623 Berlin, Germany

It is shown how a system containing two or more distinctly different types of self-assembled aggregates can be characterized by a combination of scattering methods. The system studied represents an aqueous micellar solution of a poly(paraphenylene) type oligomer in which each benzene unit of the aromatic backbone is doubly substituted with a hydrophobic and a hydrophilic group, viz., an alkyl chain ($-C_{12}H_{25}$) and an oxyethylene chain ($-CH_2O(EO)_3Me$). This oligomer, in which the boundary hydrophobic/hydrophilic is parallel to the aromatic backbone, is insoluble in pure water but soluble in micellar solutions of nonionic surfactants like C_8E_4. Cryogenic transmission electron microscopy reveals the existence of fiberlike aggregates of mean length greater 200 nm and a diameter of ca. 6 nm in an excess of small surfactant micelles. Static and dynamic light scattering and small-angle neutron scattering (SANS) were used to characterize this complex liquid, and the scattering data were analyzed by model-based and model-independent procedures.

These methods yield the mean length $<L>$, the polydispersity k, and the diameter d of the fiber aggregates. The dynamic behavior of the aggregates can be described by a model for polydisperse rigid rods, taking into account their translational and rotational diffusion. The composition of the fiber aggregates (polymer-to-surfactant number ratio) was determined by SANS experiments, using perdeuterated surfactant and a contrast variation of the solvent (H_2O/D_2O).

Introduction

Colloidal systems often contain two or more types of particles of different size and shape. In many technologically important fields (e.g. food industry, emulsion technology, etc.) and in bio-relevant systems it is necessary to characterize the individual species as they exist in the complex liquid. Systems exhibiting self-assembly can be taken as model systems of such a complex behavior. For example, several surfactant systems are known to contain vesicles coexisting with compact spherical micelles in a certain concentration range (1, 2). Aqueous solutions of diblock-copolymers may contain spherical and wormlike chain aggregates at the same time (3, 4). Scattering techniques with light, X-rays and neutrons are most frequently used for an in-situ characterization of colloidal dispersions. The determination of the properties of one specific type of particle in such complex systems is a challenge for the analysis of the scattering data. Scattering data are analyzed either with model-independent or model-based approaches. In general, model-independent approaches represent underdetermined mathematical problems. Most of these methods are based on a numerical solution of integral equations. Due to the limited number of experimental data points, the noise on the data, and the limited range of the scattering vector for which data can be obtained in a given experiment, the numerical solution of the equations is usually not unique and its quality has to be evaluated on the basis of statistical methods. The most common algorithm for analyzing dynamic light scattering data is the inverse Laplace transformation of the field autocorrelation function $g_1(\tau)$, while for static scattering experiments the indirect Fourier transformation method is most commonly used (5–7). The inverse Laplace transformation leads to a relaxation rate distribution and cannot distinguish between motions with similar diffusion coefficients. The indirect Fourier transformation method yields a one-dimensional picture of the scattering length density distribution derived from the measured q dependence of the scattering intensity I, where q is the scattering vector. Therefore, these model-independent approaches can be insufficient to analyze multiple-particle systems. In this study we present a model-based analysis of dynamic and static light scattering (DLS, SLS), and of small-angle neutron scattering data (SANS), based on results of a real-space imaging technique, the cryogenic transmission electron microscopy (cryo-TEM).

The system studied in this work represents an aqueous solution of a poly(paraphenylene) type oligomer in the presence of surfactant micelles. Poly(paraphenylene)s (abbreviated as PPP) are intrinsically stiff linear molecules. Applications of these and similar backbone polymers as photo- and electroluminescent devices arise from their peculiar electronic structure (8–13). Amphiphilically substituted PPPs are a relatively new class of materials. Their structure suggests an interesting self assembly behavior, but to our knowledge only a few systems have been studied until now (14–17). One example of such backbone polymers are PPP-sulfonate polyelectrolytes. Bockstaller et al. found elongated cylindrical micelles of constant diameter but variable length, depending on the molar mass of the PPP-sulfonates. These findings point to a lengthwise aggregation of these macromolecules (18).

In this work we study the aggregation of an amphiphilically substituted nonionic poly(paraphenylene) oligomer denoted as $PPP(n_n)$, in which each benzene ring of the aromatic backbone is substituted with an alkyl chain ($-C_{12}H_{25}$) and an oxyethylene chain ($-CH_2(OC_2H_4)_3OMe$) in para position to the alkyl substituent (see Figure 1). This amphiphilic substitution generates a boundary between the hydrophilic and hydrophobic substituents in a plane

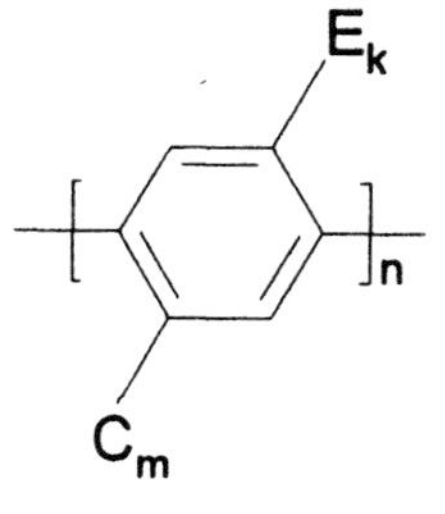

Figure 1. The poly(paraphenylene) oligomer PPP(12): $C_m = -C_{12}H_{25}$ and $E_k = -CH_2(OC_2H_4)_3OMe$ with $n \approx 12$.

which is parallel to the major axis of the PPP. The mean segment number n_n of the polymer studied in this work was 12. This oligomer PPP(12) was synthesized by Jörg Frahn on the basis of the transition metal-catalyzed Suzuki step-growth polymerization according to the procedure described elsewhere *(19–21)*. The sample studied here was characterized by gel permeation chromatography (GPC) which yielded a number and mass average molar mass of $M_n = 5200$ g/mol $(n_n = 12)$ and $M_w = 8700$ g/mol $(n_w = 21)$, corresponding to a ratio $M_w/M_n = 1.7$.

The pure polymer PPP(12) is almost insoluble in water, but it can be solubilized in aqueous surfactant solutions. Nonionic surfactants of the alkyl poly(oxyethylene) type, C_iE_j *(22, 23)*, and alkyl poly(glucoside) type, C_iG_j *(24, 25)*, were used in this work. Further information about the materials and the sample preparation is given elsewhere *(26)*.

Cryo-TEM investigations have shown that PPP(12) forms elongated fiberlike aggregates of a contour length $L_c > 200$ nm and a diameter $d = 5.9$ nm in the presence of excess surfactant micelles (see Figure 2 and ref *(26)*). The aggregates appear to be remarkably stiff, having a persistence length nearly equal to L_c.

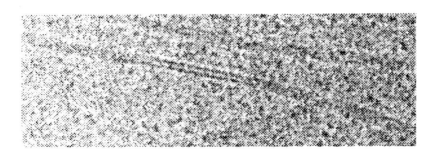

Figure 2. Cryo-TEM image of PPP(12) in a micellar solution of the surfactant C_8E_4. The small objects represent the surfactant micelles, the elongated objects represents fiber aggregates. The width of this image is equivalent to 180 nm (Reproduced from reference 26. Copyright 2003 American Chemical Society.)

Information about the experimental setup for the light scattering is given elsewhere *(4, 27)*. The SANS measurements were performed at the light water cooled and moderated swimming pool type reactor (BERII) of the Hahn-Meitner-Institut (HMI) Berlin, Germany, on the instrument V4 *(28)*, and at the high flux research reactor of the Institut-Laue-Langevin (ILL) in Grenoble, France, using the D11 instrument *(29)*. The SANS matching experiments were carried out at the Forschungszentrum Jülich (FZJ), Germany, on the instrument KWS1.

Dynamic Light Scattering

The scattering intensity I at a given scattering vector q depends on the relative positions of the scattering particles in the scattering volume. The diffusion of the particles results in an intensity fluctuation $I(t)$, with t = time. I is proportional to the square of the scattered electric field $<E>^2$. These fluctuations of I or E can be analyzed in terms of correlation functions (30). The normalized electric field autocorrelation function contains the information about the dynamics of the scatterers. For identical scattering particles (only one diffusion coefficient) it is given by

$$g_1(\tau) = \frac{\langle E_s^*(t)E_s(t+\tau)\rangle}{\langle I \rangle} = \exp(-\Gamma\tau), \qquad (1)$$

where $\Gamma = D_{tr} \cdot q^2$, with D_{tr} the translational diffusion coefficient (31) and the magnitude of the scattering vector q is given by $q = 4\pi \cdot \sin(\theta/2)/\lambda$, with θ the scattering angle and λ the wavelength in the medium. The autocorrelation function is also called the intermediate scattering function $S(q,t)$, and it represents the Fourier transform of the realspace position correlation function of the scattering particles in the system.

In polydisperse samples each particle size contributes to $g_1(\tau)$ with a specific exponential decay. Hence the correlation function is given by a weighted sum of exponentials

$$g_1(\tau) = \int_0^\infty G(\Gamma) \cdot \exp(-\Gamma\tau)d\Gamma, \qquad (2)$$

where $G(\Gamma)$ is the distribution function of relaxation rates Γ. If the system contains two or more types of particles (motions with different average diffusion coefficients), the field autocorrelation function can be expressed as the sum of these different contributions,

$$g_1(\tau) = \int_0^\infty G_1(\Gamma_1) \cdot \exp(-\Gamma_1\tau)d\Gamma_1 + \int_0^\infty G_2(\Gamma_2) \cdot \exp(-\Gamma_2\tau)d\Gamma_2 + \dots. \qquad (3)$$

$g_1(\tau)$ can be measured directly in a heterodyne light scattering experiment. However, it is more convenient to perform a homodyne experiment, which leads to the intensity time autocorrelation function $g_2(\tau)$,

$$g_2(\tau) = \frac{\langle E_s^*(t)E_s(t)E_s^*(t+\tau)E_s(t+\tau)\rangle}{\langle I \rangle^2}. \tag{4}$$

$g_1(\tau)$ and $g_2(\tau)$ are connected by the Siegert relation

$$g_2(\tau) = a \cdot |g_1(\tau)|^2 + b, \tag{5}$$

where a is the amplitude and b a baseline parameter. These two parameters have the value one in a perfect experiment.

The data for $g_1(\tau)$ can be analyzed in a model-independent way. Most common is the analysis by an inverse Laplace transformation of $g_1(\tau)$, using the FORTRAN program CONTIN (32, 33). The resulting relaxation rate distribution function $G(\Gamma)$ contains information about the number of motions with different average diffusion coefficients, the z-averaged relaxation rates $\bar{\Gamma}_i$, and the polydispersity. Different motions can only be distinguished if the average relaxation rates differ by at least half a decade and if the relative amplitudes are not too different. Otherwise only one distribution function will be obtained for two or more different motions.

In the present system we have to deal with at least two different types of polydisperse particles, fiber aggregates and surfactant micelles. Analysis of the field autocorrelation function $g_1(\tau)$ by an inverse Laplace transformation (CONTIN, 100 grid points) did not yield a consistent picture of the dynamic behavior of the system. Specifically, the number of separated contributions in the relaxation rate distribution, and the angular dependence of the average diffusion coefficients of these contributions, did not exhibit any systematic trends. Therefore, the data were analyzed by a model-based approach. Based on the observation of two different species on the cryo-TEM micrographs, the experimental correlation functions were fitted by a sum of two stretched exponential functions (Kohlrausch-Williams-Watts (34, 35)). The stretching exponents k_{WW} can be transformed into a relaxation rate distribution $G_i(\Gamma)$. With this approach a fairly good linear relation between the relaxation rate Γ and the square of the scattering vector q, as expected for pure translational motions (see eq 1), was found for the micellar contribution. The resulting values of the diffusion coefficient D and stretching exponent k_{WW} are similar to those found for pure C_8E_4 solutions ($D \approx 5\text{--}8 \cdot 10^7$ nm^2s^{-1} and $k_{WW} \approx 0.98\text{--}1$ (27)). However, the quality of the fits, especially in the time domain around the relaxation time τ of the fiber contribution, was unsatisfactory. In addition, the relaxation rate Γ attributed to the translational motion of the fiber aggregates exhibits a gradual change from a q^2 to a q^3 dependence at higher q, and the corresponding stretching exponents are extremely low (about 0.7 or down to 0.6). Such low values of kww cannot be attributed solely to a broad size distribution of the aggregates. An increase of the q dependence from q^2 to q^3 can be explained by the internal dynamics of the particles which is monitored at higher q values. For flexible wormlike particles the internal dynamics is due to a fast motion of the individual polymer chain segments in the particle. For stiff rodlike particles the rotational motion will contribute to the scattering in the high-q region.

In the present system one type of particles has a highly anisotropic shape, which leads to translational as well as rotational diffusion. Several models have been proposed to describe the dynamic behavior of cylindrical or wormlike particles. The cylindrical particles presented in this work are stiff, i.e. the persistence length L_p is of similar magnitude as the contour length L_c. To deal with these fiberlike particles we use a model of stiff rods proposed by Pecora (30, 36–37), which takes into account the rotational and translational motions in a decoupled way. In this model it is assumed that the rod consists of n identical, optically isotropic segments arranged along a line of length L. The translational motion can be described by one averaged diffusion coefficient D_{tr} which is independent of the rotational diffusion (D_{rot}). The overall motion of the particle is determined by combinations of translation and rotation. As can be seen from eq 6, the resulting dynamic structure factor is a sum of these different modes of motion. The amplitudes S_n of the modes relate to different length scales in relation to the length L, as they depend on the product of the scattering vector q and the length L.

$$g_{1,Fiber}(\tau,L) = \sum_{n=0}^{10} S_n(2n,qL) \cdot \exp\left(-\left(D_{tr}q^2 + (2n(2n+1)) \cdot D_{rot}\right) \cdot \tau\right) \qquad (6)$$

Here D_{rot} is the rotational diffusion coefficient and the amplitudes S_n are proportional to the square of the integral over a spherical Bessel function of the $2n$-th order (36). For values of $qL \leq 4$, only the pure translation (S_0) contributes to the dynamic structure factor. For values of $qL \leq 25$, the first 11 modes have to be taken into account (36).

Both the translational diffusion coefficient and the rotational diffusion coefficient are dependent on the geometry (length L, diameter d) and flexibility (L_p/L_c) of the rods. For rigid particles, Kirkwood and Riseman derived a relation between the diffusion coefficients and the particle geometry, based on the Oseen approximation (38), viz.,

$$D_{tr} = \frac{k_B T}{3\pi\eta_0 L}\ln\left(\frac{L}{d}\right) \quad \text{and} \quad D_{rot} = \frac{9D_{tr}}{L^2}, \tag{7}$$

where k_B is the Boltzmann constant, T the temperature, and η_0 the viscosity of the solvent. The model of Broersma (39, 40), even in its corrected form, does not provide a better representation than the simple Kirkwood-Riseman approach (36). However, for real systems the distribution of rod lengths L has to taken into account. In the present work, the Schulz-Zimm distribution was adopted, which allows for both an exponential and a distorted Gaussian distribution. For self assembled elongated structures like living polymers an exponential distribution is often observed (41). The measured $g_2(\tau)$ of polydisperse stiff rods can be analyzed by applying eq 6 in combination with the chosen model. This model-based analysis leads to the average length $<L>$, the diameter d, and the shape and width of the length distribution. Replacement of one stretched exponential function with the model of stiff rods, and taking into account a length distribution of the fiber length L, yielded an improved fit of the data, but a fully satisfactory representation of the experimental data was obtained only by allowing for a third contribution with a small translational diffusion coefficient ($D \approx 6\cdot10^5$ nm^{2}s^{-1}). The origin of this slow mode is not understood. In polyelectrolyte solutions and in the PPP-sulfonate/water systems, where such slow motions are also observed, they are attributed to a week cluster formation of the polymer particles. Alternatively, impurities in the polymer sample may also give rise to such a slow mode.

The intensity correlation function $g_2(\tau)$ is related to the field autocorrelation functions $g_1(\tau)$ of the three species by

$$g_2(\tau) = a \cdot \left(\frac{a_{Fiber}g_{1,Fiber}(\tau) + a_{Cluster}g_{1,Cluster}(\tau) + a_{Micelle}g_{1,Micelle}(\tau)}{a_{Fiber} + a_{Cluster} + a_{Micelle}}\right)^2 + 1, \tag{8}$$

where a is the amplitude of g_2 and a_i are the amplitudes of the individual field correlation functions $g_{1,i}$. In the data analysis the g_1 functions of the micelles and clusters were represented by a single exponential decay according to eq 1. The normalized fiber contribution $g_1(\tau)_{Fiber}$, with a polydispersity distribution for the length L according to Schulz-Zimm, is given by

$$g_{1,Fiber}(\tau) = \frac{\sum_L p(k,<L>,L) \cdot g_{1,Fiber}(\tau,L)}{\sum_L p(k,<L>,L)}, \tag{9}$$

where $p(k, <L>, L)$ is the size distribution (Schulz-Zimm, ref (41)), with $k = 1/(M_w/M_n - 1)$, and $g_{1,Fiber}(\tau, L)$ is calculated according to eq 6. The diffusion coefficients D_{tr} and D_{rot} were calculated from the length L and the diameter d using the Kirkwood-Riseman relation (eq 7). The value for the diameter, $d = 7.1$ nm, was estimated from the cryo-TEM micrographs, $d_{TEM} = 5.9$ nm, taking into account a hydration shell of two water molecules around the perimeter, with an estimated thickness of 0.3 nm (42) for each water molecule. The parameters of the model (see eq 1, 6, 7, 8, and 9) are independent of the scattering vector q, except the amplitudes a_i. Hence a parameter set for a given q must describe the dynamic behavior at other q values as well. To implement this into the data analysis, a simultaneous fit of $g_2(\tau)$ at four different q values was performed. Free adjustable parameters in the fit procedure were: the overall amplitude $a(q)$, the mean fiber length $<L>$, the polydispersity k (Schulz-Zimm), the amplitude $a(q)_{Cluster}$, the translational diffusion coefficient $D_{tr,Cluster}$, the amplitude $a(q)_{Fiber}$, and the translational diffusion coefficient $D_{tr,Fiber}$. Figure 3 displays the measured intensity autocorrelation function $g_2(\tau)$ and the respective fits according to eq 8. The amplitude $a(q)_{Fiber}$ was fixed to 1. The residuals in Figure 3 show that the model underlying eq 8 fits the data equally well in the whole domain of relaxation times. The minor deviations around $\tau = 10^{-4}$ s might be due to the single exponential decay used to describe the slightly polydisperse micellar contribution. The main fit results are listed in Table 1. In Figure 4 the field autocorrelation function $g_1(\tau)$ of each of the different contributions is shown together with the resulting overall $g_1(\tau)$ at $\theta = 60°$. The errors in the resulting amplitudes a_i and diffusion coefficients D amount to 3–8%, estimated by a parameter variation and χ^2 analysis according to (43). The individual errors for $<L>$ and k are in the same range, but the values of $<L>$ obtained in the above multiparameter fit are not independent of the polydispersity parameter k. A parameter analysis shows that different combinations of $<L>$ and k (resulting in similar length distribution functions $p(k, <L>, L)$) yield nearly equally good fits of the data. For $<L>$ values from 400 to 850 nm and k values in a range from 1.5 to 4, χ^2 values below the limit $1.1 \cdot \chi^2_{minimum}$ are obtained. The trend of an increase of $<L>$ with increasing concentration was found for all parameter sets. However, the error of $<L>$ is estimated to $\pm 20\%$, which means that the error bars cover all observed $<L>$ values for $k = 1.5$ to 4 if $<L>$ is given for a mean k of 2.5.

The data analysis outlined above shows that the mean length of the fiber aggregates, $<L>$, increases with increasing polymer concentration c (c_{PPP} in Table 1). Thermodynamic models (44) and mean-field theories (45) of reversible chain aggregation ('living polymers') predict a dependence as $<L> \propto c^\alpha$, with $\alpha = 0.5$. It is of interest to see if the present system conforms to this model. For the present system the width of the length distribution, represented by k, can be assumed to be independent on the concentration c. Therefore, we use the $<L>$ values obtained for a fixed value of k (k set to 2.5, see Table 1) to

Table 1: Results of the Simultaneous Fit of $g_2(\tau)$ with Eq 8 for Four Samples of Different Concentrations, Measured at Four Scattering Angles ($\theta_i = 30°, 60°, 120°,$ and $150°$) at 25 °C.[a]

S	c_{C8E4} [wt %]	c_{PPP} [wt %]	$<L>$ [nm]	k	$<L>_{k=2.5}$ [nm]
1	0.49	0.052	560	3.88	487
2	0.98	0.10	464	3.27	412
3	1.95	0.21	423	1.64	511
4	3.91	0.41	761	2.10	791

S	$D_{tr,Cluster}$ $[nm^2 s^{-1}]$	$R_{H,0,Cluster}$ [nm]	$D_{tr,Micelle}$ $[nm^2 s^{-1}]$	$R_{H,0,Micelle}$ [nm]
1	$7.89 \cdot 10^5$	311	$8.22 \cdot 10^7$	2.98
2	$6.45 \cdot 10^5$	380	$6.73 \cdot 10^7$	3.64
3	$6.03 \cdot 10^5$	407	$5.95 \cdot 10^7$	4.12
4	$4.34 \cdot 10^5$	567	$5.30 \cdot 10^7$	4.62

[a] The hydrodynamic radius R_H is calculated using the Stokes-Einstein equation ($\eta_0 = 0.8904$ cP).

test their concentration dependence. The concentration dependence of these adjusted values of $<L>$ is illustrated in Figure 5. One notes that the values at polymer concentrations above 0.1 wt % conform reasonably well to a power law for living polymers, but the value at the lowest polymer concentration deviates strongly from this power law. One may speculate that the growth law for living

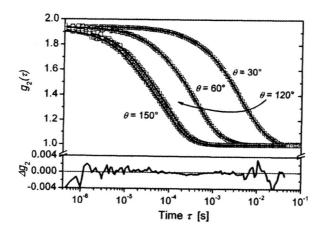

Figure 3. DLS of a sample of 0.1 wt % PPP(12) and ca. 1 wt % C_8E_4 in water at 25 °C: Intensity autocorrelation function $g_2(\tau)$ measured at four scattering angles θ (symbols) and simultaneous fit of the four curves by eq 8 (lines). The quality of the fit is indicated by $\Delta g_2 = g_2(\tau)_{fit} - g_2(\tau)_{measur.}$ at $\theta = 30°$.

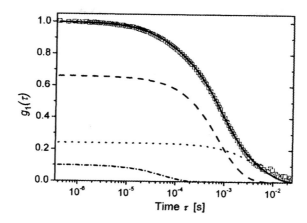

Figure 4: DLS of a sample of 0.2 wt % PPP(12) and ca. 2 wt % C_8E_4 in water (25 °C, $\theta = 60°$): Field autocorrelation functions $g_1(\tau)$ calculated from the measurements (symbols) and from the simultaneous fit (full curve) on the basis of eq 5, and the contributions to $g_1(\tau)$ from the three different species: clusters (· ·), fibers (— — —), and micelles (—·—·—).

polymers may not apply to the present fiber aggregates as these are made up of polymer and surfactant (see Section 4). However, in all scattering experiments the ratio of polymer and surfactant concentrations was small and nearly the same ($c_{PPP}/c_{C8E4} \approx 1/8$), such that the presence of the surfactant should not affect the results. A linear fit to the $<L>$ vs. c data is also shown in Figure 5. Further discussion of the influence of the polymer and its concentration will be given elsewhere (46).

In the analysis of the concentration dependence of $<L>$ we have assumed that the amount of polymer contained in clusters and not contributing to the fibers contribution can be neglected. Even if the cluster contribution is caused by impurities in the polymer sample, this approach is still valid. The scattering intensity is proportional to R_g^6, where R_g is the radius of gyration of the particle. As R_g of the clusters is by a factor 5 greater than for the fibers, one expect an intensity ratio $I_{Cluster}/I_{Fiber} \approx 15000$ for equal volume fractions of the two species. However, the static intensity $A_{Cluster}$ at $q = 0$ found for the clusters is only 10 to 100 times higher than the fiber intensity (see Table 2). Therefore, the mass concentration of the clusters must be less than 1% of the mass concentration of the fibers and can be neglected in this context.

The reliability of the data analysis outlined above was tested by comparing the results for the hydrodynamic radius of the surfactant micelles, $R_{H,0}$, in a pure surfactant solution with the respective value in the presence of the polymer. Such a comparison is shown in Figure 6 for a range of surfactant concentrations. The data for pure micellar solutions of C_8E_4 were taken from an earlier study (27). Values of $R_{H,0}$ were calculated from the diffusion coefficient $D_{tr, Micelle}$ on the basis of the Stokes-Einstein relation, using the viscosity of pure

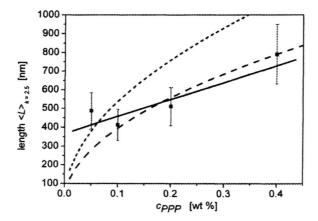

Figure 5: Dependence of the mean length <L> of the fiber aggregates on the polymer concentration c_{PPP}: values of <L> for fixed polydispersity parameter k (symbols). The solid line represents a fit by a linear function. The power law for living polymers, <L> = $L_0 \cdot c^{0.5}$ with L_0 = 1250 (— — —) and L_0 = 1700 (- - -) is also indicated.

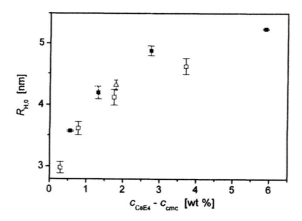

Figure 6: Hydrodynamic radius $R_{H,0,Micelles}$ of C_8E_4 at 25 °C, derived from the present measurements (open squares) and from pure aqueous solutions of C_8E_4 (open triangle, dark symbols: data taken from ref (27)).

water. Although the resulting values of $R_{H,0}$ of the micelles in the polymer system are somewhat smaller (by ca. 8%) than in the pure micellar solution, the concentration dependence of $R_{H,0}$ in the two systems is very similar. Thus the result of this comparison lends credibility to the data analysis of the polymer solutions.

As a second consistency test of the present model we can take the q dependence of the relative amplitudes a_{Fiber}, $a_{Cluster}$, and $a_{Micelle}$, which should be given by the respective static form factors. To extract the information about the fiber aggregates and surfactant micelles from the total scattered intensity, the following assumptions are made: (a) The form factor $P(q)$ of the surfactant micelles in the polymer system is the same function as in the pure aqueous surfactant solution; (b) the structure factor $S(q)$ of the polymer solution is equal to that of the pure surfactant solution, and in both cases $S(q) \approx 1$ for the relevant concentration and q range. With these assumptions the static scattering intensity can be modeled by the sum of the weighted individual form factors.

The static model must be capable to describe the q dependence measured in a static light scattering experiment as well as the q dependence of the relative amplitudes of the different types of particles found by analyzing the DLS data with the dynamic model. The following specific assumptions were made in this analysis of the relative amplitudes a_i: For the micelles and cluster aggregates a form factor of polydisperse spheres was assumed (although for the micelles the effect of polydispersity is negligible, and $P(q) \approx 1$ in the entire q range). For the fiber aggregates the form factor of monodisperse cylinders is used. (A model of polydisperse cylinders with a Schulz-Zimm distribution of cylinder length would cause only minor changes.) In the fitting procedure, the amplitudes $a_{Cluster}$ and $a_{Micelle}$ were expressed relative to a_{Fiber}, which was formally fixed to 1, i.e.

$$a_i = \frac{A_i \cdot P(q)_{Sphere,poly}}{A_{Fiber} P(q)_{Fiber,Cylinder}}. \tag{10}$$

Here, $P(q)_{Sphere,poly}$ is the form factor of homogeneous spheres with a Schulz-Zimm distribution of the sphere radius (47). Such a size distribution is often observed for microemulsions and micellar systems (48). $P(q)_{Fiber,Cylinder}$ is the form factor of a homogeneous cylinder of length L and cross-sectional diameter d (47). The cross-sectional diameter d is of no importance in this low-q range. Figure 7 displays the relative amplitudes from the fit and modeled by eq 10 for sample 3. The static amplitudes A_i are functions of the contrast factor K_i, the mass concentration c_i and the molar mass $M_{w,i}$. A_{Fiber} is set to 1, and the fiber length L (cylinder length) is set to 600 nm (see Table 1). For such large length the Guinier region is at lower q values than the experimental q range.

Table 2: The Relative Static Amplitude A_i and Radius R_i Obtained by Applying Eq 10 to the q Dependence of the Relative Amplitudes a_i. (A_{Fiber} is set to 1.)

s	$A_{Cluster}$	$R_{Cluster}$ [nm]	$A_{Micelle}$	$R_{Micelle}$ [nm]
1	9	249	0.013	2.38
2	19	304	0.033	2.91
3	25	330	0.045	3.3
4	95	453	0.046	3.54

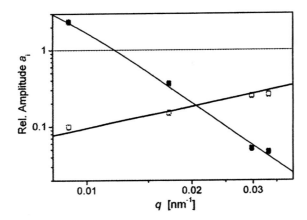

*Figure 7: Relative amplitudes a_i as a function of q for micelles (open symbols)
and cluster aggregates (dark symbols). The lines represent the modeled q
dependence according to eq 10. The amplitudes are expressed relative to the
amplitude of the fiber aggregates.*

Therefore, the deviations in the length are only of minor importance. The radii
for the cluster and micellar form factor are taken as $R_i = 0.8 \cdot R_{H,0}$. In this case
the precise value of the micellar radius is again of minor importance in view of
the low experimental q range in relation to the micellar radius. Table 2 lists the
adjusted A_i values and the values of R_i used in the analysis.

Static Light Scattering

For small particles (radius of gyration $R_g < \lambda/20$ where λ is the wave length
of light in the medium) the scattered radiation contains no information on the
particle form factor, because the whole particle acts as one single scattering
center in the q range probed in light scattering experiments. For larger particles
($R_g > \lambda/20$) and somewhat higher concentrations at which the interparticle
structure factor can be taken into account in terms of a virial expansion, the
normalized scattering intensity R_θ (Rayleigh ratio, ref (49)) is given by

$$R_\theta = Kc \cdot P(q) \cdot \left[\frac{1}{M} + 2B_2c + 3B_3c^2 + \ldots\right]^{-1}. \tag{11}$$

Here, the constant K contains all optical properties of relevance for the experiment, M is the molar mass, and c is the mass concentration. B_2 and B_3 represent the second and third virial coefficient. For dilute solutions the virial coefficients can be neglected, which is equivalent to assuming $S(q) \approx 1$ for the structure factor of the system.

The normalized static scattering intensity R_θ can be expressed as a sum of the respective form factors $P(q)$ weighted with the respective amplitudes A_i. Figure 8 shows a comparison of the measured Rayleigh-ratio R_θ and the modeled static intensity according to

$$R_\theta(q) = A \cdot \left(P(q)_{Fiber,Cylinder} \right.$$
$$\left. + A_{Cluster} \cdot P(q)_{Clus,Sphere,poly} + A_{Micelle} \cdot P(q)_{Mic,Sphere,poly} \right). \tag{12}$$

Note that the only adjustable parameter in this relation is the amplitude A. All other parameters are taken from the dynamic light scattering data analysis (Table 2). A is a function of K_{Fiber}, c_{Fiber}, and $M_{w,Fiber}$, but K_{Fiber} and c_{Fiber} are not known due to the cluster contribution. For this reason an accurate

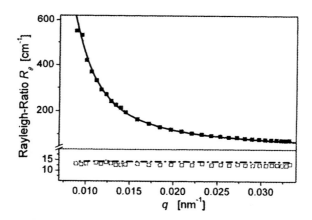

Figure 8: Top: measured total static intensity R_θ (dark squares) and model curve according to eq 12 (solid line), with values of A_i and R_i taken from Table 1, $L_{Fiber} = 600$ nm, $A_{Fiber} = 1$, $A = 352\,cm^{-1}$; bottom: measured static intensity R_θ for a pure C_8E_4 solution (open squares) and the micellar contribution from the model (dash-dotted line).

determination of the molar mass of the fiber aggregates by light scattering is not possible. (A rough estimate yields $M_{w,Fiber} \approx 10^6$ g/mol.) Also shown in Figure 8 is a comparison of the measured static intensity R_θ of a solution of C_8E_4 in water with the amplitude of the micellar contribution $A_{Micelle}$ in the polymer system. The agreement is quite good in view of the approximations inherent in this analysis.

The weighted form factors $P(q)$ of the three types of aggregates in the system are displayed in Figure 9. As to be expected, the form factor of the surfactant micelles shows no q dependence and the form factor of the fiber aggregates exhibit the q^{-1} dependence characteristic of cylindrical particles in the whole q range of the light scattering experiments. The cluster form factor follows approximately a q^{-4} dependence and contributes significantly only at low-q values.

The static model outlined above (eq 12) is expected to apply also to the analysis of small-angle neutron scattering (SANS). Due to the higher q range in SANS experiments ($0.02 < q < 2.4$ nm^{-1}) as compared to light scattering ($0.007 < q < 0.034$ nm^{-1}), the micelle size and cross section of the fiber aggregates will directly affect the measured $I(q)$. On the other hand, the cluster aggregates will not contribute to SANS in view of their low contribution at $q > 0.02$ nm^{-1} (see Figure 9).

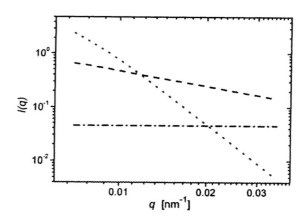

Figure 9. Simulation of the static scattering intensity for the clusters $(\cdots)$, the fibers $(- - -)$, and micelles $(-\cdot-\cdot-)$. The data for A_i and R_i are taken from Table 1, $L_{Fiber} = 600$ nm.

Small Angle Neutron Scattering (SANS)

The analysis of the SANS data is based on the same principles as outlined for the static light scattering data. Small-angle scattering curves from colloidal solutions can be represented by

$$I(q) = C \cdot \Delta\sigma^2 \cdot P(q)S(q), \tag{13}$$

where C is a concentration-dependent factor, and $\Delta\sigma$ is the excess scattering length density of the particles ($\Delta\sigma = \sigma_{particle} - \sigma_{solvent}$). The particle form factor $P(q)$ and the interparticle structure factor $S(q)$ are functions of the scattering vector q. For low concentrations and intermediate or large values of q again $S(q) \approx 1$, hence the analysis of the experiments has mainly to deal with the particle form factor (48, 50).

Contrast matching experiment

Besides the characterization of the size and shape of individual particles it is of interest to determine the composition of the aggregates in self-assembled multicomponent systems. A comparison of the static scattering intensities of one of the components measured in the absence of other components, and extracted from the total scattering intensity of a mixed system, can reveal information on the composition of mixed aggregates. However, a more elegant method is based on contrast variation in SANS experiments, as it involves a smaller number of adjustable parameters and can lead to more accurate results. In the case of two different types of scattering particles this application of contrast-variation SANS requires a sufficiently large difference in the scattering length density of the two different species contributing to the total scattering intensity. In neutron scattering experiments this is usually achieved by employing deuterated compounds. Fully deuterated surfactant, d-C_8E_4, was used to generate the contrast difference between the micelles and fiber aggregates.

The scattering intensity at a given q value (say, q') is a function of the concentration, the contrast between solvent and the particle, the form factor, and the structure factor. For two different types of particles, $I(q')$ is the sum of the two contributions. However, real particles have an inhomogeneous scattering length density. Therefore, q' has to be chosen such that one can monitor the particle as a homogeneous body with some mean scattering length density, defined by the real scattering length density distribution. For the present system at a value of $q' = 0.1$ nm^{-1}, the cluster contribution can be neglected and the form factors $P(q')$ of the micelles and fiber aggregates are causing no intraparticle interference effect on $I(q')$. For given concentrations the total scattering intensity $I(q' = 0.1 \text{ nm}^{-1})$ of the fibers and the micelles is a function of the contrast $\Delta\sigma$, i.e. of the scattering length density of the solvent $\sigma_{solvent}(x_{D2O})$, and can be written as

$$I(x_{D_2O}) = \left(C_{Fib} \cdot \Delta\sigma_{Fib}^2 \cdot P(q')_{Fib} + C_{Mic} \cdot \Delta\sigma_{Mic}^2 \cdot P(q')_{Mic} \right) \cdot S(q')_{mix}, \quad (14)$$

where x_{D2O} is the mole fraction of D_2O in the H_2O/D_2O solvent mixture, $\Delta\sigma = \sigma_{particle} - \sigma_{solvent}$, $C_{particle}$ is a concentration dependent factor, $P(q')_{particle}$ is the respective value of the form factor, and $S(q')_{mix}$ is the structure factor of the mixture (with the assumption that the particle interaction can be described by one structure factor $S(q')_{mix}$ for given concentrations c_{Fib} and $c_{Mic.}$). The scattering length density of the H_2O/D_2O solvent mixture can be calculated from the known mole fraction x_{D2O} and the scattering length densities of pure H_2O and D_2O. The scattering length density of the pure d-C_8E_4 micelles can also be calculated, and the x_{D2O}-independent factors $C_{particle}$, $P(q')_{particle}$, and $S(q')_{mix}$ can be condensed into a factor $C^*_{particle}$. The expression for $I(x_{D2O})$ then becomes

$$I(x_{D_2O}, q') = C^*_{Fib} \cdot \left(\sigma_{Fib} - \sigma_{solvent} \right)^2 + C^*_{Mic} \cdot \Delta\sigma_{Mic}^2. \quad (15)$$

By fitting eq 15 to the scattering data for a wide mole fraction range x_{D2O} of the H_2O/D_2O solvent mixture, the parameters C^*_{Fib}, σ_{Fib}, and C^*_{Mic} can be determined. The result of such a fit is shown in Figure 10. It can be seen that the measured dependence of I on the contrast is described with good accuracy by this relation. The parameter C^*_{Mic} can be determined independently from the scattering intensity of d-C_8E_4 solutions in the absence of the polymer. On the assumption that $P(q')_{Mic}$ has the same value in the absence and presence of the polymer, and that $S(q')_{mix} \approx S(q')_{pure}$, the ratio of $C^*_{Mic,mix}$ to $C^*_{Mic,pure}$ yields the amount of dissolved surfactant which is incorporated in the fiber aggregates. From the data shown in Figure 10 and complementary data for d-C_8E_4 solutions without polymer, we derive a surfactant-to-polymer number ratio of 3 : 1, corresponding to a surfactant-to-polymer mass ratio of 1 : 4. However, the error limits of this determination are high, about 100%, due to the fact that only a small fraction of the surfactant is incorporated into the fiber aggregates. Accordingly, $C^*_{Mic,mix}$ and $C^*_{Mic,pure}$ differ by no more than 4% and the error in the determination of C^*_{Mic} is 2% (from the fit for $C^*_{Mic,mix}$ and from the measurement for $C^*_{Mic,pure}$).

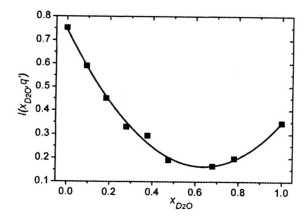

Figure 10. Total SANS intensity at q = 0.1 nm⁻¹ as a function of the mole fraction x_{D_2O} for a solution of 0.2 wt % PPP(12) and 2 wt % d-C₈E₄ in mixtures of H_2O and D_2O (symbols), and a fit according to eq 15 (line). The errors of I(q') are indicated by the size of the symbols.

Cross section of the fiber aggregates

To gain further structural information about the fiber aggregates, SANS measurements on two solutions of 0.2 and 0.3 wt % PPP(12), respectively, and 2 wt % of (nondeuterated) C_8E_4 in pure D_2O were performed and compared with the scattering curve of a sample of the same surfactant concentration in the absence of the polymer. This "blank curve" was subtracted from the scattering curve of the polymer plus surfactant solution, taking into account that in the latter a small fraction of surfactant does not exist in form of free micelles but is incorporated into the fiber aggregates. In the present case, this fraction is ca. 4% for the 0.2 wt % PPP sample, and ca. 5% for the 0.3 wt % PPP sample. Accordingly, on subtracting the scattering contribution caused by the surfactant micelles, the scattering curve of the pure surfactant solution is taken with a correction factor 0.96 and 0.95, respectively. The two resulting *difference curves* show no significant deviations in their q dependence. Figure 11 shows an example of the subtraction process. To test the justification of this procedure we compare a difference curve with the scattering curve of a mixture with perdeuterated d-C_8E_4/PPP(12) in pure D_2O. This comparison is shown in Figure 12. The scattering length density of the pure d-C_8E_4 micelles

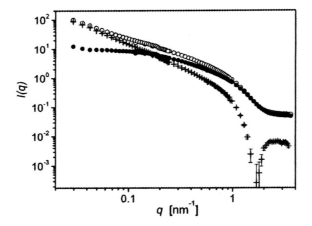

Figure 11. Absolute scattering curves I(q) of (a) a 0.3 wt % PPP(12) plus 2 wt% C_8E_4 solution (open cycles) and (b) a 2 wt % C_8E_4 solution (full cycles), both in D_2O. The crosses indicate the difference of (a) and (b) with an appropriate correction factor (see text).

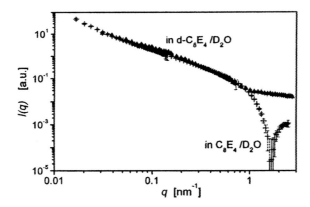

Figure 12. Comparison of the difference curve representing the fiber contribution (crosses) with the directly measured scattering curve with fully deuterated C_8E_4 as the surfactant (black symbols). Micelles of deuterated C_8E_4 are nearly completely matched by the solvent (D_2O), and thus scattering is almost solely due to the fiber aggregates.

nearly equals the scattering length density of D_2O and thus the contribution of the micelles can be neglected in this case. In the low-q region where the scattering intensity is sensitive to the inhomogeneity of the radial scattering length density distribution, the two curves in Figure 12 agree within the limits of error. The difference curve in Figure 11 and subsequent figures exhibit a pronounced minimum near $q = 1.5$ nm^{-1} as to be expected for the form factor of cylinders. However, for a real experiment such a sharp minimum is not expected even for cylindrical particles of perfectly monodisperse cross section, due to imperfect beam collimation and the wavelength dispersion of the neutron beam. The sharpness of the minimum is likely to be an artifact in the subtraction of the two scattering curves. Assuming a slightly lower correction factor for the scattering curve of the pure surfactant solution (corresponding to a somewhat higher surfactant-to-polymer ratio in the fiber aggregates) would lead to a smeared (less sharp) minimum in the difference curve. However, these details have no significant effect on the analysis of the difference curves.

Figure 13 shows a fit of the difference curve for fiber aggregates of PPP(12) in micellar solution of C_8E_4 by the form factor of a homogeneous cylinder. The cross-sectional radius of the fiber aggregates, $R_{CS,hom.}$, derived from this fit is 2.35 nm.

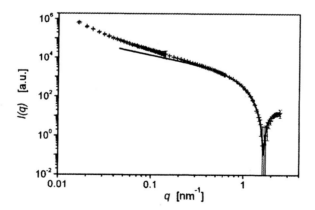

Figure 13. Difference curve representing the fiber contribution for a solution of 0.2 wt % PPP(12) and 2 wt % C_8E_4 in D_2O (symbols), and a fit by a form factor of a homogeneous cylinder (full curve).

The model-independent analysis of the fiber cross section according to the indirect Fourier transformation method by Glatter (5–7) yields the pair distance distribution function $p(r)$, and the Fourier transform of $p(r)$ represents $I(q)$ (see Figure 14). The radial scattering length density distribution function (see Figure 15) is the deconvolution of $p(r)$. This model-independent analysis yields a cross-sectional radius $R_{CS,distr.}$ of 3.0 nm. As expected, the model of a

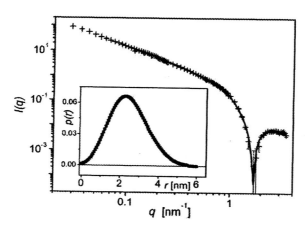

Figure 14. Difference curve representing the fiber contribution for a 0.3 wt % PPP(12) plus 2 wt % C_8E_4 solution (crosses). The line represents the Fourier transform of the pair distance distribution function p(r). Inset: Pair distance distribution function p(r) obtained by Fourier transformation of the measured data.

homogeneous cylinder yields a smaller value of R_{CS} than the model of inhomogeneous cylinders, in which the excess density $\sigma(r)$ decreases steadily. The resulting value of R_{CS} is similar to the estimated width of the PPP(12) molecules in the direction normal to the PPP backbone. Assuming an all-trans configuration of the hydrocarbon and oxyethylene side chains, the calculated length of a monomer of PPP(12) is 3.2 nm (26). On the basis of these results it is likely that the PPP(12) oligomers are arranged with their aromatic backbones oriented in the direction of the fiber axis, the aliphatic side groups pointing inward and the oxyethylene chain pointing into the aqueous medium. The radial excess scattering length density distribution $\sigma(r)$ of Figure 15 is consistent with such a core–shell cylinder model, when the alkyl side chains are forming the core, the aromatic backbone of the PPP(12) molecules are arranged in an inner shell and the hydrophilic oxyethylene chains forming the

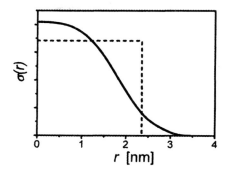

Figure 15. Radial excess scattering length density distribution function obtained by deconvolution of p(r) (see Figure 14) for the fiber aggregates (solid line), and according to a box model of a homogeneous cylinder (dashed line).

outer layer (corona): In this case, the core will exhibit the highest contrast against the solvent (D_2O), while the hydrated oxyethylene chains will exhibit a much smaller contrast. The cross-sectional structure of the fiber aggregates will be discussed in more detail elsewhere (46).

Conclusion

We have studied the aggregation of an amphiphilically substituted poly(paraphenylene) in micellar surfactant solutions by a combination of light scattering and small-angle neutron scattering techniques. As in many other colloidal systems of technical interest, this mixed polymer-surfactant system contains distinctly different particles, and the focus of the present article has been on methods to derive structural information about one of these particles in the presence of an excess of other particles of colloidal size.

Cryo-TEM images of mixed aqueous solutions of the polymer PPP(12) and the surfactant C_8E_4 indicate the existence of elongated and stiff (fiberlike) aggregates of a contour length $L_c > 200$ nm and a cross-sectional diameter of ca. 6 nm in the presence of an excess of surfactant micelles. The dynamic behavior of this system is studied by dynamic light scattering. The measured intensity autocorrelation function is made up of three contributions, two of which are attributed to the surfactant micelles and the fiber aggregates. The nature of the third contribution is not clear. From the observation that the hydrodynamic radius R_H of these entities increases with increasing polymer concentration, it is speculated that the third type of particles may represent clusters of fiber aggregates, as known from polyelectrolyte type poly(paraphenylene)s. The contribution of the fiber aggregates is analyzed in terms of a dynamic model of stiff rods, which shows that the persistence length

L_p of the aggregates is nearly equal to their contour length $<L_c>$, which is about 500 nm, somewhat increasing with increasing polymer concentration. The polydispersity of the contour length can be represented by a Schulz-Zimm distribution. Based on these results a static model is presented which describes the relative amplitudes for both the dynamic and static light scattering data.

The cross-sectional radius of the fiber aggregates was determined by SANS. We show that the contribution of the surfactant micelles to the scattering intensity can be subtracted from the overall scattering curve $I(q)$, while the contribution of the cluster aggregates does not contribute to the scattering in the q range of the SANS experiments. The $I(q)$ difference curve containing the information about the fiber aggregates was analyzed by two different methods, viz., using a model of homogeneous cylinders, and by the indirect Fourier transformation method via the pair distance distribution function. As to be expected, the latter method yields somewhat higher values of the cross-sectional radius R_{CS} than the former, which is based on a step-function of the scattering length density while the latter yields a smooth radial scattering length density profile. The results for R_{CS} are consistent with the diameter of the fiber aggregates derived from cryo-TEM and hint at a lengthwise arrangement of the PPP backbone in the fiber aggregates, such that the hydrophobic and hydrophilic side chains are pointing inward and outward, respectively, reminiscent of a cylindrical micelle.

The amount of surfactant contained in the fiber aggregates was determined by SANS contrast matching experiments, using fully deuterated surfactant and a set of H_2O/D_2O mixtures of different relative content of D_2O, to attain different contrasts of the surfactant micelles and fiber aggregates against the solvent. For the aggregates of PPP(12) with the surfactant C_8E_4 studied in this work a number ratio of surfactant to polymer molecules of 3 : 1 was found, though the limits of experimental error are rather high.

In conclusion, we have discussed strategies how scattering techniques can be used to determine properties of colloidal particles in the presence of other scattering particles. Although of models used in the analysis are specific for the present system, which consists of elongated fiber aggregates (and perhaps clusters of such fiber aggregates) in an excess of surfactant micelles, most of the strategies can be adapted to account for other types of systems as well.

Acknowledgement

The authors wish to thank A. D. Schlüter (FU Berlin) for providing the sample of the polymer PPP(12) and to Ch. Böttcher (FU Berlin) for producing cryo-TEM micrographs of the polymer-surfactant system. The authors are grateful to A. Brandt (HMI Berlin), C. Dewhurst (ILL Grenoble) and W. Pickout-Hintzen (FZ Jülich) for help with the SANS measurement at these institutions. Financial support from the Deutsche Forschungsgemeinschaft

(DFG) through SFB 448, and complementary support through the Fonds der Chemischen Industrie, is also gratefully acknowledged.

References

1. Schönfelder, E.; Hoffmann, H. *Ber. Bunsenges. Phys. Chem.* **1994**, *98*, 842-852.
2. Würtz, J.; Hoffmann, H. *J. Coll. Inter. Sci.* **1995**, *175*, 304-317.
3. Won, Y.-Y.; Brannan, A. K.; Davis, H. T.; Bates, F. S. *J. Phys. Chem. B* **2002**, *106*, 3354-3364.
4. Nordskog, A.; Egger, H.; Findenegg, G. H.; Hellweg, T.; Schlaad, H.; Berlepsch, H. v.; Böttcher, C. *Phys. Rev. E.*, submitted.
5. Glatter, O. *Acta Physica Austriaca* **1977**, *47*, 83-102.
6. Glatter, O. *J. Appl. Crystallogr.* **1980**, *13*, 7-11.
7. Glatter, O.; Strey, R.; Schubert, K.-V.; Kaler, E. W. *Ber. Bunsenges. Phys. Chem.* **1996**, *100*, 323-335.
8. Kranzelbinder, G.; Byrne, H. J.; Hallstein, S.; Roth, S.; Leising, G.; Scherf, U. *Phys. Rev. B* **1997**, *56*, 1632-1636.
9. Savvateev, V. N.; Yakimov, A.; Davidov, D. *Adv. Mater.* **1999**, *11*, 519-531.
10. Grice, A. W.; Bradley, D. D. C.; Bernius, M. T.; Inbasekaran, M.; Wu, W. W.; Woo, E. P. *Appl. Phys. Lett.* **1998**, *73*, 629-631.
11. Harrison, B. S.; Foley, T. J.; Bouguettaya, M.; Boncella, J. M.; Reynolds, J. R.; Schanze, K. S.; Shim, J.; Holloway, P. H.; Padmanaban, G.; Ramakrishnan, S. *Appl. Phys. Lett.* **2001**, *79*, 3770-3772.
12. Pan, J.; Scherf, U.; Schreiber, A.; Haarer, D. *J. Chem. Phys.* **2000**, *112*, 4305-4309.
13. Neher, D. *Adv. Mater.* **1995**, *7*, 691-702.
14. Bo, Z.; Zhang, Ch.; Severin, N.; Rabe, J. P.; Schlüter, A. D. *Macromolecules* **2000**, *33*, 2688-2694.
15. Engelking, J.; Wittmann, M.; Rehahn, M.; Menzel, H. *Langmuir* **2000**, *16*, 3407-3413.
16. Bo, Z.; Rabe, J. P.; Schlüter, A. D. *Angew. Chem., Int. Ed.* **1999**, *38*, 2370-2372.
17. Liu, T.; Rulkens, R.; Wegner, G.; Chu, B. *Macromolecules* **1998**, *31*, 6119-6128.
18. Bockstaller, M.; Köhler, W.; Wegner, G.; Fytas, G. *Macromolecules* **2001**, *34*, 6353-6358.
19. Frahn, J.; Schlüter, A. D. *Synthesis* **1997**, *11*, 1301-1304.
20. Frahn, J. Ph.D. thesis, Freie Universität Berlin, Berlin, Germany, 1999.
21. Schlüter, A. D. *J. Polym. Sci., Part A:, Polym. Chem.* **2001**, *39*, 1533-1556.

142

22. Strey, R. *Cur. Opin. Colloid Interface Sci.* **1996**, *1*, 402-410.
23. Kahlweit, M.; Strey, R. *Angew. Chem.* **1985**, *97*, 655-669.
24. Stubenrauch, C. *Cur. Opin. Colloid Interface Sci.* **2001**, *6*, 160-170.
25. Stubenrauch, C.; Findenegg, G. H. *Langmuir* **1998**, *14*, 6005-6012.
26. Fütterer, T.; Hellweg, T.; Findenegg, G. H.; Frahn, J.; Schlüter, A. D.; Böttcher, C. *Langmuir* **2003**, accepted, Intended Vol. *19*.
27. Strunk, H. Ph.D. thesis, Technische Universität Berlin, Berlin, Germany, 1995.
28. Keiderling, U.; Wiedenmann, A. *Physica B* **1995**, *213&214*, 895-.
29. Lindner, P. *Physica B* **1992**, *180&181*, 967-972.
30. Zwanzig, R. *Ann. Rev. Phys. Chem.* **1965**, *16*, 67-102.
31. Berne, B. J.; Pecora, R. *Dynamic Light scattering*; John Wiley & sons Inc.: New York, NY, 1976.
32. Provencher, S. W. *Computer Phys. Comm.* **1982**, *27*, 213-217.
33. Provencher, S. W. *Computer Phys. Comm.* **1982**, *27*, 229-242.
34. Williams, G.; Watts, D. C.; Dev, S. B.; North, A. M. *Trans. Faraday Soc.* **1971**, *67*, 1323-1335.
35. Lindsey, C. P.; Patterson, G. D. *J. Chem. Phys.* **1980**, 73, 3348-3357.
36. Russo, P. S. In *Dynamic Light scattering*; Brown, W., Ed.; Clarendon Press: Oxford, UK, 1993; Cha. 12.
37. Pecora, R. *J. Chem. Phys.* **1968**, *49*, 1036-1043.
38. Doi, M.; Edwards, S. F. *The Theory of polymer dynamics*, Cha. 7, Oxford University Press: Oxford, UK, 1986.
39. Broersma, S. *J. Chem. Phys.* **1960**, *32*, 1626-1631.
40. Broersma, S. *J. Chem. Phys.* **1960**, *32*, 1632-1635.
41. Drögemeier, J.; Hinssen, H.; Eimer, W. *Macromolecules* **1994**, *27*, 87-95.
42. Hellweg, T.; Eimer, W.; Krahn, E.; Schneider, K.; Müller, A. *Biochim. Biophys. Acta* **1997**, 1337 , 311-318.
43. Lösche, M.; Schmitt, J.; Decher, G.; Bouwman, W.G.; Kjaer, K. *Macromolecules* **1998**, *31*, 8893-8906.
44. Israelachvili, J. *Intermolecular and Surface Forces*, 2. Auflage, Academic Press: London, UK, 1991; Cha. 17.
45. Cates, M. E. *Macromolecules* **1987**, *20*, 2289-2296.
46 Fütterer, T.; Hellweg, T.; Frahn, J.; Schlüter, A. D.; Findenegg, G. H., to be published.
47. Pedersen, J. S. *Adv. Coll. Inter. Sci.* **1997**, *70*, 171-210.
48. Gradzielski, M.; Langevin, D.; Farago, B. *Phys. Rev. E* **1996**, *53*, 3900-3919.
49. Bender, T. M.; Lewis, R. J.; Pecora, R. *Macromolecules* **1986**, *19*, 244-245.
50. Hellweg, T.; Langevin, D. *Physica A* **1999**, *264*, 370-387.

Chapter 9

Aggregation and Gelation in Colloidal Suspensions: Time-Resolved Light and Neutron Scattering Experiments

Peter Schurtenberger[1], Hugo Bissig[1], Luis Rojas[1], Ronny Vavrin[1], Anna Stradner[1], Sara Romer[1,2], Frank Scheffold[1], and Veronique Trappe[1]

[1]Department of Physics, University of Fribourg, CH–1700 Fribourg, Switzerland
[2]Current address: Swiss Federal Lab for Materials Testing and Research, CH–8600 Duebendorf, Switzerland

We present a time-resolved study of the aggregation and sol-gel transition in concentrated colloidal suspensions. We use diffusing wave spectroscopy (DWS) to obtain quantitative information about the microscopic dynamics all the way from an aggregating suspension to the final gel, thereby covering the whole sol-gel transition. In order to obtain additional information on the corresponding structural changes we have designed a combined SANS-DWS experiment. This allows us to simultaneously measure both the time evolution of the local dynamics as well as the microstructure as the aggregation and gelation proceeds.

144

Understanding the structure and dynamics of colloidal particle suspensions and gels is of significant interest both for research and industry (1). Moreover colloidal systems serve as convenient models to address fundamental issues such as liquid-like ordering, crystal and glass formation, fractal growth and structural properties of random networks. This is primarily due to the fact that colloidal particles can be produced with well defined properties such as shape, size or surface charge density, and that the strength as well as the range of the interaction potential can be tuned easily. We can for example use highly charged polystyrene particles at very low ionic strength, which leads to an effective pair potential that is well described by a so-called Yukawa potential, i.e. a long range exponential decay with the Debye length as the characteristic decay length of the potential. On the other hand we can add salt or use colloidal particles stabilized by a polymer layer, which then leads to a typical hard sphere interaction potential. If we add even higher quantities of salt to charge stabilized colloids or remove the stabilizing polymer layer, the attractive van der Waals interactions completely dominate and the particles start to aggregate irreversibly. These idealized cases are shown in figure 1.

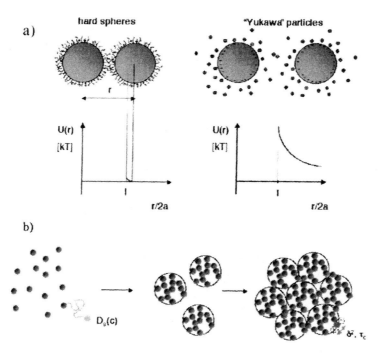

Figure 1: (a) Interaction potential for sterically and charge stabilized particles as a function of the distance r normalized with the particle diameter 2a.(b) From stable suspensions to cluster formation and gelation.

Aggregation and gelation in complex fluids has been for a long time a field of intense research where both fundamental as well as applied questions are equally important. Applications of gels and sol-gel processing include such different areas as ceramics processing, cosmetics and consumer products, food technology, to name only a few. Despite the randomness and the complexity of the sol-gel transition it has always attracted fundamental researchers due to the unique features of gels and the strong similarities between very different gels. The sol-gel transition displays a very rich behavior of different physical properties that can be characterized by distinct scaling laws (2). Irreversible aggregation of colloidal particles is known to result in the formation of space filling gels, where the motion of the constituent particles becomes constrained and non-ergodic behavior is observed in dynamic light scattering. These gels show solid-like properties and exhibit for example a low frequency plateau modulus in oscillatory shear experiments. Considerable efforts have been made to obtain a more detailed understanding of the relation between microscopic and macroscopic properties, and for gels formed at low particle concentrations this was successfully done using fractal concepts.[3, 4] However, until now only few attempts have been made to study and understand aggregation and gelation at high particle concentrations, despite the fact that the high concentration regime is of crucial interest in many technical applications as for instance in the processing of inorganic particles for the production of ceramics.

In this chapter we shall present results from a study of aggregation and gelation in concentrated colloidal suspensions using a combination of small-angle neutron scattering (SANS) and DWS. We in particular investigate whether the concepts that have been so successful in understanding the behavior of dilute systems can also be applied to concentrated suspensions and gels. At low volume fractions and in the absence of gravitational forces colloidal particles with attractive interactions aggregate into large clusters and finally form a macroscopic gel. In the gel network the individual clusters show a fractal structure that leads to a power law dependence of the structure factor $S(q) \sim q^{-d_F}$ for $a < 1/q < R_C$, where the fractal dimension is a measure for the compactness of the individual cluster. For diffusion limited cluster-cluster aggregation (DLCCA) $d_F = 1.8$, while for reaction limited cluster-cluster aggregation (RLCCA) $d_F = 2.1$ is expected (5-7). However higher values are possible if the cluster become more compact due to internal restructuring or cluster-cluster interactions (e.g. at elevated densities) (for a more detailed description see e.g. (6, 7) and references therein). In the most simple picture the individual clusters grow at the same rate until they fill up the whole accessible volume, with a critical cluster radius given by

$$R_c \approx a\Phi^{-1/(3-d_F)} \tag{1}$$

Most of the fundamental research has been carried out on the macroscopic properties of gels, e.g. their viscoelastic behavior, and attempts have been made to interpret them based on fractal models for the gel structure. However, it is much more difficult to access information about the microstructural properties and furthermore link them to the macroscopic properties of the gel. New developments both in light scattering techniques and in theory now allow us for the first time to bridge this gap even for very concentrated and turbid systems (*8-11*).

Structure and dynamics of stable suspensions

Before turning to aggregation and gelation, we characterize the structure and dynamics of the initial stable suspensions. The relevant length (several hundred nanometers to micrometers) and time (microseconds to hours) scales in colloidal suspensions are typically much larger than in atomic or molecular systems and are thus easy to access in experimental studies. This is one of the main reasons why colloidal suspensions are ideal model systems to investigate phase transitions or the structure and dynamics of metastable states such as glasses or gels. The interaction between colloidal particles is of fundamental importance for the phase behavior of colloidal suspensions, as well as for their mechanical properties and, most importantly for applications in the paint and food industry, their stability against aggregation induced by van-der-Waals forces or depletion attraction. Interaction effects and the correspondingly strongly influenced positional correlations between the particles will also strongly influence the optical properties of dense systems. These specific features of concentrated and strongly correlated colloidal systems have recently attracted considerable attention (*12,13*).

We can investigate the structure and dynamics of stable suspensions using scattering experiments. For charge stabilized particles at low ionic strength, the Debye length is large and we observe the formation of highly ordered suspensions ("supercooled liquids") already at relatively low volume fractions. The corresponding static structure factor $S(q)$ measured in a scattering experiment exhibits a pronounced peak at relatively low values of the scattering vector q, which can only be seen with light but not with neutron or X-ray scattering due to the accessible range of q values in these measurements. However, the large scattering contrast of these particles in aqueous suspension leads to strong multiple scattering even at very low volume fractions, and static (SLS) and dynamic (DLS) light scattering experiments thus become very difficult or impossible. One can overcome this problem and suppress contributions from multiple scattering in SLS and DLS using different cross correlation schemes. In particular the so-called 3D cross correlation (3DDLS) experiments has been very successful (*8, 9, 14*). This is demonstrated in figure 2, where the resulting structure factors are shown for two deionized suspensions of polystyrene latex particles (radius a=58.7 nm) at two different volume fractions ($\Phi = 0.0042$ and $\Phi = 0.0105$, for details see ref. 14). Two photographs of the suspensions are shown as an inset and demonstrate the considerable turbidity of these samples and the power of the 3DDLS technique that allows measuremenzs of the static and dynamic structure factor even under these extreme conditions.

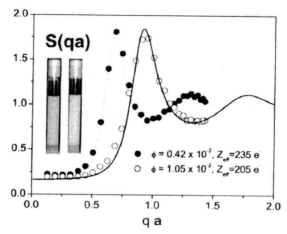

Figure 2: Static structure factors S(qa) of deionized suspensions of polystyrene latex particles (radius a=58.7 nm) at two different volume fractions (Φ = 0.0042 and Φ = 0.0105) measured with 3DDLS. Solid lines: best-fit polydisperse HNC calculations using a Yukawa interaction potential with adjustable volume fraction and effective charge Z_{eff} (see ref (15)). Also shown are pictures of the two samples that illustrate the considerable turbidity.

It is important to point out that previous studies of suspensions of highly charged systems at these volume fractions have often been hampered by difficulties to find systems that do not readily crystallize. In fully deionized aqueous suspensions of monodisperse spheres normally crystallization occurs already at very low volume fractions ($\Phi < 10^{-3}$) *(16,17)*. If electrolyte is added it is often found difficult to control, maintain and determine the ionic strength with the required precision, though carefully performed experiments of this kind have been reported *(16, 18, 19)*. We have chosen a different way to avoid these complications by using a mixture of alcohol and water as a solvent which efficiently avoids crystallization by modifying the solvent dielectric constant. Thus we are able to work at full deionized conditions thereby providing a well controlled model system of charged spheres with volume fractions well above $\Phi = 10^{-3}$.

If we now add salt to these suspensions, the electrostatic repulsion will be screened and we can tune the interaction potential such that we first obtain an effective hard sphere potential. Under these conditions the structure factor is much less pronounced, and we have to go to much higher concentrations in order to observe a measureable peak. Moreover, for this particle size the peak is shifted to higher values of q (it occurs at a value that corresponds to the characteristic distance for the next nearest neighbor shell that for hard spheres is $q \approx 2\pi/2a$) and is almost independent of concentration. Therefore we now have to go to small-angle neutron scattering in order to investigate S(q). A typical example for the resulting q-dependence of the scattering intensity and the structure factor as obtained using a theoretical calculation based on integral equation theories with the Percus-Yevick closure relation is shown in figure 3 for latex particles (a = 80 nm, polydispersity approximately 3%) at a volume fraction of about $\Phi = 0.3$ (*20*). It is important to point out that in this case I(q) contains contributions both from the effective structure factor S(q) as well as from the particle form factor P(q), and that instrumental resolution effects result in a considerable smearing of I(q) that needs to be taken into account in any attempt to quantitatively analyze SANS data.

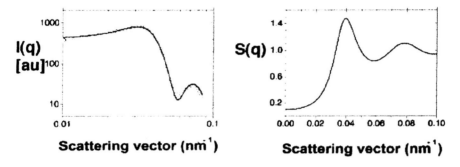

Figure 3: Neutron scattering intensity I(q) (left) and resulting structure factor S(q) (right) for a polystyrene suspension with particle radius a = 80 nm, polydispersity 3% and effective volume fraction Φ = 0.289 (see ref. (20) for details)

In addition to the structural properties we can also investigate the dynamic behavior as a function of particle concentration and salt content. Information about the dynamic properties on lengths scales comparable to the size of the particles is usually obtained from dynamic light scattering in the single scattering regime (*21, 22*). However, we have already seen that in dense suspensions DLS cannot be applied in most cases due to strong multiple scattering of light. In addition the q-range and hence the length scales probed by standard light scattering techniques ($\lambda/n \sim 400nm$) are limited to $q < 0.03$ nm^{-1} which is not sufficient for many concentrated systems. An elegant way to overcome this limitation is to either use 3DDLS for systems where we look at highly charged particles at low ionic strength, or then take advantage of the multiple scattering process rather than avoiding it by using Diffusing Wave Spectroscopy (DWS). DWS works in the limit of very strong multiple scattering, where a diffusion model can be used in order to describe the propagation of the light across the sample (*23-25*). Using such a diffusion approximation, one can then determine the distribution of scattering paths and calculate the temporal autocorrelation of the intensity fluctuations analogous to DLS. It is thus still possible to study the dynamics of a colloidal suspension by measuring the intensity fluctuations of the scattered light observed either in transmission or reflection. The fluctuations of the scattered light measured in transmission result from the variation of the total path length by a wavelength of light. However, since the light is scattered from a large number of particles, each individual particle must move only a small fraction of a wavelength for the cumulative change in path length to be a full wavelength. Therefore, despite the fact that DWS does not yield explicit information on the q-dependence of the so-called dynamic structure factor $S(q,t)$, it is capable in providing unique information about the local dynamical properties, i.e. the mean square displacement $<\Delta r^2(t)>$ of the individual particles. It probes particle motion on very short length scales and can for example measure motions of particles of order $1\mu m$ in diameter on length scales of less than $1nm$ (10). We have in fact used both 3DDLS and DWS/SANS in order to investigate the dynamics of strongly interacting colloidal suspensions, and in particular investigated the role of hydrodynamic interactions in such systems (14,15,20). However, our general approach of combining SANS and DWS is not limited to stable particle suspensions, but can be equally well applied to other colloidal systems, such as aggregating and gelling suspensions. It is this particular feature that we shall concentrate on in the remainder of this chapter.

Aggregation and gelation in destabilized suspensions

Strongly aggregated colloidal particle gels have been extensively investigated in the past, and have been found to possess fractal character over a range of length scales whose width primarily depends on the particle volume fraction, Φ. Because of the fascinating properties of fractals most investigations were restricted to low volume fractions, where the fractal character extends over a wide range of length scales, and where the structural and dynamic properties of the system are determined by the fractal structure. We extended the range of volume fractions studied to larger Φ. In our investigations, the aggregation and gelation of charge stabilized and buoancy-matched concentrated latex dispersions with volume fractions typically in the range of $0.01 \leq \Phi \leq 0.3$ is induced through an increase of the ionic strength. However, for highly concentrated samples it is very difficult to achieve a homogeneous, reproducible destabilization by simply adding salt. Recently a novel method based on an in-situ variation of the ionic strength has been introduced (26). The destabilization of the colloidal suspensions is induced with an internal chemical reaction which allows to slowly produce ions (the urease catalyzed hydrolysis of urea). We have subsequently extended this approach to concentrated suspensions of polystyrene particles (27). Figure 4 shows a typical example of the temporal evolution of the rheological properties (storage (G') and loss (G") moduli) of such a destabilized suspension of polystyrene particles (a = 85 nm, polydispersity 4%) at a volume fraction of $\Phi = 0.045$. Initially the suspension has typical Newtonian fluid properties and G" dominates. Due to the formation of larger clusters both G' and G" then increase. At approximately 300 seconds

150

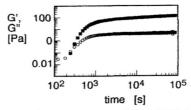

Figure 4: Time evolution of the storage modulus G'(t) (solid squares) and the loss modulus G"(t) (open circles) for a suspension of destabilized latex particles with radius a = 85 nm and a volume fraction of Φ = 0.045. Measurements were made with a MCR 300 rheometer from Paar Physica using a double gap couette system at a constant frequency (ω = 10 rad/s).

we then observe a steep increase of G', which is now larger than G". This is the classical signature of the gel point. After about 1 hour the elastic modulus changes only gradually with time.

While such experiments are capable of yielding an estimate of the gel point and of the viscoelastic properties of the suspension and the final gel, they do not provide us with any information on the structural evolution of the system. Moreover, despite the fact that one generally tries to work at low strain values, rheological measurements are not really non-invasive, which is particularly important for weak gels formed close to the gel point and at lower volume fractions, and they require a considerable amount of material. We have thus started to use DWS and SANS as non-invasive tools for a true in-situ and real-time investigation of aggregation and gel formation in concentrated suspensions.

Figure 5a shows typical snapshots out of a sequence of correlation functions $g(\tau)$-1 during aggregation and gelation of a latex particle suspension (particle radius a = 85 nm, volume fraction Φ = 0.11). For ergodic systems, the mean square displacement of the correlated Brownian particles can be successfully modeled by means of an averaged short-time diffusion coefficient. This leads to a correlation function well approximated by a single exponential decay (10). During the initial period the formation of large aggregates and clusters leads to a dramatic slowing down of the single particle diffusion and a corresponding shift of the characteristic relaxation time τ_c of the intensity autocorrelation function to longer decay times. However, at later stages in the aggregation process the clusters fill the entire sample volume and gelation occurs. Because of the high volume fraction the network is very stiff, and as a consequence the trapped particles can only execute limited motions about their fixed averaged positions. After about 10 days the gel changes only very little and we take that time as the terminal state. The correlation function now exhibits a distinctly different decay which is described by a stretched exponential with an arrested decay leading to a plateau.

After the sol-gel transition, we are confronted with an additional difficulty due to the non-ergodicity of these systems. In solid-like media, such as gels, the

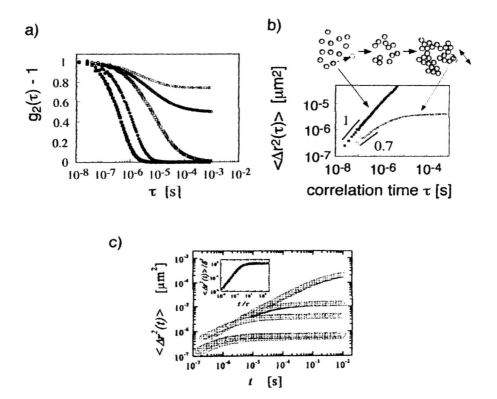

Figure 5: Sol-gel transition of a colloidal suspensions (radius a = 85 nm, volume fraction Φ = 0.11) measured over 10 days with a two-cell setup. A) The destabilized system shows a transition from a liquid state (solid points), characterized by an almost exponential decay of the correlation function, to a solid state (open symbols) after about 80 min. In the gel-state a continuously increasing plateau builds up in the correlation function, $g_2(\tau)$-1, characteristic for the finite storage modulus of a solidlike system. B) Particle mean square displacement of a colloidal system in the sol and the gel. C) Mean square displacement as function of time for gels composed of polystyrene spheres (a = 85nm) at different volume fractions: Φ = 0.25 [triangle down], Φ = 0.20 [hexagon], Φ = 0.11 [circle], Φ = 0.045 [triangle up], Φ = 0.010 [square]. The inset shows the master curve obtained by normalizing t by τ_c and $<\Delta r^2(t)>$ by δ^2. The solid line represents the calculation according to Eq. 2 with p = 0.65.

scatterers are only able to make limited Brownian excursions. As a consequence, the time-averaged intensity correlation function of the scattered light is different from the ensemble-averaged correlation function. For dilute, nonergodic samples different approaches have been proposed in order of properly average the signal (28). The extension to turbid systems is however not always practicable. We were able to develop a completely new and simple experimental scheme in order to investigate nonergodic samples with DWS (27, 29). We overcome the nonergodicity limitation of light scattering in concentrated systems by using a combination of two cuvettes. The first cuvette contains the gelling sample that can be either ergodic or nonergodic, whereas the second one is a turbid ergodic system that leads to an additional decay of the correlated signal at long delay times. We thus obtain the true gel contribution over a wide lag-time interval from 10 ns $\leq t \leq 1$ ms.

The dramatic change in the local particle dynamics becomes even more clearly visible when looking at the time dependence of the corresponding mean square displacement $<\Delta r^2(t)>$. The particle dynamics in the initial stable suspensions as well as in the aggregating suspensions prior to the gel point (solid symbols) exhibit the typical characteristics of free particle diffusion due to Brownian motion. This is reflected by an almost exponential decay of the correlation functions and leads to a linear dependence of $<\Delta r^2(t)>$ on time (indicated with a line with slope one in figure 5b). During the initial period the formation of large aggregates and clusters leads to a dramatic slowdown of the single particle diffusion. Later the long time behavior of $<\Delta r^2(t)>$ becomes more and more constrained. In this regime, before gelation, the particle shows simple diffusive motion only on a length scale of $<\Delta r^2(t)> \approx 1\text{-}2$nm, representing only a small fraction of the particle diameter (27).

At the gel point a quite dramatic change in the particle dynamics occurs, and the short time behavior changes from Brownian to a subdiffusive motion (open symbols) well described by a power law $<\Delta r^2(t)> \sim t^p$. We find that in the gel state the average mean square displacement is well described for all t by a stretched exponential

$$\left\langle \Delta r^2(t) \right\rangle = \delta^2 \left[1 - e^{-(t/\tau_c)^p} \right] \qquad (2)$$

leading to a plateau at long times (with $p = 0.7 \pm 0.05$).

We find, within our time resolution, that the exponent for diffusion $p = 1$ drops at the gel point and takes a value of $p \approx 0.7$ for all times $t > t_{gel}$. This indicates that already at the gel point t_{gel} almost all particles are connected to the gel network. A comparison between time-resolved rheological measurements and the DWS experiments demonstrates that the qualitative change in microscopic particle dynamics indeed coincides with a dramatic change in the macroscopic viscoelastic properties of the samples at the gel point. This is shown in figure 6, where the storage (G') and loss (G") moduli measured at a single oscillation frequency and the exponents p obtained from DWS are plotted

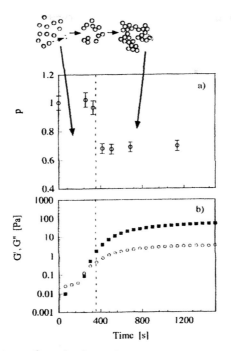

Figure 6: Comparison of results from classical rheology and DWS during the aggregation and sol-gel transition for a suspension of latex particles (radius a = 85 nm, volume fraction $\Phi = 0.045$) (a): Time evolution of the exponent p obtained from the short-time mean square displacement ;(b): Time evolution of the storage modulus G'(t) (solid squares) and the loss modulus G"(t) (open circles) obtained from an oscillating rheological measurement.

as a function of time. We observe the steep increase of G', indicating the transition from a sol to a gel, at the same time where the exponent p drops from 1 to about 0.7.

Clearly, the volume fraction also has a profound effect on the dynamic properties of the gels. At the lowest volume fraction, $\Phi = 0.01$, the mean square displacement first rises sub-linearly at short times, and then curves towards a constant plateau value, indicating that the particle excursion is restricted to a maximum mean square displacement, δ^2. As the volume fraction increases δ^2 systematically decreases to lower values, reflecting the increased restriction to particle motion. Moreover, the cross-over time, τ_c, at which the time dependence of $<\Delta r^2(t)>$ changes from sub-diffusive to a time independent plateau behavior, clearly shifts towards lower values between $\Phi = 0.01$ and $\Phi = 0.11$, while remaining roughly constant for $\Phi = 0.11\text{-}0.25$. Remarkably, all data can be scaled onto a single master-curve by normalizing t with the characteristic crossover time and $<\Delta r^2(t)>$ with the maximum mean square displacement, as shown in the inset of Figure 5c. This mastercurve is well described by eqn. 2, as denoted by the solid line in the inset of Fig. 5c.

The short time dynamics (eqn. 2) observed for the concentrated colloidal gels is qualitatively similar to what has previously been found in dilute fractal gels. However, for concentrated gels we obtain reduced values for τ_c and δ^2, reflecting the compactness of the dense gel (3). This suggests that the exponent $p = 0.7$ is a common feature of colloidal gels even when the development of a loose fractal structure is suppressed due to the high space filling. Based on their data obtained at low volume fraction gels, Krall and Weitz (3) suggested to describe the short time fluctuations of the gel as a series of overlayed damped oscillators with a upper cut-off linked to the characteristic size, R_c, of the gel. Thus, the maximum mean-square displacement and the characteristic relaxation time are both related to a characteristic spring constant $\kappa_c(R_c) = \kappa_0(a/R_c)^\beta$ of the system (given by the slowest mode), which then sets the scale for the macroscopic properties. Here κ_0 is the effective spring constant for bond bending between two particles, a is the particle radius and $\beta \approx 3.1$ an elasticity exponent. The elastic modulus of the gel is then related to the characteristic time τ_c as $G_0 = \kappa_c/R_c = 6\pi\eta/\tau_c$, where η is the solvent viscosity (3). Thus it is possible to extract G_0 for different particle sizes and volume fractions from light scattering measurements and compare them to the moduli measured with classical rheology. We indeed find good agreement between the light scattering data and the rheological measurements up to quite high particle volume fractions $\Phi \approx 0.1$. However, at higher values of Φ we observe significant deviations, and we are currently investigating possible reasons for this discrepancy.

The combination of DWS and rheological measurements thus demonstrates that at the sol-gel transition a dramatic change of the local dynamic behavior of the particles occurs which is directly related to the build-up of solid-like elastic properties. However, both types of experiments do not provide any information on the corresponding changes in the microstructure that would be important in any attempt to better understand this important ergodic-nonergodic transition. Therefore we have designed a combined SANS-DWS experiment at the SANS instrument of the Swiss neutron scattering facility SINQ which allows us to simultaneously measure both the local dynamics as well as the microstructure as the aggregation and gelation proceeds. For highly concentrated colloidal suspensions and gels a combination of SANS and diffusing wave spectroscopy (DWS) is of particular use. DWS provides information on the local dynamic properties of the individual particles, whereas SANS gives access to the strutural properties on similar length scales. The combination of both methods thus allows us to obtain structural and dynamic information on a very large range of length and time scales that could not be obtained otherwise. For the chosen instrument configuration we can in fact observe the aggregation and sol-gel transition with a temporal resolution of 10 - 15 minutes, which is sufficiently fast when compared to the aggregation kinetics chosen for these experiments.

In a first set of experiments using particles with radius a = 80 nm (polydispersity approximately 6%), we performed simultaneous SANS and DWS measurements at relatively high volume fractions between $0.19 \leq \Phi \leq 0.26$ (30). A typical set of data that characterizes the structural and dynamic evolution as a function of time is shown in figure 7. We observed that the structure factor peak which is typical for strongly correlated particles disappears as soon as the ionic strength increases due to the screening of the repulsive electrostatic interactions. However, while the DWS measurements shown as an inset in figure 7 indicate that dramatic changes in the local dynamics occur for a long time, the SANS pattern quickly reaches its final appearance. The q-dependence of the intensity is in fact in quite good agreement with the structures found in Monte Carlo simulations of colloidal suspensions with attractive particles (30). Our experiments thus indicate that a fluid-like structure is arrested in the course of the gel formation and that the changes in the gel dynamics are due to an increasing bond stiffness in the network and not caused by a structural evolution on length scales accessible in these experiments after the gel point.

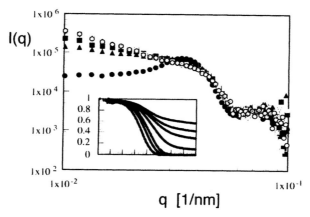

Figure 7: Time evolution of the q-dependence of the scattered intensity of a particle suspension with $\Phi = 0.26$ undergoing a sol-gel transition. The corresponding DWS results from the in-situ backscattering experiments are given in the inset. Shown are data (from bottom to top) for the stable suspension at $t = 0$ s and time evolution after destabilization for $t = 15, 30, 55, 80, 105,$ and 210 min (full symbols before the gel point and open symbols after the gel point).

However, these initial experiments suffered from several shortcomings. First of all, for the chosen monomer size the critical cluster is such that we cannot resolve the fractal structure with SANS. Moreover, in the DWS experiments we had to obtain ensemble averaged correlation functions using an approach in which we recorded a sufficient number of statistically independent time averaged intensity autocorrelation functions (for details see ref. (31)). We have thus modified our DWS-SANS experiment as shown in figure 8. A He-Ne laser beam is reflected by a mirror and illuminates a sample which is located in a multi-sample cell holder. The mirror is positioned in the incoming neutron beam

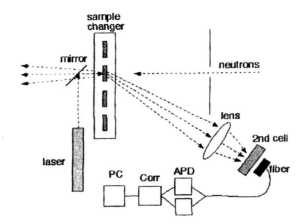

Figure 8: Schematic top view of the set-up for simultaneous light and small-angle neutron scattering..

under 45°. The multiple scattered light is then collected in transmission geometry using a modified two cell set-up. The two cells are separated by a distance of several centimeters. Light transmitted diffusively from the sample cell is imaged via a lens onto a 2nd cell (containing a highly viscous colloidal suspension of moderate optical density or a solid second cell (such as a glass frit) which is translated mechanically). Subsequently the light is detected with a single mode fiber, guided to two avalanche photodiodes (APD) and analyzed with a digital correlator performing a pseudo cross-correlation function. In order to increase the resolution of the SANS experiments we have also used a set of focusing neutron lenses.

In addition to the instrumental improvement, we have also moved to smaller particles, which do give us access to the fractal regime in SANS experiments and thus allow for a detailed and quantitative determination of the time evolution of the suspension and aggregate structure during the sol-gel process. This is illustrated with first experiments using latex particles with radius a = 9.5 nm, where we looked at volume fractions between $0.01 \leq \Phi \leq 0.1$. An example of the time dependence of the scattering experiment in a SANS experiment with a destabilized latex suspensions at $\Phi = 0.038$ is shown in Figure 9. At this concentration and for the given particle and solvent (mixture of H_2O/D_2O in order to obtain buoancy-match for latex particles (27)) we have sufficient scattering contrast to perform measurements with a sampling period of 15 minutes. We find that, for the chosen conditions, the aggregation process and the scattering curves rapidly evolve during the first 1-4 hours while afterwards no important changes could be detected anymore, similarly to what we have found in the experiments with the larger particles shown in Figure 7. However, in contrast to the previous data we clearly seen that even for the case of quite concentrated gels a power law regime can be identified. This is demonstrated in the inset of Figure 9, where data for the scattered intensity of the "final" gels from measurements at different volume fractions are plotted. From our data we consistently obtain a fractal dimensions of approx. 2.34 ± 0.06, which is somewhat higher than the theoretical predictions for DLCCA and RLCCA (*15*). At the highest concentrations we find clear deviations from the simple fractal power law due to the finite cluster size (Eqn. 1). This allows us to determine R_c from the SANS experiments using appropriate expressions for the structure factor of fractal aggregates and gels.

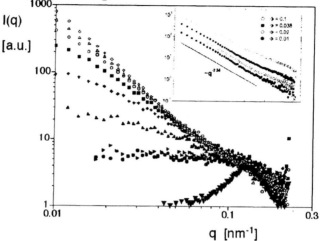

Figure 9: Time evolution of the scattered intensity from SANS experiments with destabilized latex suspensions (radius a = 9.5 nm, volume fraction Φ = 0.038). From bottom to top:Stable suspension and after 15 min, 1h, 2h, 3h, 4h, 5h, 6h, and 10 h. Inset: Intensity of the final gels for different volume fractions: Φ = 0.01, 0.02, 0.038, 0.1. Solid line: Power law $\sim q^{-2.34}$

At the same time we have also made an attempt to use DWS for the characterization of the dynamics of these gels. Our previous experiments have been mostly restricted to mesoscopic particles in the range 200-1000 nm. Application to smaller nanostructures represents a significant challenge both concerning the experimental techniques as well as the fundamental physical properties. DWS works best in the the limit of highly turbid systems where light is scattered from individual particles with a size comparable or larger than the wavelength, $ka \geq 1$. For $ka > 1$, DWS measures the (average) single particle mean square displacement while for smaller sizes contributions from collective modes become increasingly important (*25*). No attempt has been made however, to our knowledge, to analyze the dynamics of systems consisting of nanosized particles, i.e. where $ka \ll 1$. In a stable suspension such systems appear almost transparent at all densities and are not accessible to DWS. However, the dense colloidal gels become strongly turbid and hence in principle it should be possible to obtain valuable information about the local gel-dynamics from DWS. We were in fact able to derive a functional form for the correlation function

158

measured in a DWS experiment in transmission geometry from a colloidal gel with small primary particles which is based on the original work of Krall and Weitz and test it with our data obtained with the final particle gel shown in figure 9 (see ref (15) for details). We obtained a quantitative description of the measured data, and the resulting values for the key parameters l^* and R_c were in quantitative agreement with independent estimates. However, it is clear that in order to obtain a more complete picture of the scale dependent gel properties it will be necessary to study in more detail the concentration dependence of both the dynamic and static gel properties.

Conclusions

Our results show that a combination of SANS, 3D dynamic light scattering and DWS provides ample opportunities to access all relevant parameters necessary to characterize dense colloidal systems. Contrary to previous studies of concentrated suspensions these techniques are fully non-invasive and do not require additional modifications of the system properties such as index matching. A particularly interesting feature is the emerging link between the structural and dynamic behavior of the gels on microscopic length scales and their resulting macroscopic mechanical properties such as the gel modulus and its frequency dependence (31). This aspect of micro-rheological measurements may yield valuable information also in the case of numerous applied systems. This requires, however, a thorough (and further improved) understanding of how thermal fluctuations couple to a gel network at even higher densities. Another very interesting and rapidly progressing area is the investigation of the aging behavior of such dynamically arrested non-equilibrium systems. Here we have recently seen that new experimental approaches have been presented based on DWS measurements of colloidal gels (32,33).

Acknowledgement

We gratefully acknowledge financial support from the Swiss National Science Foundation, COST (Action P1) and the Swiss Federal Office for Education and Research. We thank the PSI, Villigen, Switzerland for providing the neutron research facilities and we gratefully acknowledge the expert help of our local contact Joachim Kohlbrecher.

References

1. Evans, D. F. and Wennerström, H., The Colloidal Domain: Where Physics, Chemistry, Biology and Technology Meet, VCH Publishers Inc.: New York, 1994.

2. Brinker, C. J. and Scherer, G. W., Sol-Gel Science: the Physics and Chemistry of Sol-Gel Processing, Academic Press: San Diego, 1990.
3. Krall, A. and Weitz, D. A., Phys. Rev. Lett. **1998**, 80, 778.
4. Gisler, T., Ball, R. C., and Weitz, D. A., Phys. Rev. Lett. **1999**, 82, 1064.
5. Weitz, D.A., and Oliveria, M., Phys. Rev. Lett. **1984**, 52, 1433; Weitz, D.A., Huang, J.S., Lin, M.Y., and Sung, J., Phys. Rev. Lett. **1984**, 53, 1657; Dimon, P., Sinha, S.K., Weitz, D.A., Safinya, C.R., Smith, G.S., Varady, W.A., and Lindsay, H.M., Phys. Rev. Lett. **1986**, 57, 595; Klein, R., Weitz, D.A., Lin, M.Y., Lindsay, H.M. Ball, R.C., and Meakin, P., Prog. Coll. Int. Sci. **1990**, 81, 161.
6. Dickinson, E., J. Coll. Int. Sci. **2000**, 225, 2 and references therein.
7. Sorensen, C.M., Aerosol Sci. Tech. **2001**, 35, 648.
8. Overbeck, E., Sinn, C., and Palberg, T., *Progr. Colloid Polym. Sci.* **1997**, 104, 117.
9. Urban, C., Romer, S., Scheffold, F., and Schurtenberger, P., *Macromol. Symp.* **2000**, 162, 235.
10. Weitz, D. A., Zhu, J. X., Durian D. J., and Pine D. J., In *Structure and dynamics of strongly interacting colloids and supramolecular aggregates in solution*; edited by S.-H. Chen, J. S. Huang and P. Tartaglia, Kluwer Academic Publishers: Dordrecht, 1992; p 731 .
11. Dawson, K., *Curr. Opin. Colloid Interface Sci.* **2002**, 7, 218.
12 Scheffold, F., Härtl, W., Maret, G., and Matijevic, E., *Phys. Rev. B* **1997**, 56, 10942.
13 F. J. P. Schuurmans, D. Vanmaekelbergh, J. van de Lagemaat and A. Lagendijk, *Science* **1999**, 284, 141.
14 Rojas, L F., Urban, C., Schurtenberger, P., Gisler, T., and Grünberg, H.H., *Europhys. Lett.* **2002**, 60, 802.
15 Rojas, L F., Vavrin, R., Urban, C., Kohlbrecher, J., Stradner, A., Scheffold, F., and Schurtenberger, P., *Faraday Discuss.* **2003**, 123, 385-400.
16 Russel, W. B., Saville D. A., Schowalter, W. R, *Colloidal Dispersions*, Cambridge University Press: Cambridge, 1989.
17 Härtl, W., Beck, Ch., and Hempelmann, R., *J. Chem. Phys.*, **1999**, 110, 7070.
18 Nilsen, S.J., and Gast, A.P., *J. Chem. Phys.*, **1984**, 101, 4975.
19 Overbeck, E., Sinn, C., and Watzlawek, M., *Phys. Rev. E* **1999**, 60, 1936.
20 Rojas, L F., Romer, S., Scheffold, F., and Schurtenberger, P., *Phys. Rev. E* **2002**, 65, 051403.
21 Pusey, P. N. in Neutrons, *X-Rays and Light: Scattering Methods Applied to Soft Condensed Matter*, edited by P. Lidner and T. Zemb; North Holland, Elsevier: Amsterdam (2002).
22 Schurtenberger, P., and Newman, M. E., in *Environmental Particles*, edited by J. Buffle and H. P. van Leeuwen, Lewis Publishers: Boca Raton, (1993), Chap. 2, pp. 37-115.
23 Maret, G., and Wolf, P. E., *Z. Phys. B* **1987**, 65, 409.
24 Pine, D. J., Weitz, D. A., Chaikin, P. M., and Herbolzheimer, E., *Phys. Rev. Lett.* **1988**, 60, 1134.

160

25 Weitz, D. A., and Pine, D. J., in *Dynamic Light Scattering*, edited by W. Brown, Oxford U. Press: New York, (1993), Chap. 16, pp. 652-720.

26 Gauckler, L. J., Graule, T. J., Baader, F. H., and Will, J., *Key Eng. Mater.* **1999**, 159, 135.

27 Romer, S., Scheffold, F., and Schurtenberger, P., *Phys. Rev. Lett.* **2000**, 85, 4980.

28 Xue, J.-Z., Pine, D. J., and Milner, S. T., *Phys. Rev. A* **1992**, 46, 6550.

29 Scheffold, F., Skipetrov, S. E., Romer, S., and Schurtenberger, P., *Phys. Rev. E* **2001**, 63, 61404.

30 Romer, S., Urban, C., Lobaskin, V., Scheffold, F., Stradner, A., Kohlbrecher, J., and Schurtenberger, P., *J. Appl. Cryst.* **2003**, 36, 1.

31 Scheffold, F., and Schurtenberger, P., *Soft Materials* (2003, to appear)

32 Bissig, H., Trappe, V., Romer, S., and Cipelletti, L., submitted to *Phys. Rev. Lett.*

33 Cipelletti, L., Bissig, H. Trappe, V., Ballesta, P., Mazoyer, S., *Phys. Cond. Mat.* **2003**, 15, 257.

Chapter 10

Aging of Soft Glassy Materials Probed by Rheology and Light Scattering

Eugene Pashkovski[1], Luca Cipelletti[2], Suliana Manley[3], and David Weitz[3]

[1]Colgate Palmolive Company, R&D, Technology Center, 909 River Road, Piscataway, NJ 08855
[2]GDPC UMR 5681, CNRS and Université Montpellier II, Montpellier, France
[3]DEAS, Harvard University, 40 Oxford Street, Cambridge, MA 02138

Aging of soft glassy materials studied by rheology, small-angle Dynamic Light Scattering (DLS) and Diffusing Wave Spectroscopy (DWS) is presented. We use these methods to obtain the dynamic signature of aging for various colloidal systems such as compressed water-in-oil emulsions, mixtures of surfactant, oil and water, and xanthan pastes. In all cases, we observe a significant reduction in the colloidal mobility with sample age as derived from measured intensity correlation functions. For all systems, an unusual shape of the age-dependent part of the correlation functions was found, ruling out conventional diffusive mobility. Using an analogy between molecular and colloidal glasses, we analyze a violation of the Generalized Stokes-Einstein Relation using a combination of DWS and rheology.

Rheological properties of soft glassy materials reflect their metastable structure and structural disorder (1). Slow structural relaxation (i.e. aging) of such materials causes rheological responses to depend on the age of the sample, t_w, defined as the time elapsed since the sample was quenched in the glassy state. Slow dynamics in such systems was recently discussed in terms of jamming (2); the drastic decrease in mobility occurs in the jammed state due to high concentration and/ or inter-particle interactions. Experimentally, slow dynamics can be probed by studying the response of the system to external perturbations or by directly measuring the dynamic structure factor. Experimental evidence of slow relaxations and aging was observed by Struik for polymer glasses (3). His approach was based on mechanical creep measurements that provide information on system's mobility evolution versus its age. Remarkably, very similar aging behavior was observed for many different glassy amorphous polymers as well as for molecular glasses. Molecular mobility that defines the age-dependent relaxation process of these systems diminishes with age t_w according to the simple power law, $M(t_w) \propto t_w^{-\mu}$, where μ is the aging exponent.

This general trend was also found for magnetic glasses; recently, aging of these systems has been the subject of intense theoretical and experimental interest (4). In these experiments, the sample is cooled below the glass transition temperature, T_g, while an external magnetic field is applied. The age is defined as the "waiting" time t_w spent in the glass phase, before cutting the field. At zero field, a rapid decrease of the magnetization is followed by a slow age-dependent relaxation. Remarkably, the age-dependent part of the magnetization shows the same scaling behavior as the creep compliance of polymeric glasses. In both cases, the response functions may be scaled as a function of t / t_w^{μ} implying that the relaxation time $\tau \propto t_w^{\mu}$.

The aging process has been studied theoretically using several approaches, which can be applied for broad classes of systems that display non-stationary dynamics (5-7). One of the main features of non-stationary dynamics is the violation of the fluctuation-dissipation theorem (FDT), which relates the response and correlation functions of the system at equilibrium:

$$\chi(t,t_w) = \frac{1}{T}[C(0,t_w) - C(t,t_w)] \qquad (1)$$

where $\chi(t,t_w)$ is the response at time t to a constant magnetic field applied at t_w, $C(t,t_w)$ is the correlation function, and T is the temperature (6). At equilibrium, a parametric plot of $\chi(C)$ vs. C yields a straight line, whose slope is $-1/T$. When the system is cooled below T_g, it stays out of equilibrium, and the dependence of $\chi(C)$ is non-linear. Thus, the slope of the curve $\chi(C)$ defines the FDT violation factor $X(C)$:

$$\frac{d\chi(C)}{dC} = -\frac{X(C)}{T} \qquad (2)$$

The concept of "effective temperature", $T_{\text{eff}} = T/X(C)$, defines the physical meaning of $X(C)$ for systems with slow dynamics. As the violation occurs for $X(C)<1$, the effective temperature is higher than the bath temperature, $T_{eff} > T$. During aging, the system "cools", i.e. $X(C) \rightarrow 1$ for $t_w \rightarrow \infty$. Violations of FDT were analyzed for spin (7) and molecular glasses such as glycerol below the glass transition (8,9). In these measurements, the weak violation of FDT was observed such that $X(C)$ increased and T_{eff} decreased with aging time. Thus, FDT- violation is certainly a signature of aging behavior of broad classes of glassy systems.

Jammed colloidal systems as "soft glasses"

Jammed colloidal systems have many common features with polymer, spin or structural glasses. This similarity was recognized by Sollich (10). His theory on the rheology of "soft glassy materials" (SGM) allows one to incorporate the so called "trap model" (11) developed for glasses. The rheological response of SGM depends on age as a result of inherent metastability and restricted mobility. Experimental studies of soft microgels (12) and concentrated silica suspensions (13) have shown that the low frequency data can be scaled similarly to the magnetization curves for spin glasses, or the creep curves for polymers. These analogies suggest that aging phenomena in colloids may be studied by investigating the violation of the FDT, as it is done for magnetic and molecular glasses. For colloidal systems, a convenient way to do so is to study the violation of the Generalized Stokes-Einstein Relation (GSER).

At equilibrium, the GSER provides the relationship between the mean square displacement of particles of radius a and the viscoelastic properties of the complex fluid (14). The viscoelastic spectrum $\tilde{G}(s)$ as a function of Laplace frequency (15) s is given by:

$$\tilde{G}(s) = \frac{k_B T}{\pi a s < \Delta \tilde{r}^2(s) >} \qquad (3)$$

Here, k_B is the Boltzmann constant, the tilde denotes the unilateral Laplace transform. The time-domain creep compliance can be obtained from eq. (3) and the simple relationship between the shear modulus and the shear creep compliance, $\tilde{J} = 1/(s\tilde{G})$ (15):

$$J(t) = \frac{\pi a}{k_B T} < \Delta r^2(t) > \qquad (4)$$

Measurements of the mean square displacements of monodisperse tracer particles allows viscoelastic parameters of complex liquids to be calculated (14-17). In these experiments, the value of $< \Delta r^2(t) >$ is obtained by dynamic light scattering (DLS) techniques, either diffusing wave spectroscopy (DWS) (16) or single scattering (17). This approach, if used for such systems as semi-dilute polymer solutions or moderately concentrated colloidal dispersions, allows one to calculate the viscoelastic properties, as the GSER is not violated.

Violation of Generalized Stokes-Einstein Relation using light scattering study

Jammed systems such as foams, colloidal gels and concentrated pastes are metastable systems for which GSER is generally violated. The violation factor depends on the observation timescale, and increases with time (decreases with frequency). Despite the metastability, the system slowly evolves towards the equilibrium state and therefore its rheological properties depend on age. As a result, there are two different time scales to be considered, the observation time t and the aging time t_w. Typically, for $t \ll t_w$ GSER is not violated and therefore high-frequency responses may be age-independent. For instance, a good estimate of the high-frequency rheological response was obtained from DWS measurements of the mean square displacement of monodisperse droplets in concentrated emulsions (18). To investigate the violation of GSER in jammed systems, one should extend the measurements of $< \Delta r^2(t) >$ to much longer times, $t \geq t_w$. For traditional detection schemes that use a single detector (19), the value of $g_2(t)$ must be extensively averaged over time; the measurement time is typically several orders of magnitude longer than the system relaxation time. The single-detection scheme becomes impractical for $t \geq t_w$; the dynamics may be age-dependent for aging systems. To probe the slow dynamics, so-called multi-speckle detection schemes (20-22) based on a 2D CCD camera detector have been developed. These techniques allow one to decrease the measurement time by several orders of magnitude, down to the order of the system relaxation time. Aging effects probed by multi-speckle techniques can be compared with age-dependent rheology. This gives information on aging at different time and length scales. The displacement that is resolved in our rheological experiments is 0.5-1 microns, whereas DWS can resolve displacements of particles as low as a few nanometers. Comparing the dynamic behaviors on meso and micro scales allows investigating violation of GSER for colloidal systems, in analogy with FDT violation for magnetic and molecular glasses.

Aging of jammed systems

One of the important features discovered for colloids, is the unusual shape of the intensity correlation function which decays *faster* than an exponential function, $g_1(q,t) \propto \exp[-(t/\tau)^p]$ with $p>1$. This unusual shape was discovered using multi-speckle DLS for low-volume fraction colloidal gels formed by 10-nm polystyrene particles (21). The parameter p that defines the shape of the "compressed" exponential was $p=1.5$. Furthermore, the relaxation time was found to depend on the scattering vector as $\tau \propto q^{-1}$. The latter rules out the diffusive motion of particles that sets $\tau \propto q^{-2}$. A phenomenological model explaining this behavior introduces local stress sources that appear as a result of shrinking of the gel (21). A more rigorous model developed by Bouchaud and Pitard (23) computes the dynamic structure factor based on random appearance of micro-collapses in a gel, resulting in motions of particles that contribute to the decorrelation of the scattered light intensity. Bouchaud and Pitard also found that $p=1.5$ in an early-time regime; however, for intermediate-time regime $p\sim1.25$ and for long-time regime, the "compressing" exponent reaches $p \to 1$. Though both models ($21,23$) were developed for particulate gels, the same unusual shape of the correlation function was observed for widely different colloidal systems, such as compressed emulsions, "onion" phases, and micellar cubic phase (24). For these systems, DLS measurements have reproduced the "compressed" exponential shape of the dynamic structure factor $f(q,t) \approx g_1(q,t) \propto \exp[-(t/\tau)^p]$. Note, that the DWS measurements for Laponite gels also reproduce this behavior (25). In this case, the tracer micro particles were dispersed in the gel to induce multiple scattering. For a given q in DLS, the relaxation time increases with sample's age as $\tau \propto t_w^\mu$. The aging exponent μ appears to be system-dependent and, in some cases, it changes with t_w. All this observations reveal the universality of slow dynamics of jammed colloidal systems having different structural organization. In this Chapter, we discuss aging behavior of compressed emulsions, multiphase oil-in water emulsions, and microgel suspensions studied by DLS, DWS, and by rheological methods. DLS has been a very useful tool to characterize slow dynamics but it is limited to the single-scattering regime. Multispeckle DWS may be applied to a much wider class of systems, including compressed emulsions with non-matching refractive indices in their continuous and dispersed phases and concentrated colloidal suspensions and gels. However, characterization of aging behavior using DWS is still poorly understood. As observed in DLS experiments, DWS measurements reveal similar aging effects that occur for very different systems. To demonstrate this, we compare transparent compressed emulsions with two turbid systems, a mixture of Pluronic F108, water and mineral oil and a xanthan paste with dispersed tracer particles.

Compressed emulsions

When the volume fraction of droplets ϕ exceeds the critical packing volume fraction ϕ_c an emulsion system becomes compressed with elasticity $G'(\phi) \propto (\sigma/a) \cdot (\phi/\phi_c - 1)$ (26). The Laplace pressure (σ/a), the ratio of surface tension to the droplet radius, depends on preparation conditions and the nature and the concentration of the surfactant which stabilizes the droplets. In our study, the water phase (ϕ=0.77) was dispersed in the oil phase (cyclomethicone), which contained a silicon-based polymeric surfactant, (Copolyol, Dow Corning 522C, c=0.25 wt. %). Refractive indices of water and oil phases were matched by adding PEG-600 and a salt (NaCl) to the water phase. The two phases were separately prepared, than mixed using high-speed lab mixer. The speed was selected such that the average particle size was 1.0 ± 0.2 μm. Confocal microscopy shows no coalescence of droplets during aging for this polydisperse emulsion. The emulsion was centrifuged for 15 min at rate between 1500 and 1900 rpm, to avoid scattering from air bubbles entrapped in the optical cell. The end of the centrifugation was set as $t_w = 0$.

Figure 1a shows the dynamic structure factors measured simultaneously for scattering vectors parallel to the acceleration imposed by centrifugation for t_w=300s. The dynamic structure factor decays as $f(q,t) \approx \exp[-(t/\tau)^p]$ with $p \approx 1.5 \pm 0.08$ for all values of scattering vectors q. The q-dependence of the relaxation time has a very unusual form, $\tau \propto q^{-0.9 \pm 0.3}$ (see inset), ruling out the diffusive motion of droplets as a mechanism responsible for the decay of structure factor (for diffusive motion, $\tau \propto q^{-2}$). Figure 1b shows the age dependence of the relaxation time measured at q=0.22 μm^{-1}. As can be seen, the relaxation time τ dramatically increases with age, but decreases with centrifugation speed suggesting the role of stress relaxation; the build-in stress is likely to be higher for the samples centrifuged at higher rate (24). The dependence of mobility on age indicates similarity between aging of colloids and molecular and spin glasses. Moreover, slow non-diffusive dynamics implies that GSER must be violated in the low frequency range. Analysis of this violation is given below for xanthan paste using combination of DWS and rheology.

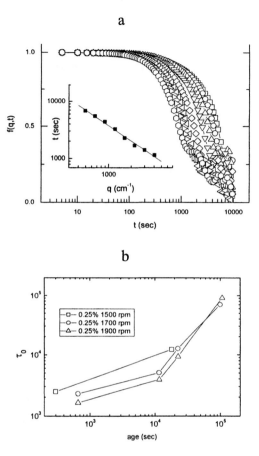

Figure 1   a: Final decay of the dynamic structure factor measured by multispeckle DLS for a concentrated emulsion of age $t_w \approx 600$ sec. Lines are "compressed" exponential fits to the data (see text). Inset: q dependence of the characteristic relaxation time of the final decay of f(q,t). The line is a power law fit to the data, yielding an exponent 0.90 ± 0.03, indicative of "ballistic" motion on long time scales.

b: Age dependence of the characteristic time of the final decay of the dynamic structure factor for a compressed emulsion. The different symbols refer to samples centrifuged at different rates for the same time, 15 minutes (Reproduced with permission from Ref. 24. Copyright 2003 Royal Society of Chemistry).

The multiphase mixture of Pluronic, water and oil

The multiphase mixture of Pluronic F108, HO- $(CH_2CH_2O)_{127}$- $[CH_2CH(CH_3)O]_{48}$-$(CH_2CH_2O)_{127}OH$, water, and mineral oil (27.8/44.4/27.8%) was prepared using moderate mixing speed. The size of oil droplets was found to be in the range of 1-5 μm. Similar systems are well described in ref. (27). The emulsion was heated and kept at 73^0C in a Linkam cell (CSS 450, sample thickness 2 mm) for 20 min; then, the sample was cooled down to 23 ^0C; this moment defined the sample age t_w. The field correlation functions are presented in Figure 2a for different t_w. All curves decay as "compressed exponential" functions with the relaxation time increasing with age (Fig.2b); initially, $\tau \propto t_w^{1.3}$ but with increasing age this dependence becomes weaker, and finally, the aging stops. Interestingly, the "compression" exponent p slightly increases with age (Fig. 2c). The result that the aging "stops" after several hours is confirmed by the stress relaxation measurements for this system with identical thermal history (Fig. 2d). The relaxation times estimated by fitting the curves to the two-element stress relaxation model ($G(t) = G_0 + G_1 e^{-t/\tau_1} + G_2 e^{-t/\tau_2}$) do not depend on sample age for t_w>2000s. Moreover, $G(t)$ does not completely relax approaching a constant value G_0 of the relaxation modulus, thus showing that the system is very rigid ($G_0 \approx 10^4$ dyn/cm^2). It is possible that a complete relaxation would occur at times longer than the time of the relaxation experiments, 10^3s; this may be the case as DLS shows the complete decay of the dynamic structure factor to occur at t>10^4s for similar systems (24). It should be noted that DWS probes much shorter length- and time-scales in comparison to rheology. The fact that the microscopic mobility (estimated by DWS) differs so dramatically from macroscopic dynamics (obtained from stress relaxation data) may be one of the fundamental features of the non-stationary dynamics. The difference between mesoscale dynamics and microscopic mobility probed by DWS and rheology respectively, indicates that GSER is violated. The non-diffusive mobility of colloid particles in soft glasses and GSER-violation underline the universality of aging behavior of these systems, as well as their similarity to spin glasses (4).

Xanthan pastes

To further investigate the relationship between DWS and rheology for systems that display aging phenomena, a xanthan paste was selected. Suspensions of xanthan microgel particles in (1:1) water-glycerin mixture were prepared. The elastic modulus of the solutions changes linearly with concentration, $G' \propto (c-c_0)$ with $c_0 = 0.08\%$. Above this concentration, the microgel particles form an elastic gel with the storage modulus higher than the loss modulus in a wide frequency range (28), similar to the compressed

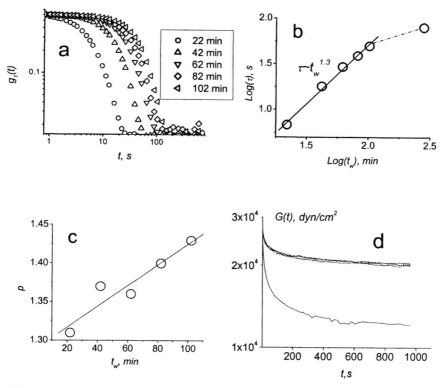

*Figure 2. Mixture of Pluronic F108, water and mineral oil (27.8/44.4/27.8%).
a- correlation functions $g_1(t,t_w)$ collected for $t_w =22$ -102 min that can be fitted
to $g_1(t,t_w) \propto \exp[-(\tau/t)^p]$; b- relaxation time τ vs. aging time t_w; c- the
"stretching" exponent p vs. t_w; d- The relaxation modulus is the step-strain
experiment vs. aging time, the curves collected for $t_w>1500s$ are identical.*

emulsions (26). A paste with $c = 1.45\%$ was utilized in this study. The condition of $c \gg c_0$ indicates that the individual xanthan microgel particles are strongly deformed. Using regular DWS for 0.4% xanthan paste (29), the mean square displacement of probe particle was found to decrease exponentially with particle size. Furthermore, strongly subdiffusive mobility of particles with the mean square displacement $< \Delta r^2(t) > \propto t^{0.45}$ suggests that the probe particles are "caged" between xanthan microgels. Thus, any rearrangements of the xanthan microgels must be reflected in the dynamics of the probe particles. These particles are much bigger than the microgels mesh size $\xi \sim 70 nm$, as measured by AFM (29). The dimensions of microgel blobs and the tracers are approximately similar.

Creep measurements

Rheologically, aging behavior of the xanthan paste was studied using a strain-controlled rheometer (ARES, Rheometric Scientific) in the creep mode. The initial preshear sets the sample age. The creep measurements are made using a very small stress (1 dyn/cm^2) to ensure conditions close to free aging. This condition is fulfilled when the applied constant stress is much less than the yield stress which was found to be ~ 15 dyn/cm^2. The preshear (100 strain units) is applied and a number of consecutive retardation and relaxation runs are performed such that each new retardation experiment starts when the sample is $\sim 90\%$ relaxed from the previous run. Creep response curves clearly depend on sample age (Fig.3a). The curves can be scaled onto a mastercurve by horizontal shifting with the shift factor depending on the sample age (see insert in Fig.3a); $J(t', t_w) \propto t' / t_w^{0.86}$, i.e. the system is close to full aging ($\mu = 1$). Even for $t' / t_w^{0.86} \gg 1$ no steady flow was observed and $J(t) \propto t^{0.88}$. Thus, the age-dependent relaxation time can be found by fitting the data to the single-element Voight model with no flow term ($J(t) = J_0 + const \cdot [1 - \exp(-t / \tau)]$). We found that for $t_w > 700s$, the relaxation time changes almost linearly with aging time, $\tau \propto t_w^{0.95 \pm 0.1}$ (Fig. 3b).

DWS Measurements

The comparison between rheological and DWS measurements for the xanthan paste can further explore the analogy between colloids and spin and molecular glasses. We use a camera-based DWS setup (20) in the transmission geometry, that allows for studying slow mobility of gels due to the multi-speckle detection. To ensure that we are in the multiple scattering regime, 1.1 μm latex tracers were added; the mean free path length $l^* \sim 200$ μm was small compared to the cell thickness (2 mm). A crossed polarizer was used to avoid the detection of the transmitted light. The intercept of the correlation function $g_2(0, t_w)$ depends on the optical setup; we normalized this value using the regular DWS setup for which $g_2(0, t_w) = 1$. In our experiment, the value of $g_2(0, t_w)$ was found to increase with t_w from $g_2 \sim 0.87$ to 0.99 (Fig. 3c). This is equivalent to an increase of the elastic modulus $G(0, t_w) = k_B T / 2\pi a < \Delta r_0^2(0, t_w) >$ upon aging. The mean square displacement of particles $<\Delta r^2(t)>$ calculated using standard DWS formalism (16),

$$g_1(t) = \sqrt{g_2(t) - 1} = \int_0^\infty P(S) \exp\left(-\frac{1}{3} k_0^2 < \Delta r^2(t) > \frac{S}{l^*}\right) dS, \qquad (5)$$

is unusually low, i.e. several orders of magnitude lower than the particle radius. In this equation, $P(S)$ represents a distribution function of the light path length S through the sample, l^* is the mean free path of photons, and $k_0 = 2\pi n / \lambda$ is the wave vector of the incident light ($\lambda = 488$ nm) in the medium with refractive index n. This formalism was developed for dynamically "homogeneous" systems with continuous motion of all scatterers. There are systems where the dynamic events are localized both in space and time (30); it is suggested that temporal inhomogeneities are a fundamental feature of dynamics of jammed systems such as foams and colloidal gels. In this case, the use of eq. (5) for calculating the mean square displacement of particles is problematic.

The intensity correlation function curves can be scaled horizontally using the reduced time $t'/t_w^{1.5}$ (Fig. 3d) leaving short-time data unscaled (additional vertical scaling, not used in Fig 3d, would perfectly collapse all curves). The correlation functions can be fitted to the exponential $g_2(t', t_w) = g_2(0, t_w) \cdot \exp\{-[t'/\tau]^p\}$ with the "stretching" exponent values $1.45 < p < 1.65$. The relaxation time increases with t_w as $\tau \propto t_w^{1.5}$, i.e. faster, than in the creep compliance experiment ($\mu = 0.86$). In addition, the relaxation time is shorter for DWS. The high value of the aging exponent obtained by DWS may be due to aggregation of the tracer particles; the increase of the plateau values $g_2(0, t_w)$ with t_w demonstrates some kind of structural evolution. In contrast to this, the creep compliance curves scale perfectly together using only horizontal shifting showing no structural changes but purely aging effects.

Fluctuation-dissipation plot

The high-frequency rheological parameters are in reasonably good agreement with those calculated from DWS using eq. (5), $J(0, t_w = 3h) \approx 2.5 \cdot 10^{-3}$ cm/dyn^2, whereas the value calculated from DWS is $2\pi a < \Delta r_0^2(t_w = 3h) > / k_B T$ $\approx 2.2 \cdot 10^{-3}$ cm/dyn^2. This indicates that the GSER holds for $t' = t - t_w \ll t_w$ but

172

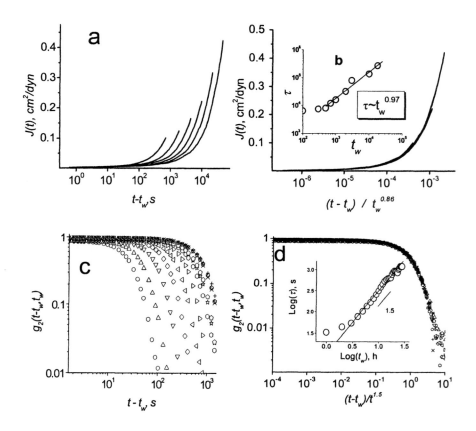

Figure 3. Xanthan paste (1.5 wt. % in 1:1 water-glycerin mixture). a - creep compliance curves recorded at different t_w using a small stress (1 dyn/cm²); b- the scaled creep compliance curves and the dependence of creep relaxation time vs. aging time (the insert); c-the intensity correlation functions recorded at different t_w ; d- the scaled curves and the dependence of DWS relaxation time vs. aging time (the insert).

is violated for $t \approx t_w$. One can compare mechanical and DWS data to analyze the violation of GSER (see example for t_w=3h in Fig. 5). In this plot, called a fluctuation-dissipation (FDT) plot, rheologically measured creep compliance is given as a function of $< \Delta r^2(t) > \pi a / k_B T$. The straight line in this plot represents the GSER, and it is violated for long observation times. Note that the standard formalism to calculate $< \Delta r^2(t) >$ is applied here, assuming temporally and spatially homogeneous dynamics; if this is not the case, the analysis of GSER violation is problematic. Indeed, preliminary tests using the Time Resolved Correlation analysis proposed in *(30)* show no heterogeneous dynamics, at least for t_w<5 hrs.

The FDT plot for t_w=3hrs (Fig. 4) may be interpreted similarly to FDT plot for spin glasses. At short times, the slope of the creep compliance vs. $< \Delta r^2(t) > \pi a / k_B T$ is close to 1. The deviation from the straight line for longer times is a result of the GSER violation, which is, by analogy with spin and molecular glasses, interpreted as due to an increase of the effective temperature *(6)*. The compressed exponential decay of the correlation function implies that $< \Delta r^2(t) > \propto t^{1.6}$ for long time. The creep compliance, on the other hand, changes as $J(t) \approx [1 - \exp(-t/\tau)] + t/\eta \propto t$. This discrepancy between the time dependences of $J(t)$ and $< \Delta r^2(t) >$ is a signature of the GSER violation. This may explain why an unusual "compressed" exponential decay of the correlation function is observed for a broad class of colloidal systems.

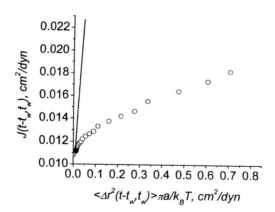

Figure 4. Fluctuation-dissipation plot for t_w=hr: dependence of $J(t - t_w, t_w)$ vs. $< \Delta r^2(t - t_w, t_w) > \cdot \pi a / k_B T$

Conclusion

Light scattering experiments show that jammed colloidal systems such as water-in-oil emulsions, concentrated mixtures of water, oil and surfactants, and microgel pastes display slow non-diffusive dynamics. The combination of light scattering and rheology allows one to analyze the relationship between the mesoscopic and microscopic mobility; this analysis shows the low-frequency violation of GSER, in analogy with FDT-violation for molecular and spin glasses. Thus, the non-diffusive dynamics of jammed colloids may be interpreted in terms of age-dependent FDT- violation, as is done for spin and molecular glasses. According to this concept, the degree of FDT-violation decreases with the sample age which is equivalent to "cooling" of the system during aging. These methods provide a new approach to characterizing aging phenomena in colloidal systems.

Acknowledgements

We gratefully acknowledge financial support from Colgate-Palmolive Co. and from CNRS and Région Languedoc Roussillon. We thank M. Johansson for preparation of compressed emulsions.

References

1. Solich, P.; Lequeux, F.; Hebraud, P.; Cates, M.E.; *Phys. Rev. Lett.,* **1997**, *78*, 2020.
2. *Jamming and Rheology. Constrained Dynamics on Microscopic and Macroscopic Scales.* Liu, A. J.; and Nagel, S. R., Eds. Taylor & Francis, London and New York 2001.
3. Struik, L.C.E.; *Physical Aging in Amorphous Polymers and Other Materials.* Elsevier, NY 1978.
4. Vincent, E. ; Hammann, J. ; Ocio, M. ; Bouchaud, J.-P. ; Cugliandolo, L. F.; Slow *dynamics and Aging in Spin Glasses. In: Complex Behavior of Glassy Systems,* Rubí, M.; Pérez-Vicente, C.; Eds., Springer, 1997, p.184.
5. Cugliandolo, L. F.; Dean, D. S.; Kurchan, J.; *Phys. Rev. Lett.,* **1997**, *79*, 2168 .
6. Cugliandolo, L. F.; Kurchan, J.; Peliti, L.; *Phys. Rev. E,* **1997**, *55*, 3898.
7. Cugliandolo, L. F.; Grempel, D. R.; Kurchan, J.; Vincent, E. ; *Europhys. Lett.,* **1999**, *48*, 699.

8. Grigera, T. S.; Israelov N.E.; *Cond-mat/9904351*.
9. Leheny, R. L.; Nagel, S.R.; *Phys. Rev. B*, **1998**, *57*, 5154.
10. Sollich, P.; *Phys. Rev. E*, **1998**, *58*, 738.
11. Bouchaud, J.-P.; *J. Phys. A*, **1996**, *29*,3847.
12. Cloitre, M.; Borrega, R.; Leibler, L.; *Phys. Rev. Lett.*, **2000**, *85*, 4819.
13. Derec, C.; Ajdari, A.; Ducouret, G.; Lequeux, F.; *C. R. Acad. Sci. Paris*, *1(ser IV)*, 1115.
14. Schnurr, B.; Gittes, F.; MacKintosh, F.C.; Schmidt, C.F.; *Macromolecules*, **1997**, 30, 7781.
15. Mason, T. G., Gisler, T., Kroy, K., Frey, E., Weitz, D. *J. Rheol.*, **2000**, *44*, 917.
16. Weitz D. A., Pine, D. J. , *Dynamic Light Scattering*, Edited by W. Brown, Oxford, 1992, Chapter 16.
17. Dasgupta, B. R., Tee, S.-Y., Crocker, J. C., Frisken, B.J., and Weitz, D.A. *Phys. Rev. E*, **2002**, *65*, 051505.
18. Mason, T. G., Gang, Hu, Weitz, D.A. *J. Opt. Soc. Am. A*, **1997**, *14*, 139.
19. Nobbmann, U. ; *Characterization down to nanometers: Light scattering from proteins and micelles* In: "Mesoscale Phenomena in Fluid Systems", Eds. Case, F and Alexandris, P., ACS 2003.
20. Cipelletti, L.; Ramos, L.; *Current Opinion in Colloid & Interfacial. Science.*, **2002**, *7*, 228.
21. Cipelletti, L.; Manley, S.; Ball, R.C.; Weitz, D.A.; *Phys. Rev. Lett.*, **2000**, *84*, 2275.
22. Schturtenberger, P. , Bissig, H., Vavrin, R., Stradner, A., Scheffold F., Trappe, V..; Aggregation and gelation in colloidal suspensions - time-resolved light and neutron scattering experiments. In: "Mesoscale Phenomena in Fluid Systems", Eds. Case, F and Alexandris, P., ACS 2003.
23. Bouchaud, J.-P.; Pitard, E.; *Eur. Phys. J.*,**2001**, *E6*, 231.
24. Cipelletti, L.; Manley, S.; Weitz, D.; Pashkovski, E.; Johansson, M.; *Faraday Discuss.*, **2003**, *123*, 237.
25. Knaebel , A., Bellour, M., Munch , J.P., Viasnoff, V., Lequeux, F., Harden, J.L. ; *Europhys.Lett.*, **2000**, *52*, 73.
26. Mason, T. G. , Bibette, J., Weitz, D. A. *Phys. Rev. Lett.* **1995**, *75*, 2051.
27. Alexandris, P.; *Small-angle scattering characterization of block copolymer micelles and lyotropic liquid crystals*. In: "Mesoscale Phenomena in Fluid Systems", Eds. Case, F and Alexandris, P., ACS 2003.

28. Pashkovski, E.E.; Masters J.G.; Mehreteab, A.; *Langmuir*, **2003**, *19*, 3589.
29. Kirby, A.R., Gunning, A. P., Morris, F. G.; *Carbohydrate Research*, **1995**, *267*, 161.
30. Cipelletti, L.; Bissig, H.; Trappe, V.; Ballesta, P.; Mazoyer, S.; *J. Phys. Condens. Matter*, **2002**, *15*, S257; Bissig, H., Trappe V., Romer S., Cipelletti, L.; preprint no. cond-mat/0301265.

Chapter 11

Nanoscale versus Macroscale Friction in Polymers and Small-Molecule Liquids: Anthracene Rotation in PIB and PDMS

Mark M. Somoza[1,2] and Mark A. Berg[1,*]

[1]Department of Chemistry and Biochemistry, University of South Carolina, Columbia, SC 29208
[2]Current address: Institut für Physikalische und Theoretische Chemie, Technische Universität München, Lichtenbergstrasse 4, 85748 Garching, Germany

As an object becomes smaller than the size of a polymer molecule, the friction it experiences in a polymer melt may be altered from the friction on a macroscopic object. These mesoscopic dynamics are unique to polymeric solvents. The mechanisms causing this effect and the transition from small-molecule to polymer-like behavior have been explored by using the rotation time of anthracene as a measure of nanoscale friction in liquids. The viscosity experienced by anthracene molecules is determined as a function of polymer chain length over the range from the dimer to the entangled polymer. The nanoviscosities in poly(dimethylsiloxane) (PDMS) and poly(isobutylene) (PIB) develop very differently as a function of chain length, despite similar static structures of the polymers. These results are attributed to higher torsional barriers in PIB than in PDMS. We suggest that a dynamic length scale is important in determining the friction experienced by mesoscopic particles in polymers.

Introduction

For macroscopic objects moving in simple fluids, continuum hydrodynamics is highly successful. With the recognition of viscoelastic and high shear effects, motion in polymeric fluids can also be treated with hydrodynamics. However, as the size of a moving object becomes sub-micron, the continuum assumption of hydrodynamics comes into question. Nevertheless, the Stokes–Einstein–Debye (SED) model, which applies continuum hydrodynamics to molecular-sized objects, has considerable success in small-molecule fluids [see (1) and references in (2)]. However, for large-molecule, polymeric fluids, hydrodynamics is expected to breakdown for nanoscale objects.

In the studies summarized here (2-5), we look at the breakdown of the SED model as the molecular size of the fluid increases from the small-molecule to the polymeric regime. The aim is to identify the length scale associated with the breakdown of hydrodynamics and to understand the mechanisms governing the motion of nanoscale objects in polymers. Based on our results, we suggest that torsional relaxations of the polymer backbone are the dominant mechanism for facilitating the motion of small objects. The important length scale defining "small" is a dynamic one determined by the average distance between torsional relaxations that occur within the characteristic time of the object's motion.

As our probe of small object motion, we measure the rotation time of a solute molecule. In general, hydrodynamics predicts that the rotation time τ_r of a particle due to Brownian motion is determined by the ordinary viscosity η_∞ of the solvent,

$$\tau_r = \frac{\lambda V_h}{6kT} \eta_\infty + \tau_0,\qquad(1)$$

where k is the Boltzmann constant, T is the temperature and V_h is the hydrodynamic volume of the particle (1). The particle shape and boundary conditions determine the parameter λ. The time τ_0 is added phenomenologically to correct for deviations at very low viscosity (6).

For a nanometer-sized object, this equation may not hold. However, the object still experiences a friction, which can be expressed as a viscosity by inverting eq 1,

$$\eta_{nm} = \frac{6kT}{\lambda V_h}(\tau_r - \tau_0).\qquad(2)$$

The concept of generalizing hydrodynamics to molecular levels by including length dependent properties is well established (7). In polymers, a number of important lengths are conceivable: the radius of gyration the polymer chain, the spacing of entanglements, the length of an effective freely jointed segment, the width of the chain. These lengths may define a variety of different of size ranges where different mechanisms control the effective viscosity on an object.

We work with an anthracene molecule as the probe object. It has a maximum dimension of ~1 nm (Figure 1), so we call the effective viscosity on it the "nanoviscosity" and differentiate it from the ordinary viscosity on a macroscopic object—the "macroviscosity". This object size is larger than the chain width of the polymers studied here, but can be smaller than other important lengths characterizing the polymer. Our terminology leaves open the possibility that a different viscosity may apply at other length scales, e.g., an "angstrom-viscosity" for objects smaller than the chain width (e.g., He, O_2) or a "100-nm-viscosity" for objects approaching the spacing of entanglements. With this terminology, the SED model is equivalent to $\eta_\infty = \eta_{nm}$. This relationship is expected to hold if the solvent molecules are sufficiently small relative to the object.

Solute rotation in either small-molecule solvents or polymers is a well-studied subject. More extensive references to this literature can be found in references 2, 8 and 9. The new features of our studies are: (1) data over an extensive molecular-weight range that connect studies in small-molecule solvents to those in polymeric solvents, (2) the detailed comparison of two polymers with similar static structure, but different torsional dynamics, and (3) the careful calibration of the hydrodynamic behavior of the probe object, a feature that allows the measurement of the absolute values of the nanoviscosity. Most of the

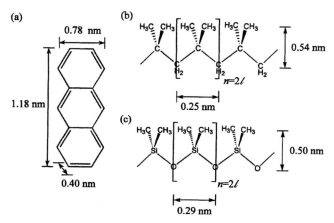

Figure 1. Structures and van der Waals dimensions of: (a) anthracene, (b) poly(isobutylene) (PIB), (c) poly(dimethylsiloxane) (PDMS). (Reproduced from reference 2. Copyright 2002 American Chemical Society.)

existing studies of solvent-size effects have been limited to small molecules. In polymers, studies of temperature dependence at a single molecular weight are most common. Solute rotation at temperatures near the glass transition yield a variety of important phenomena (9,10). However, our studies are far from the glass transition and are not likely to be influenced by these effects.

Other methods can also be used to study polymer motion at short length scales including neutron scattering and computer simulation. Particularly relevant examples for this work are comparisons of neutron scattering in PIB and PDMS by Arbe et al. (11) and simulations of the effect of torsional barriers by Krushev et al. (12). These and many related studies focus on the dynamics of pure polymers. In contrast, this study of solute motion involves both the dynamics of the polymer itself and the mechanism of coupling a small, foreign object to those motions.

We approach this material in the spirit of a review, emphasizing the major results and the interpretation suggested by them. Many important experimental details, supporting data and arguments, discussions of less successful interpretations and overlap with other approaches are omitted. The interested reader is referred to the original papers (2-5).

Calibration and Validation of a "Nanoviscometer"

Anthracene was chosen as nanometer-scaled probe because it is a near ideal hydrodynamic rotor. Anthracene follows the SED relationship (eq 1) closely as long as the solvent molecule is similar in size or smaller than the anthracene molecule. Figure 2 shows the rotation times of anthracene in various small-molecule solvents as a function of the solvent viscosity along with a fit to eq 2. The resulting value of $\lambda V_h/6kT = 8.68$ ps/cP corresponds closely with the predicted hydrodynamic parameters of a $1.18 \times 0.75 \times 0.42$ nm ellipsoid with slip boundary conditions. (Slip boundary conditions are often found for molecular rotation (1).) These hydrodynamic dimensions agree well with the van der Waals dimensions of anthracene (Figure 1).

This study is important both because it provides the calibration parameters for our definition of nanovisosity (eq 2), and because it validates the concept of using anthracene as a viscometer. Deviations from simple hydrodynamic rotation occur for many solutes due to strong solvent-solute interactions or a non-rigid structure. However, none of these problems occurs in anthracene. Solvent specific deviations from the SED relationship are present and complicate fits over a small viscosity range (inset, Figure 2). However, the dominant role of normal hydrodynamics becomes clear over a the large viscosity range used in this study.

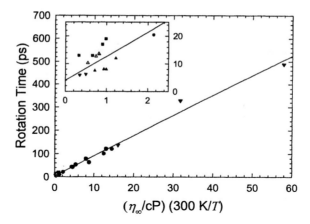

Figure 2. Rotation times of anthracene dissolved in various small-molecule solvents and the fit of the data points to eq 1. (■) – CCl_4, C_2Cl_4, C_6D_6, C_3D_6O (13); (▲) – cyclohexane (14); (△) – toluene; (λ) – benzyl alcohol; (▼) – ethylene glycol, cyclohexanol, isooctane, 162 g/mol PDMS. (Reproduced from reference 2. Copyright 2002 American Chemical Society.)

In the end, our use of anthracene rotation as a viscometer is very similar to the use of a capillary viscometer. In both cases, a hydrodynamic problem is solved to generate an equation that predicts an experimentally measurable time from the viscosity. The validity of the hydrodynamic model is verified on known solutions. These known solutions also provide the calibration factors for the viscometer. The hydrodynamic equation is then inverted to give the viscosity of unknown samples from measured times.

The essential difference between a capillary viscometer and the anthracene viscometer is the size of the apparatus. The fact that the anthracene viscometer can yield the same viscosity as the capillary viscometer in small-molecule fluids is an interesting result in itself. Here we are more interested in expanding to the case where the solvent molecules are much larger than the anthracene, i.e., cases where the continuum assumption of hydrodynamics is strongly violated. In polymers, we expect the friction on a nanoscopic object to be inherently different from the friction on a macroscopic object.

It is possible that in this mesoscopic regime, the structure of hydrodynamics could be so strongly violated that the concept of viscosity would become useless. However, extending the concept of viscosity to nanoscale objects proves to be a consistent and useful way to analyze our results.

Small-Molecule to Polymer Transition in Poly(isobutylene)

The results of our studies are summarized in Figure 3, which shows both the macroviscosity derived from ordinary viscometry and the nanoviscosity determined from molecular rotation times in three similar polymers. The poly(isobutylene) (PIB) and poly(dimethylsiloxane) (PDMS) results are from our measurements of anthracene rotation (2,5). The n-alkane [poly(ethylene), PE] results are from measurements of biphenyl rotation by Benzler and Luther (15). To facilitate comparisons between different polymers, the polymer length has been characterized by the number of backbone bonds $\ell = \sigma M/M_r$, where M is the molecular weight of the polymer, M_r is the molecular weight of the repeat unit and σ is the number of backbone bonds per repeat unit ($\sigma = 2$ for PIB and PDMS, $\sigma = 1$ for PE). Samples of variable polydispersity and microstructure were used, but they did not show effects that were discernable against the large molecular-weight effects.

As many as four stages can be identified in the evolution from small-molecule to high polymer. In this section, we focus on one polymer, PIB (Figure 3a), where these stages are most clearly defined.

For the smallest molecules ($\ell < 7$), the viscosity is a weak function of chain length. For somewhat longer molecules, the viscosity increases rapidly with chain length. We attribute this transition to the change from a compact molecular shape, i.e., quasi-spherical, to one with a large aspect ratio, i.e., rod-like. For the packing of spheres, the number of contacts is strictly invariant to the size of the spheres. The same is roughly true for small, compact molecules. The number of intermolecular contacts is roughly constant so long as the chain length does not greatly exceed the chain width. Once the chain becomes rod-like, the number of intermolecular contacts increases rapidly with length. The contacts impede motion, and the viscosity increases.

The second transition occurs when the macro- and nanoviscosities become different, near $\ell = 20$. This is the primary effect sought in these studies and represents the breakdown of the SED model in polymers. The transition length will be denoted ℓ_{SED}. Above ℓ_{SED}, the macroviscosity continues to increase with chain length.

The surprising aspect of this length is that it is quite long. The SED model holds even when the polymer molecules are 2.5-3 times larger than the anthracene molecule, judging either by length or by volume. Because the SED model assumes a continuous fluid, the simple expectation is that this model would already be failing when the polymer molecules are the same size as the anthracene molecule.

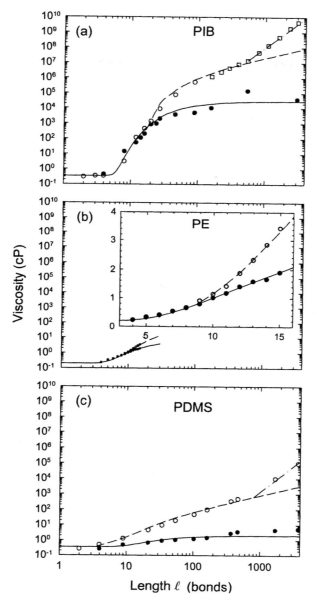

Figure 3. Macroviscosity (open) and nanoviscosity (solid) of three polymers versus the number of backbone bonds. See text for fits. (Reproduced from reference 5. Copyright 2003 American Chemical Society.)

Above ℓ_{SED}, the nanoviscosity also increases with chain length. However, the increase is much weaker than for the macroviscosity, and it levels off for long polymers. Because the rotation of a solute molecule must be a property of a local region near the solute, we expected the nanoviscosity to reach an asymptotic value in the infinite polymer limit. What was not anticipated, was that the limit would only be reached long after ℓ_{SED}.

The reason for the slow approach to the limiting value is suggested by empirical fits to the nanoviscosity data

$$\eta_{nm}(\ell) = \eta_{nm}^0 + (\eta_{nm}^\infty - \eta_{nm}^0)e^{-A/\ell}, \tag{3}$$

which are shown as solid curves in Figure 3. Because the density of chain ends scales as $1/\ell$, the $1/\ell$ dependence suggests that the slow growth to the asymptotic value is related to the loss of end effects.

In principle, the nanoviscosity is a more fundamental property than the macroviscosity, and knowledge of the nanoviscosity should allow the macroviscosity to be predicted. Polymer viscosity below the entanglement length has generally been described by the Rouse model (16). In this model, the macroviscosity is predicted from the segmental friction ζ. The nonlinear behavior of PIB macroviscosity has been attributed to the length dependence of this segmental friction. A natural hypothesis is that the nanoviscosity and segmental friction are closely related. We assume that they are proportional to one another, $\zeta(\ell) \propto \eta_{nm}(\ell)$. Coupling this idea with the Rouse model's linear dependence of the macroviscosity on the chain length gives

$$\eta_\infty(\ell) \propto \ell \times \eta_{nm}(\ell) . \tag{4}$$

As normally formulated, the Rouse model does not attempt to treat the small-molecule limit properly. We know that in this limit, the SED model holds

$$\eta_\infty(\ell) = \eta_{nm}(\ell) . \tag{5}$$

We have made a simple interpolation between these two regimes

$$\frac{\eta_\infty(\ell)}{\eta_{nm}(\ell)} = 1 + \begin{cases} B(\ell - \ell_{SED}) & \ell > \ell_{SED} \\ 0 & \ell < \ell_{SED} \end{cases} . \tag{6}$$

This formula yields the dashed fits in Figure 3. This simple model accounts for the rise of the macroviscosity before chain entanglements become the dominant mechanism. The good fits suggest that the nanoviscosity measured by solute rotation is closely related to the segmental viscosity, and that the macroviscosity follows from the nanoviscosity through a Rouse-like mechanism.

Equation 6 also provides a specific method to define the value of ℓ_{SED}. Although the values from these fits are slightly shorter than obtained by judging by eye ($\ell_{SED} = 17$ for PIB), these values are a consistent and objective means to make comparisons between polymers.

The final stage in the development of the viscosities is the rapid increase in macroviscosity when entanglement becomes important. In Figure 3a, this region occurs above $\ell_e = 607$, where the macroviscosities are fit to a power law with an exponent of 3.4 (dot-dashed line). No entanglement effect is seen in the nanoviscosity. This result is reasonable, because the nanoviscosity should be determined by effects on a length scale smaller than the distance between entanglements.

Poly(dimethylsiloxane) and Poly(ethylene)

The data on PIB demonstrate the dramatic development of mesoscale dynamics in polymers. In small-molecule solvents, the viscosity on millimeter and nanometer sized objects is the same; in entangled polymers, they differ by as much as five orders-of-magnitude. This dramatic change suggests a fundamental shift in the mechanisms causing friction on a small object.

An important clue to the nature of these mechanisms comes from a comparison to PDMS. In terms of structure, PDMS and PIB are quite similar. In PDMS, the carbon chain of PIB is replaced by a silicon-oxygen chain (Figure 1). The bond lengths and angles of the backbone differ slightly between PMDS and PIB (17). However, many key measures of structure—chain width, characteristic ratio (C_∞), self-diffusion coefficient and entanglement length—do not change significantly (11).

Despite these similarities, the macroviscosity of PDMS is much lower than that of PIB of the same chain length. Figure 3c shows that the same is true of the nanoviscosity. For the longest chains, the difference is four orders-of-magnitude. Figure 3 also shows that the origin of this difference lies in the breakdown of the SED model in the oligomeric region. The breakdown length for PDMS is much shorter ($\ell_{SED} = 3.5$) than for PIB ($\ell_{SED} = 17$). The entire region of rapid growth of both macro- and nanoviscosity ($\ell = 7-17$ in PIB) is missing in PDMS.

Once this difference is recognized, the other features in PDMS follow the same pattern as in PIB. In the small-molecule limit, the macro- and nanoviscosities are the same. After ℓ_{SED}, the nanoviscosity increases weakly toward the long polymer value. The nanoviscosity data can be fit with the same model used for PIB (eq 3, solid curve in Figure 3c), suggesting that the elimination of end effects is again the reason for the slow convergence to the asymptotic value. Below the entanglement length, the macroviscosity can be predicted from the nanoviscosity by our Rouse-like model (eq 6, dashed curve in Figure 3c). The nanoviscosity is unaffected by entanglement.

Although the static structures of PIB and PDMS are quite similar, the torsional kinetics are very different. In PIB, there is substantial steric interaction between the methyl groups during torsional transitions. As a result, the energy barrier between gauche and trans conformations is high compared with thermal energies (6 kcal/mol = 10 kT at 300 K). PDMS has more open bond angles and longer bond lengths, which combine to almost eliminate the steric interaction between methyl groups. Consequently, torsional barriers in PDMS are negligible (<0.5 kcal/mol = 0.8 kT at 300 K).

The results of solute rotation in n-alkanes [oligo-poly(ethylene), PE] support the role of torsional barriers as the critical feature in determining ℓ_{SED}. The torsional barrier height in PE (2.9 kcal/mol = 4.9 kT at 300 K) is intermediate between those of PIB and PDMS.

Numerous studies of solute rotation in alkanes have been reported in the literature [see references in (5)]. We focus on reanalyzing the results of Benzler and Luther on biphenyl rotation in n-alkanes extending up to hexadecane (15). (Higher n-alkanes crystallize.) An independent calibration of the hydrodynamic factors for biphenyl is not available. We follow Benzler and Luther's conclusion that the SED equation works below $\ell = 9$ and use these points for calibration. The n-alkane data can then be plotted in the same format used for PIB and PDMS (Figure 3b).

As predicted, the point of divergence of the macro- and nanoviscosities follows the torsional barrier height (PIB > PE > PDMS). In the next section, we will outline a schematic model that explains how the torsional barrier height is connected to the breakdown of the SED model and to the development of distinct hydrodynamic properties for mesoscale objects.

Before presenting that model, we note that previous models for the breakdown of the SED model fail to describe our data, even at a qualitative level (1,18). For small-molecule liquids, most theories and most interpretations of experimental data have emphasized the ratio of solute and solvent sizes and have predicted that the breakdown will be underway when the sizes are equal. As already discussed, the PIB results show that the breakdown of the SED model can be postponed until the solvent is several times larger than the solute. The comparison between different polymers shows that the breakdown can occur at different solvent sizes even for the same size solute. Although some dependence of the SED breakdown length on solute size has been demonstrated (15), we believe that this is a secondary effect. We argue that the primary effect is determined by a dynamic length scale related to torsional transitions, not to a simple ratio of solute and solvent sizes.

A Model Based on Torsional Relaxations

In thinking about why the viscosity on a large and a small object can be different, it is useful to think in terms of a viscoelastic material. At sufficiently short times, all materials behave elastically and can resist stresses, including the stress of an object trying to undergo thermally induced rotation. With time, the material relaxes and allows the object to move. In viscoelastic theory, this relaxation is expressed a drop in the shear modulus $G(t)$ with time. The viscosity is the integral of this modulus (19)

$$\eta = \int_0^\infty G(t)\, dt \,. \tag{7}$$

It is also useful to think of the compliance, $J(t) = 1/G(t)$, which is the response of the material to a constant force.

Figure 4 shows schematic representations of several motions that allow for compliance in a polymer. The polymer moves in response to a force from a rotating object, thereby opening up space for the rotation to continue. Figure 4a shows a rigid-body motion of a polymer molecule, i.e., rotation or translation of the molecule without changing its conformation. Figure 4b shows an "end flip." In this motion, a single torsion near an end of the chain can move one end of the chain, without moving the other end. Figure 4c shows a crankshaft motion, i.e., concerted torsions at two points on the chain that allow the intermediate portion of the chain to move without moving either end.

The motion of either a macroscopic or a nanoscopic object can be accommodated by rigid-body rotations or translations of the polymer molecules. Both crankshaft and end-flip motions produce enough space to accommodate the motion of a small object, but not a large one. Roughly speaking, the modulus for a large object results from only the rigid-body compliance, $G_\infty(t) = 1/J_{rigid}(t)$, whereas the modulus for a small object results from the sum of rigid-body, end-flip and crankshaft compliances, $G_{nm}(t) = 1/(J_{rigid}(t) + J_{end}(t) + J_{crank}(t))$. Thus, the nanoviscosity can be smaller than the macroviscosity because extra relaxation channels are open for a small object.

Consider the behavior of PDMS as a function of chain length. For very short chains, there are no conformational changes available, or at least none large enough to accommodate the rotation of anthracene. Rigid-body motions dominate both the macro- and nanoviscosities, and so they are the same. In PDMS, torsional transitions are nearly barrierless, so as soon as the chain becomes long enough to have different conformations, end-flip and crankshaft relaxations become effective. With these torsional relaxation channels open, the nanoviscosity becomes smaller than the macroviscosity. Thus, ℓ_{SED} is quite short in PDMS.

For moderately long chains, end-flips will be the most important relaxation mechanism, because only one torsion is required and most positions on the chain are near a chain end. As the chain length increases and density of chain ends drops, the end-flip mechanism becomes ineffective. In the long polymer limit, only the crankshaft mechanism remains to accommodate small-object motion. As the end-flip relaxation channel closes, there is a slow rise in the nanoviscosity toward the long polymer limit that is governed by the decreasing density of chain ends.

In PIB, the presence of significant torsional barriers adds a new feature. Even though PIB can ultimately access as many conformations as PDMS can, PIB cannot access them as quickly. For example, if the mean time between

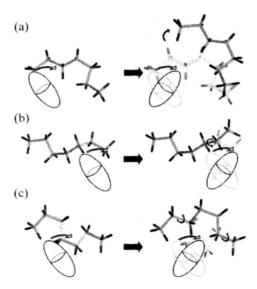

Figure 4. Schematic representation of the three types of polymer motion that can accommodate rotation of a small object (ellipsoid). (a) The polymer molecule can move as a rigid body. (b) A single torsional transition can flip one end of the polymer chain. (c) Two concerted torsional transitions lead to a crankshaft motion that leaves the ends of the chain in place. (Reproduced reference 5. Copyright 2003 American Chemical Society.)

torsional transitions on a chain is 10 ns, but the rotation time is 500 ps, torsional relaxation will not be effective in allowing rotational motion and will not contribute to the nanoviscosity. We can say that PIB is *dynamically inflexible*, even though is seen as flexible in static measurements of structure.

The consequences of this dynamic inflexibility can be seen in Figure 3a. For short chains, only rigid-body motions are effective and the macro- and nanoviscosities are the same. As the chains become longer ($7 < \ell < 17$), the molecules become rod-like, and the number of contact of a given molecule with other polymer molecules increase rapidly. As a consequence, rigid-body motions become increasingly difficult, causing the viscosity to rise rapidly. However, torsional transitions are too slow to provide an effective relaxation channel, so the macro- and nanoviscosities remain the same during this increase.

Three features conspire to bring this situation to an end. First, as just mentioned, rigid-body motions become more difficult with increasing chain length, decreasing their contribution to the total compliance. At the same time, as the chain becomes longer, the chance that a torsional transition will occur somewhere on the chain rises, increasing the contribution of the torsional (end-flip and crankshaft) mechanisms. Finally, as the viscosity increases and the rotation slows, the chance of having a torsional transition within the rotation time goes up.

At some chain length, the torsional relaxation mechanisms become important relative to the rigid-body mechanism, and the macro- and nanoviscosities diverge. This length is ℓ_{SED}. In PIB, this crossover is at $\ell_{SED} = 17$. In *n*-alkanes (PE), the lower torsional barriers shift the balance of these factors, so the breakdown of the SED model occurs for shorter chains, $\ell_{SED} = 8$.

Finally, we comment on the length scale that determines the breakdown of the SED model ℓ_{SED}. This length does not correspond to any conventional measure of molecular or polymer structure. Conventional length scales correspond to *static* correlation functions—averages over an ensemble, or equivalently, averages of a single molecule over a long time. The breakdown of the SED model is related to a *dynamic* correlation function—one that depends on the time scale of the measurement. For example, we can define a dynamic length scale $\ell(\tau)$ as the mean distance between torsional transitions that have occurred within the time τ on an infinitely long chain. For short times, $\ell(\tau)$ will be very long. With time, $\ell(\tau)$ decreases. At long times, when a transition has occurred at every possible position, it reaches $\ell(\infty)$, which is equal to the bond length, a static measure of structure. The SED model breaks down when it is probable to have a torsional transition somewhere along a finite chain. Roughly speaking, $\ell_{SED} \sim \ell(\tau_{rot})$. Beyond this length, the chain must be regarded as dynamically flexible. Models that do not account for this flexibility, such as the SED model, will no longer be valid.

Acknowledgements

This work was supported by Office of Naval Research through Grant No. N00014-97-1-0806 and by National Science Foundation through Grants No. CHE-9809719 and CHE-0210986.

References

1. Fleming, G. R. In *Chemical Applications of Ultrafast Spectroscopy*; Oxford University Press: Oxford, 1986.
2. Sluch, M. I.; Somoza, M. M.; Berg, M. A. *J. Phys. Chem. B* **2002**, *106*, 7385.
3. Zhang, Y.; Sluch, M. I.; Somoza, M. M.; Berg, M. A. *J. Chem. Phys.* **2001**, *115*, 4212.
4. Somoza, M. M.; Sluch, M. I.; Berg, M. A. *Femtochemistry and Femtobiology: Ultrafast Dynamics in Molecular Science*; Douhal, A. and Santamaria, J., Ed.; World Scientific Publishing: Singapore, 2002, pp 289.
5. Somoza, M. M.; Sluch, M. I.; Berg, M. A. *Macromolecules* **2003**, in press.
6. Bauer, D. R.; Alms, G. R.; Brauman, J. I.; Pecora, R. *J. Chem. Phys.* **1974**, *61*, 2255.
7. Boon, J. P.; Yip, S. In *Molecular Hydrodynamics*; Dover Publications, Inc.: New York, 1980.
8. Ediger, M. D. *Annu. Rev. Phys. Chem.* **1991**, *42*, 225.
9. Ediger, M. D. *Annu. Rev. Phys. Chem.* **2000**, *51*, 99.
10. Sillescu, H. *J. Non-Cryst. Solids* **1999**, *243*, 81.
11. Arbe, A.; Monkenbusch, M.; Stellbrink, J.; Richter, D.; Farago, B.; Almdal, K.; Faust, R. *Macromolecules* **2001**, *34*, 1281.
12. Krushev, S.; Paul, W.; Smith, G. D. *Macromolecules* **2002**, *35*, 4198.
13. Lettenberger, M.; Emmerling, F.; Gottfried, N. H.; Laubereau, A. *Chem. Phys. Lett.* **1995**, *240*, 324.
14. Jas, G. S.; Wang, Y.; Pauls, S. W.; Johnson, C. K.; Kuczera, K. *J. Chem. Phys.* **1997**, *107*, 8800.
15. Benzler, J.; Luther, K. *Chem. Phys. Lett.* **1997**, *279*, 333.
16. Ferry, J. D. In *Viscoelastic Properties of Polymers*; Third ed.; John Wiley & Sons: New York, 1980.
17. Flory, P. J. In *Statistical Mechanics of Chain Molecules*; Hanser Publishers: Munich, 1989.
18. Dote, J. L.; Kivelson, D.; Schwartz, R. N. *J. Phys. Chem.* **1981**, *85*, 2169.
19. Tschoegl, N. W. In *The Phenomenological Theory of Linear Viscoelastic Behavior*; Springer: Berlin, 1989.

Chapter 12

Sculpting Nanoscale Liquid Interfaces

**Richard C. Bell, Hanfu Wang, Martin J. Iedema,
and James P. Cowin[*]**

Pacific Northwest National Laboratory, Box 999, Richland, WA 99352

Mesoscale-structured liquids are difficult to understand because their geometry is often poorly known. We create "structured liquids" that have well defined, sharply modulated structures using thin film molecular beam epitaxy at low temperature. Upon warming, the film regains its fluidity, which the ions probe before the initial structures can disperse. Using ion mobility to probe the spatially varying flow properties with 0.5 nm resolution, we show how nanometer films of glassy 3-methylpentane (3MP) are much less viscous at the vacuum-liquid interface. In addition, we are able to get a better understanding of liquid-liquid interfaces and even map the solvation potential of an ion near the oil/water interface.

It has been demonstrated that fluids confined to nanometer-sized spaces have properties that can be dramatically different from those of the bulk (*1*). Interfacial fluidity is of great importance to understanding biological systems (e.g. ion migration through cell membranes), heterogeneous chemistry (e.g. nanometer powders or zeolite-like structures), or in nanoscale devices. Bulk fluids can be altered in the nanometer region of the interface, having a viscosity, molecular density and ordering different than that of the bulk. These perturbations to the interfacial properties can profoundly alter the fluids' interfacial transfer kinetics or its success as a lubricant or adhesive (*2*). We devised a non-perturbative probe of the interfacial fluidity and solvation near the free-, liquid- and solid-liquid interfaces of nanometer fluid films, with 0.5 nm spatial resolution. The method relies on both molecular beam epitaxy of "structured liquids" with sub-nanometer tailored layers, and the use of a soft-landing ion source (< 1 eV) to precisely insert ions into the layers to monitor the fluidity and solvation.

Experimental

The details of the experimental technique used in these studies have been reported elsewhere (*3,4*). Briefly, the experimental set-up consists of three main steps: 1) deposition of the film followed by the 2) soft-landing of the ions on the surface (next, step 1 may be repeated) and finally 3) the film is heated and the motion of the ions monitored electrostatically. The ions collectively generate a substantial electric field across the film, which provides the driving force to test the ions' mobility. This will be discussed in detail later. The entire process takes place in an UHV chamber with a base pressure of $\leq 2 \times 10^{-10}$ Torr.

The films are epitaxially deposited with submonolayer accuracy onto a Pt(111) substrate maintained at less than 30 K using a molecular beam. This is below the glass transition temperature T_g of 77 K for 3MP (*5*) where it passes from the solid to liquid state, so no fluid motion is possible, allowing for very well defined initial film geometries. The thickness of the films are expressed in nominal monolayers dosed onto the surface, with one monolayer (ML) defined as the amount of the solvent molecules required to just saturate the most strongly bound first layer of molecules. The film thickness of a θ-monolayer film is estimated to be $\theta \times 3.45$ Å for water (*6*) on Pt(111) and $\theta \times 4.2$ Å for 3MP (*3*). For the solvation studies, 0 to 10 monolayers of water act as a solvation trap for ions, which are deposited on top of or between two 3MP films.

Next, D_3O^+ or Cs^+ ions are gently deposited on the 30 K film with a kinetic energy of 0 to 1 eV. We saw similar results using either hydronium or Cs^+ ions (*3*). The sample bias is controlled during deposition to compensate for the effect of film charging on the ion deposition energy. The total ion coverage in this study was small, not more than 8×10^{11} ions/cm^2. In many of the experiments, a final step was to deposit additional 3MP or H_2O on top of the ions.

The ions create a net voltage ΔV (the "film voltage") across the film that depends on the ions position. We measure this voltage to monitor the ion motion. The insulating film creates a planar capacitor where the deposited ions act to charge the capacitor. The voltage ΔV across the film (the film voltage) should thus equal $QL/\varepsilon_0 \varepsilon A$, where Q is the deposited ion charge, L is the film thickness, ε_0 the vacuum permittivity, ε the dielectric constant of the film, and A the area of the film. Integration of the Poisson equation by parts (7) yields the result that when the ions are spread out over the distribution $\rho(z)$, the usual capacitor equation is modified by replacing L with the average height $<z>$ of the ions above the metal substrate, within the film. Since no ion neutralization occurs, Q, ε_0 and A are constant. And since ε for hydrocarbons is nearly constant (8), the potential difference across the film ΔV is directly proportional to the average height (vertical displacement) of the ions within the film:

$$\Delta V = \frac{Q\langle z \rangle}{\varepsilon_0 \varepsilon A} \tag{1}$$

The film voltage ΔV can be measured to 10 mV precision with a non-contact Kelvin probe. The Kelvin probe voltage is insensitive to the presence of the water layer, since it has a much higher dielectric constant than the hydrocarbon, and is thin compared to the hydrocarbon. Thus we have a very sensitive probe of the ion motion within our sculpted films.

In our experiments we report the ion motion (via the Kelvin probe) as the composite film is warmed slowly, at 0.2 K/s. As the temperature rises above the glass temperature of 3MP, the viscosity should rapidly decrease, and the ions should begin to move in the liquid. Because $q_e \Delta V$ is much larger than kT, the field-driven ion motion is much faster than neutral molecule diffusion. Thus the ions experience the initially created structures, before the latter can diffusively blur. In addition, a mass spectrometer monitors gaseous species evolved, as the 3MP film eventually evaporates (>130 K).

Results and Discussion

Interfacial Fluidity

An important issue concerning confined liquids is whether the interfaces cause changes uniformly across the liquid, or whether the fluid is most perturbed right at the interface, with the perturbations decaying in some fashion

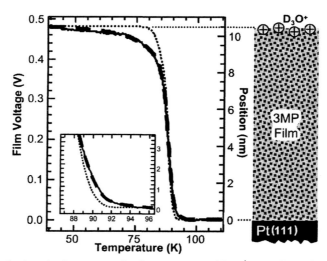

Figure 1. A typical spectrum for the transport of D_3O^+ ions through a 25 ML (10.5 nm) 3MP film. As the temperature of the film is ramped at 0.2 K/s, the ions begin to move quickly through the film near 80 to 95 K as the viscosity of the 3MP quickly decreases (5 orders of magnitude in this range). This is evidenced by a drop in the film voltage (left axis), which is directly proportional to the average height of the ions within the film (right axis). For comparison, the simulated curves use the simple continuum-based model calculated for a position-independent (bulk) viscosity (dotted), and for a position-dependent viscosity (dashed). Reproduced/Adapted with permission from reference 21. Copyright 2003, American Chemical Society.

as one moves away from the interfaces (9). Most probes used to date cannot resolve the spatial dependence of fluidity changes, or simultaneously perturb the system strongly. Free standing and supported polymer films, when heated to or above their glass transition temperatures (T_g's), have shown much evidence for interfacial effects (10-12). Typically, apparent T_g's have been measured averaged across the films using a variety of probe methods (usually optical). The free interface was found to make the film more fluid (lowering the T_g), while the polymer-solid substrate interface typically stiffens the adjacent film (with notable exceptions (12)). These experiments lacked spatial resolution, but in general concluded that the interfacial perturbations only persist approximately 10 nm into the film for these polymers (see Kawana and Jones (10)).

The transport of D_3O^+ ions through a 25 ML (10.5 nm) 3MP film is shown as the solid curve in Figure 1 (13). As the temperature of the film is increased, the viscosity of the 3MP rapidly decreases, and the ions move through the film largely between 80 and 95 K. This is evidenced by a drop in the film voltage, according to equation 1, caused by the change of $<z>$. The average ion height in the film will be a simple linear function of the film voltage (4), as indicated by the right hand axis of Figure 1.

To correctly predict the properties of these systems, it is necessary to understand and accurately model the processes responsible for fluidity and ion motion. The Stokes-Einstein relation, with minor modifications (3,4,13), works surprising well to describe the transport properties of charged particles through thin films. However, deviations from it have been observed for ultra-thin or confined liquids (1).

The ion motion through the film for an ion distribution $\rho(z)$ of charge nq_e over time should be predictable using a diffusion equation with an added term for ion mobility (4). The diffusion coefficient D deals with the thermal random-walk motion of the ions, while the ion mobility μ describes the forced ion motion caused by the electric field E (which happens to be generated by the ions themselves). The ratio of D to μ is strictly determined by the Einstein relation. We estimate the ion mobility with the Stokes formula as follows:

$$\mu(t) = \mu(T(t)) \approx \frac{nq_e}{6\pi\eta(T(t))r_i} \quad ; \quad D(t) = \mu(t)\cdot\left(\frac{kT}{q_e}\right) \qquad (2)$$

where $\eta(T)$ is the temperature-dependent viscosity of the organic film, with $T(t) = \beta t$ (β is the temperature-ramping rate) and r_i is the ion hydrodynamic radius. This mobility also will predict the ion motion under any steady-state force, including that due to chemical potential gradients such as a solvation potential near a water layer. For computational purposes the film is divided into 10000 slabs, and the change in ion density is numerically propagated in time. Time steps of 0.05 to 0.5 seconds are used.

The viscosity of bulk 3MP has been measured many times (5,14-16). Over the many orders of magnitude change in viscosity from 77.5 to 120 K, about 1 order of magnitude spread is observed for the different data sets. We use an effective bulk viscosity, expressed via a Vogel-Tamman-Fucher equation (equation 3) with $T_S = 0$. This represents ion motion in the interior of the films well, though some small, subtle long-range fluidity changes are included in it (3).

$$\eta_{eff}(T,z) = (3.72\times10^{-16} \ kg \ m^{-2} \ s^{-1})e^{-3920 \ K/(T-8.3K-T_s)} \qquad (3)$$

A simulation of the film voltage versus temperature using the ion motion from a numerical solution of the diffusion equation was calculated for an assumed hydrodynamic radius of 6 Å (approximately one complete solvation shell) and our effective bulk viscosity of 3MP from equation 3. Although the simulation resembles the experimental spectrum, severe deviations are observed at both low and high temperatures. The bulk viscosity of 3MP is much too high to permit any measurable ion motion at or below T_g (77 K), yet the experimental data shows substantial ion motion well below this temperature. Also theory predicts that all the ions should be at the bottom of the film by 95 K due to the low viscosity of the bulk film, but it is clear from the experimental data that this is not the case. Thus, the vacuum interface displays a distinct loosening of the surface layers while a stiffening of 3MP at the metal interface is observed.

This surface-enhanced fluidity can be explored by initially and precisely placing the ions at increasing distances from the free interface. This is demonstrated in Plate 1 for 3MP films where the ions are at a constant initial height of 25 ML from the Pt as increasing layers of 3MP are placed on top of the ions, shielding them from the interface. All films were created with a collective electric field of approximately 7×10^7 V/m and all curves are normalized to directly read the ion height along the vertical axis, to permit accurate comparison of the ions' average position within the different films. The bottom curve is for a film with no shielding layers. The mobility of the ions is observed to approach that expected for a position-independent bulk film viscosity as the number of shielding layers of 3MP is increased. Finally, with the ions 7 ML (2.9 nm) or more away from the interface, the mobility is extremely close to that predicted using the bulk film viscosity.

The overall thickness of the film is increased when 3MP is added on top of the ions while maintaining a base of 25 ML of 3MP. This means that the data of Plate 1 does not directly probe the spatial dependence of the fluidity of a constant-thickness film: the added 3MP layers on top of the ions changes the total film thickness. In principle, the properties of the assembly might be varying as the total film thickness is changed. To avoid this possible effect, a second set of experiments (not shown here) was performed maintaining a constant film thickness while the position of the ions within the film was changed. To avoid any far-reaching bottom interface effects, the thickness of the film was increased to 100 ML. The shielding results for the 100 ML film are nearly identical to those observed for the 25 ML case (3). This demonstrates that this perturbation propagates in a decaying fashion away from the free interface.

To numerically parameterize the interfacial effects on fluidity, we modeled them as if T_g was a function of distance from the top and bottom interfaces. Equation 3 was modified to include a z-dependent glass temperature shift, $T_S(z)$ given by:

$$T_S(z) = \left(A_1 e^{(-z/b_1)} + A_2 e^{(-z/b_2)}\right) - \left(A_3 e^{(-(L-z)/b_3)} + A_4 e^{(-(L-z)/b_4)}\right) \quad (4)$$

The thickness of the film is L and the distance from the ions to the bottom of the film is z. The positive first term gives stiffening near the Pt, and the negative term a loosening near the vacuum interface. This function was used to fit the ion motion seen for a variety of data and film thicknesses using the parameters 16.72, 6.67, 19.66, and 21.77 K for A_1 through A_4 and 2.34, 0.02, 1.45, and 0.41 ML for b_1 to b_4, respectively. As shown in Figure 1 and Plate 1 for the 25 ML 3MP films, we get a very good fit to the observed ion motion. This results in an apparent decrease in T_g as a function of distance from the free-interface and an increase in T_g near the liquid-solid interface. The viscosity calculated at 85 K is shown versus z in Plate 2. This spatial dependence of the viscosity at 85 K is inferred from the ion movement, using the shift prescribed by equation 4. This position-dependent viscosity is found to reproduce the interfacial behavior of the ions either as they traverse the whole of a given film, or are initially placed progressively further into the film.

A possible cause of the increased fluidity of the free surface may be due to an increase in the vibrational motion of the interfacial molecules leading to a decrease in T_g. The amplitude of the mean square displacements of atoms in a solid typically increases near the free interfaces. It is usually understood as being due to the decrease in the number of nearest neighbors as compared to the bulk and, will therefore produce (within the context of a simple harmonic oscillator model and pairwise forces between atoms) less restoring force per unit displacement. Such increased vibrational amplitudes have been observed both experimentally and theoretically at the free-interface of many bulk solids (*17*), and have been shown to persist over 5 molecular layers into the surface (*18*). Our results display a similar length scale. Another consequence of the free surface is that solvent molecules have room to diffuse laterally on the surface of the film. This may permit an increased effective fluidity.

Our studies clearly indicate that ion mobility in thin 3MP films changes dramatically near the interfaces, over a length scale of about 7 monolayers (2.9 nm), at the free and solid interfaces. Interestingly, confined liquids should display a tensor like viscosity, where "longitudinal" transport perpendicular to the interfaces could be very different than the "transverse" transport orthogonal to this, at a fixed distance from the interface. Mica-mica sheet experiments (*19*) and shear modulation force microscopy experiments (*9*) probe the transverse viscosity while we are sensitive to the longitudinal viscosity.

Solvation Near Interfaces

Ions in liquids near interfaces will be perturbed by their sensing of the local properties of the interfacially perturbed liquid. Additionally they are subjected to long-range coulombic interactions with the interface – essentially a solvation

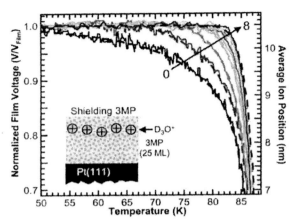

Plate 1. Shielding ions from the interfacial fluidity perturbations: Comparison
of the experimental (solid) and simulated (dashed) normalized film voltage
versus T for the top 3.1 nm of the 25 ML 3MP films with 0 to 8 ML of additional
3MP shielding the ions from the free interface. As the ions are displaced more
than 6 ML from the interface, the mobility of the ions resembles that predicted
by the position-independent bulk film viscosity. Additional shielding layers (not
shown) have no further effect. Reproduced/Adapted with permission from
reference 21. Copyright 2003, American Chemical Society.

(See Page 1 of color insert.)

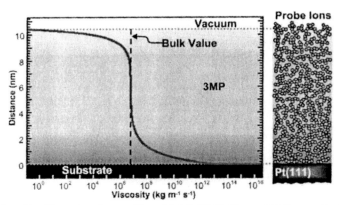

Plate 2.   Viscosity change for a 25 ML 3MP film at 85 K using D_3O^+ as the
probe ions. Notice the uniform change in viscosity from either interface to the
bulk value in the middle of film. Reproduced/Adapted with permission from
reference 21. Copyright 2003, American Chemical Society.

(See Page 1 of color insert.)

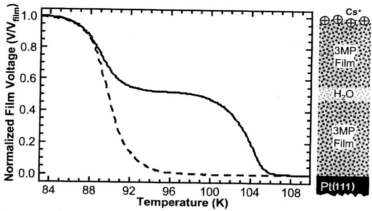

Figure 2. The temperature evolution of the film voltage for Cs⁺ migration through 200 ML 3MP / Pt (dashed curve). The ions are clearly trapped by the addition of water to the center of the film as is seen for the 100 ML 3MP / 4 ML H₂O / 100 ML 3MP / Pt (solid curve) system.

effect. This affects transport and populations of ions near interfaces. This is particularly important for interfaces between water and organic liquids. Understanding ion transport across this type of interface is crucial for biology (ion transport across cell membranes) (*20*), chemical processes (transport across or reaction at the interface of aqueous-organic solutions) (*21*), and fuel cells (*22*) to name a few. Ion motion involves leaving the aqueous region with its high solvation power, to enter an organic environment, which involves surmounting a large potential barrier.

Since we can precisely control the geometry of complex-layered systems and can precisely place and monitor ions within them, we can easily answer questions concerning the amount of water required to fully solvate an ion at the oil-water interface and to determine the shape of the solvation potential near the interface (*4,23,24*).

Figure 2 shows the effect that water has on the motion of ions across the oil-water-oil interface. The dashed curve in this figure represents ion motion through a pure 200 ML 3MP film as the temperature of the film is raised at 0.2 K/s. The solid curve is for a 200 ML 3MP film with 4 ML of water placed directly in the center of the film creating a reverse micelle. It is obvious from these curves that the ions linger in the water layer, due to the favorable solvation forces in the aqueous phase. The initial structure, shown on the right of Figure 2, does not noticeably interdiffuse before being sampled by the ions, because the ions are being pulled by the collective electric field on the order of 10^7 to 10^8 V/m and, therefore, typically move much faster than the neutral solvent (*7*).

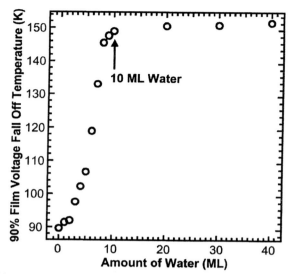

Figure 3. The experimental 90% film voltage fall-off temperature versus water coverage for 3MP films. The apparent bulk solvation limit is abruptly reached at 10 ML of water.

To determine the amount of water required to fully solvate an ion, we construct films with an initial structure of $H_2O/D_3O^+/3MP/Pt$ (23). We found no difference if the ions were placed above or below the thin water layers in these studies. More recently, we used Cs^+ to further examine the exact amount of water required to fully solvate the ions (D_3O^+ displays identical behavior, not shown, due to the similar size of both these ions). Ions remain trapped by the water layer at increasingly higher temperatures, as the thickness of H_2O is increased. Figure 3 displays the temperature at which 90% of the ions have escaped the water layer, versus the water layer thickness. This figure clearly shows that the ions require nearly 10 ML of water to become fully solvated.

The collective electric field of the ions can be easily adjusted over the range of 10^6 to 10^9 V/m, just by changing the amount of the ions, as it is proportional to $\Delta V/L$, and thus Q. The high end of this range is strong enough to equal or exceed the gradient of the solvation potential. This will strongly perturb the effect of the solvation trap. Figure 4 displays the results for a series of reverse micelle-like films with a constant physical trap and increasing electric fields. The trap consists of 3 ML of water between two 100 ML films of 3MP with widely varying amounts of ions gently deposited on top of these films all at less than 30 K. This figure displays the ΔV vs. T curves (0.2 K/s heating rate) where the initial voltages have been normalized to unit height so that all curves are proportional to the average ion height (average vertical ion displacement) in the

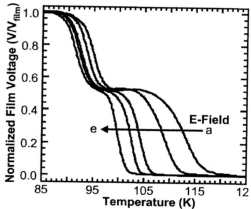

Figure 4. Electric field effect on the Cs⁺ trapping/escape from the 100 ML 3MP/ 3 ML H₂O / 100 ML 3MP / Pt at a ramping rate of 0.2 K/s. All curves are normalized to unit height so that all curves are proportional to the average ion height in the film. The electric field strengths are (a) 0.09, (b) 0.24, (c) 0.65, (d) 1.0, and (e) 1.6 × 10⁸ V/m. The arrow indicates increasing electric field E.

film. Upon heating to 90 K, all the curves display a drop to about half the initial film voltage due to ion migration down through the top 3MP layer. As expected, the drift velocity is observed to increase with increasing electric field. The ions pause in the water layer where they remain until they are able to escape the trap at higher temperatures and time (overcome the solvation potential). It is obvious how the increase in electric field greatly affects ion trapping. The trapping is nearly eliminated for an electric field of 1.6×10^8 V/m.

An increase in the hydrodynamic radius due to hydration of the ion as it passes through the water layer alone cannot account for the tens of degrees delay of ion motion in the aqueous phase (4). For example, five orders of magnitude change in the ionic radius would be required to force the ions to fall at a 10 K higher temperature than they would were they unsolvated. However, there is not enough water within the films discussed here to allow for such large hydration shells, nor would such large hydration shells be expected to be stable. Therefore, it must be the energy change of the ions moving from the aqueous to the organic phase in the form of a trap that results in the delayed ion motion.

To extract quantitative information about the solvation potential the ions feel near the oil-water interface, we need to model the solvation potential. Max Born first derived a continuum fluid equation describing the energy change for a point charge between air and another medium (25). The over-simplifications of the Born equation have been modified (26) to model the solvation an ion experiences as it passes from an aqueous environment to a hydrophobic one (see reference 4, and references therein, for details and corrections).

By modeling the results for the ion motion in the presence of a trap, we are able to deduce the depth and slope of the trap. We used a simple parameterized potential (a Lorentzian), that resembles the Born potential shape (23,24), to model the ion escape from the trap, versus temperature and electric field. We simulated the ion motion with the gradient of the trap potential added to the collective E-field term in the ion diffusion equation (7). The particles random-walk out of the potential trap, using kT to climb the potential barrier. It is necessary to know the solvent viscosity in the region near the potential barrier maximum for escaping the trap. This will be in the 3MP layer, but within a few nanometers of its interface with the water. Experiments in progress show that there is some stiffening of the 3MP near the 3MP-water interface, but for simplicity here we will assume the 3MP is bulk-like (3, 4). We found that a 166 ML 3MP film with 4 ML of water placed in the middle creates a solvation potential trap resembling 0.26 eV / $(1 + 4*(z-z_0)^2 / (6Å)^2)$ (4).

Work is in progress to probe the solvation potentials near the 3MP-water interface, as a function of water thickness. Included in that study are various experiments in which the ions are initially placed 0, 1, 2... monolayers away from the water layer, to map in even more detail the reach of the solvation potential and to determine the fluidity perturbations near the "oil-water" interface.

Conclusion

Soft-landed ions on or within sculpted films provide a remarkably versatile approach to studying both the nano-scale perturbations of the fluidity induced by their interfaces, and charge transport and solvation at interfaces. Using molecular beam techniques, one can fabricate structured liquids or interfaces that closely resemble membranes or interfaces in various materials, biological systems, fuel cells, etc. We are able to address many questions with molecular accuracy using ions as probes. Here we have shown that the interfacial properties of liquids are greatly altered at the vacuum-liquid and liquid-solid interface, displaying a continuous, orders of magnitude change in viscosity before reaching the bulk viscosity of the liquid. For the case of 3MP, this results in a loosening of the free interface and a stiffening of the solid-liquid interface with each penetrating approximately 3 nm into the film. A brief synopsis of our ongoing studies of the solvation potential at the organic-water interface displays the importance of these types of investigations. This work provides unrivaled detail of this potential, which will greatly aid in the understanding and modeling of the interactions that occur at this ubiquitous, and ever important interface.

References

1. Granick, S. *Phys. Today* **1999**, *52*, 26-31.
2. Bhushan, B.; Israelachvili, J. N.; Landman, U. *Nature* **1995**, *374*, 607-616.
3. Bell, R. C.; Wu, K.; Iedema, M. J.; Cowin, J. P. "Hydronium Ion Motion in Nanometer 3-methylpentane Films," in preparation.
4. Wu, K.; Iedema, M. J.; Schenter, G. K.; Cowin, J. P. *J. Phys. Chem. B* 2001, 105, 2483-2498.
5. Ling, A. C; Willard, J. E. *J. Phys. Chem.* **1968**, *72*, 1918-1923.
6. Wu, K.; Iedema, M. J.; Tsekouras, A. A.; Cowin, J. P. *Nucl. Instr. Meth. Phys. Res. B* **1999**, *157*, 259-269.
7. Tsekouras, A. A.; Iedema, M. J.; Cowin, J. P. *J. Chem. Phys.* **1999**, *111*, 2222-2234.
8. *Handbook of Chemistry and Physics*; Lide, D. R., Ed.; CRC: Boca Raton, 1995.
9. S. Ge, Y. Pu, W. Zhang, M. Rafailovich, J. Sokolov, C. Buenviaje, R. Buckmaster, R. M. Overney, *Phys. Rev. Lett.* **2000**, *85*, 2340-2343.
10. Kawana, S.; Jones, R. A. L. *Phys. Rev. E* **2001**, *63*, 012501.
11. Jackson, C. L.; McKenna, G. B. *J. Non-Cryst. Solids* **1991**, *131*, 221-224.
12. Keddie, J. L.; Jones, R. A. L.; Cory, R. A. *Europhys. Lett.* **1994**, *27*, 59-64.
13. Bell, R. C.; Wang, H.; Iedema, M. J.; Cowin, J. P. *J. Amer. Chem. Soc.* **2003**, *125*, 5176-5185.
14. von Salis, G. A.; Labhart, H. *J. Phys. Chem.* **1968**, *72*, 752-754.
15. Baranek, M; Breslin, M.; Berberian, J. G. *J. Non-Cryst. Solids* **1994**, *172-174*, 223-228.
16. Ruth, A. A.; Nickel, B.; Lesche, H. *Z. Phys. Chem.* **1992**, *175*, 91-108.
17. The Chemical Physics of Solid Surfaces and Heterogeneous Catalysis; King, D. A.; Woodruff, D. P. Eds.; Elsevier: New York, NY, 1981; Vol. 1, pp 145.
18. Allen, R. E.; de Wette, F. W.; Rahman, A. *Phys. Rev.* **1969**, *179*, 887-892.
19. Raviv, U.; Laurat, P.; Klein, J. *Nature* **2001**, *413*, 51-54.
20. Eisenberg, B. *Acc. Chem. Res.* **1998**, *31*, 117-123.
21. Herriott, A. W.; Picker, D. *J. Am. Chem. Soc.* **1975**, *97*, 2345-2349.
22. Srinivasan, S.; Mosdale, R.; Stevens, P.; Yang, C. *Annu. Rev. Energy Environ.* **1999**, *24*, 281-328.
23. Wu, K.; Iedema, M. J.; Cowin, J. P. *Science* **1999**, 286, 2482-2485.
24. Wu, K.; Iedema, M. J.; Cowin, J. P. *Langmuir*, **2000**, 16, 4259-4265.
25. Born, M. *Z. Phys.* **1920**, *1*, 45-48.
26. Benjamin, I. *Annu. Rev. Phys. Chem.* **1997**, *48*, 407.

Predicting Mesoscale Structure and Phenomena

Chapter 13

Modeling Structure and Properties in Mesoscale Fluid Systems

Peter V. Coveney

Centre for Computational Science, Department of Chemistry, University College London, 20 Gordon Street, London WC1H 0AJ, United Kingdom

We review some modern approaches to the study of the properties of amphiphilic fluid systems and the mesoscopic structures they can form. Lattice gas, lattice-Boltzmann, and dissipative particle dynamics methods aim at reproducing the complexity of macroscopic amphiphilic immiscible hydrodynamics by evolving very simple mesoscale particles representing fluid and surfactant molecules. These methods have proved capable of modeling a wide range of experimentally observed phenomena, and they constitute an active area of development in mesoscale fluid modeling. The power of mesoscale descriptions resides in the fact that, by contrast with the vast number of microscopic degrees of freedom available to a physical system, only a very small set of variables determines its macroscopic behavior. We show that these crucial degrees of freedom can be systematically extracted from the underlying microdynamics by coarse-graining, renormalization group procedures. We also apply these procedures to the intrinsically non-equilibrium processes of nucleation and crystal growth, and compare our results with experimental data.

Mesoscale structure and mesoscale behavior at length scales from nanometers to microns determine the properties of a wide range of materials. However, mesoscale phenomena are complex; they emerge from the collective interactions of large numbers of smaller interacting units. A dramatic example is provided by the fluid dynamics of surfactant-containing fluids, modeled with cellular automata(*1*).

Surfactants, also termed amphiphiles, are often small molecules containing both *hydrophilic* and *hydrophobic* parts; see Figure 1. Surfactants include detergents, alcohols and block copolymers, as well as phospholipids, which, in the form of self-assembled bilayer structures, are the main components of cell membranes.

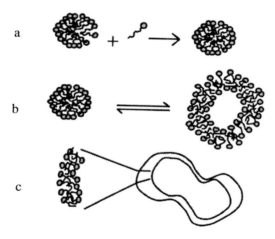

Figure 1- Schematic representation of (a) stepwise formation of spherical micelles; (b) the micelle-vesicle transition; (c) a cell membrane showing a close-up detail of the cell wall comprising lipid bilayer

Surfactants are strongly attracted to oil-water interfaces, at which their head groups reside within the aqueous phase, while their tails protrude into the oil phase. Deprived of an oil phase in solution in water, surfactants have a propensity to associate into clusters known as micelles. The simplest micelles are spherical in shape, and relatively monodisperse, typically containing between 50 to 100 monomers. But many other so-called micellar mesophases are possible, depending on various physicochemical parameters, such as *pH*, ionic strength, temperature, etc. *(2,3)*.

Micelles form lyotropic liquid crystalline phases, and can appear as lamellar phases, hexagonal-packed cylinders, body-centered cubic and face-centered

Figure 2 - Artist's depiction of surfactant mesophases: the hexagonal tubular, lamellar, and cubic mesophases.

(Reproduced with permission from reference 1. Copyright 2002 Royal Society of London.)

cubic arrays, and bicontinuous cubic phases. Their size can range from mesoscopic to macroscopic. Figure 2 displays examples of such mesophases. More random assemblies of intertwined worm-like micelles produce viscoelastic fluids *(2-6)*.

These amphiphilic fluid systems provide an example of a much broader phenomenology in complex fluids. Other complex fluids include polymers, colloids, and multiphase flows. The large-scale flow properties of these fluids are non-Newtonian and are of considerable practical interest. Every day examples include drilling fluids, tomato ketchup, and paints.

One would like to understand, model and predict the fluid flow properties of these complex systems. However, in many of these cases, in which several different length and time scales are involved, a hydrodynamic description is not known and indeed may not even exist. As a consequence, macroscopic hydrodynamic descriptions in the form of continuum partial differential equations (essentially, the Navier-Stokes equations with constitutive equations shoehorned in), the cornerstone of continuum fluid dynamics for well over a century, are of doubtful validity. Instead, a more fundamental approach is called for, in which the flow properties are computed as emergent features of a particulate ("molecular") description of the fluid.

Discrete models of amphiphilic fluid dynamics

Instead of working at the continuum level, we could return to the microscopic description of matter based on (classical) molecular dynamics (MD), but this remains today computationally too demanding. Over the past ten to fifteen years, a whole new set of modeling and simulation methods have been developed, which we can refer to collectively as *mesoscopic;* that is, they provide an essentially particulate description of matter on length and time scales that are intermediate with respect to the molecular and macroscopic levels. Figure 3 shows the location of these modeling methods within Boltzmann's program for statistical mechanics (see Figure 4).

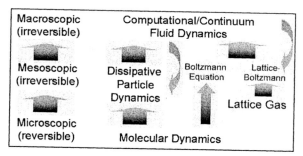

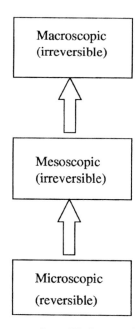

Figure 3 - Diagram showing the location of existing modeling and simulation methods across the length- and time-scales set by Boltzmann's program for statistical mechanics.
(Reproduced with permission from reference 1. Copyright 2002 Royal Society of London.)

Figure 4 - A modern interpretation of Boltzmann's program for statistical mechanics

We shall briefly describe the elements of three of these methods here, namely lattice gas, lattice-Boltzmann and dissipative particle dynamics. All follow the complexity paradigm defined earlier, in being based on very simple interacting mesoscale particle units, which together produce the complexity of macroscopic hydrodynamics. These three, however, are not the only modeling approaches to mesoscale phenomena; descriptions of other approaches are to be found in the following chapters.

(i) Lattice gas automaton methods

Lattice gas automaton models are cellular automata that conserve mass and momentum *(7)*. A regular lattice with the correct geometry to ensure isotropy of flow (see Figure 5 for the 2D case, where a triangular geometry is required), is populated with particles which collide and propagate to neighboring nodes in discrete time with discrete velocities. The variables in these models are Boolean: that is, they are comprised of binary digits, or bits (0s and 1s), denoting the presence or absence of particles on a lattice link, the particle velocities, and so on. The resulting computer codes are algorithmically efficient, stable, and highly parallel. As first shown by Frisch *et al. (8)* and by Wolfram *(9)*, hydrodynamic equations for incompressible flows (that is, the Navier-Stokes equations) are emergent when the automaton dynamics are averaged over large enough regions of space and time.

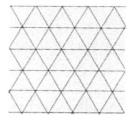

Figure 5- A portion of a two-dimensional triangular lattice, possessing the necessary symmetry for mesoscopic lattice-gas automata to reproduce Navier-Stokes fluid dynamics at the macroscopic level

A lattice-gas model for immiscible mixtures, say of oil and water, was realized by Rothman and Keller *(10)* by adopting ideas taken from electrostatics: simple massive point particles of different species are assigned different color charges (say, blue and red for water and oil, respectively), and phase separation is achieved by choosing the post-collisional order parameter flux for which its work against the color field, a vector field defined by the order parameter at surrounding lattice sites, is a minimum (the order parameter being the difference in the two color densities); see Color Plate 1 for a diagram showing the definition of color flux, J, and color field, E.

Color Plate 2 shows the phase separation process of a binary mixture of two immiscible lattice gas fluids. Collisions conserve the mass of each species and the total momentum, and the desired macroscopic hydrodynamic properties of single-phase fluids are reproduced in the bulk of each phase. Note the complexity of the interfaces that result from these models, as emergent properties of inter-particle interactions. Even such multiphase flow presents a major problem for continuum fluid dynamics, which lacks any simple way to describe or follow complex interfacial behavior over time.

We have extended this model to describe amphiphilic fluid dynamics (11). Surfactant molecules ("green") are represented by point color dipoles (one end blue, the other red) with continuous angle orientations. Now, the lattice-gas automaton dynamics conserves water, oil, and surfactant mass, as well as the total momentum. Interactions are incorporated in a manner similar to the Rothman-Keller model: a local Hamiltonian includes interactions with and amongst dipoles, and dipoles carry their orientation when they propagate (11). As a result, the much more complex and richer interfacial self-assembly phenomena associated with amphiphilic fluids can be reproduced Color Plate 3 shows how the addition of surfactant to a mixture of oil and water leads to the arrest of domain growth and the formation of a microemulsion phase.

Lattice gases exhibit the extremely attractive computational feature of unconditional numerical stability (because the state of the automaton at each time is represented by Boolean rather than floating-point numbers), and give correct hydrodynamics, immiscibility, and interfacial and self-assembly behavior (12,13), at least for low Reynolds numbers. Boolean quantities allow for algorithm efficiency and exact momentum conservation, and the particulate nature of the model ensures the presence of fluctuations. For pure hydrodynamic applications, however, lattice-gas methods need extensive ensemble averaging to remove these fluctuations. Moreover, owing to the complexity of the collisional look-up tables and the required symmetry properties of the lattice (a projected four-dimensional face-centered hypercube), the study of these models in three dimensions is a major supercomputing challenge, although still entirely viable (14).

(ii) Lattice-Boltzmann methods

Lattice-Boltzmann methods were introduced in an attempt to overcome some of the computational drawbacks of hydrodynamic lattice gases (15).

212

Binary immiscible lattice gas

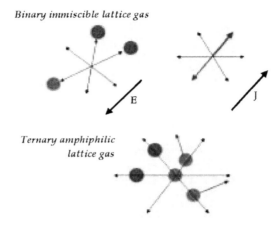

E

J

Ternary amphiphilic
lattice gas

*Plate 1 - Directions of color flux and color field in a lattice-gas automaton
model for immiscible fluids*
(See page 2 of color insert.)

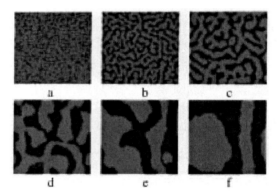

a b c

d e f

**Plate 2 - Phase separation kinetics in a binary immiscible fluid mixture (oil and
water) as simulated with a two-dimensional lattice-gas automaton model. Time
steps shown are (a) 0, (b) 50, (c) 150, (d) 500, (e) 1000, (f) 6000. The lattice size
is 128^2.**
*(Reproduced with permission from reference 11. Copyright 1996 Royal Society
of London.) (See page 2 of color insert.)*

Initially introduced by applying the Boltzmann approximation to a lattice gas automaton, lattice-Boltzmann methods are seen today by many within the continuum fluid dynamics community as a novel form of numerical finite-difference scheme for solving the Navier-Stokes equations. Unlike the particulate lattice-gas method, now each velocity at each site of the lattice is populated with a single-particle distribution function, which relaxes towards a tailored, pre-defined local Maxwellian equilibrium state, conserving mass and momentum (given infinite machine precision). With judicious choice of the equilibrium state it is possible to achieve desired hydrodynamic behavior in these models. Note, however, that these models are susceptible at the very least to numerical round-off errors owing to the use of a floating-point, rather than Boolean, arithmetic.

Although many authors have sought to construct lattice-Boltzmann models of complex fluids by an 'anti-complexity' or 'top-down' route —namely, by shoehorning into their scheme some known or guessed information about the macroscopic, equilibrium behavior— our own approach is more consistent with the notion of complexity as defined earlier. Thus, we build 'bottom-up' models in which the particles enjoy specified interactions with one another. The macroscopic fluid properties then emerge from the microscopic dynamics. In this way, we can reproduce two phase flow phenomena in a similar fashion to that done with lattice gases *(16-18)*. Moreover, we have also extended these models to include surfactants *(19,20)*, where we have been able to simulate the self-assembly of the cubic Schwarz "P" and the lamellar mesophases (Color Plates 4 and 5), and more recently, the beautiful cubic double gyroid mesophase *(21)*.

In general, lattice-Boltzmann algorithms are simple and efficient, and readily parallelizable. However, the widespread lack of numerical stability exhibited by these models is a serious problem. These instabilities arise because almost all such model developments have completely overlooked the underlying statistical mechanical basis of the method. The original Boltzmann equation had associated with it a very important property, namely an "*H*-theorem" guaranteeing the existence of a Lyapunov function *H* which evolves in a monotonically increasing manner to its minimum value at equilibrium*(22,23)*. This *H*-function is simply the negative of Boltzmann's entropy.

Motivated by this microscopic worldview, we and others have developed *entropic* lattice-Boltzmann methods which re-inject into the lattice-Boltzmann scheme some of the jettisoned remnants of statistical physics and kinetic theory *(24)*. The equilibrium distribution function is no longer specified in an arbitrary manner. Instead, an entropy (*H*-) function is first defined which can only increase in magnitude as time passes and reaches its external value at equilibrium. Indeed, for incompressible fluid flow, we have recently shown that the requirement that these models be Galilean invariant specifies the *H*-function essentially uniquely *(25)*. Such unconditionally stable lattice-Boltzmann schemes permit the fluid

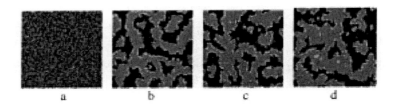

Plate 3 - *Arrest of domain growth in a binary immiscible mixture (black and red for water and oil, respectively) when surfactant (green) is added, as simulated with a two-dimensional lattice-gas automaton model. Time steps shown are (a) 0, (b) 4000, (c) 16000, and (d) 40000. The lattice size is 128^2.*

(Reproduced with permission from reference 11. Copyright 1996 Royal Society of London.) (See page 3 of color insert.)

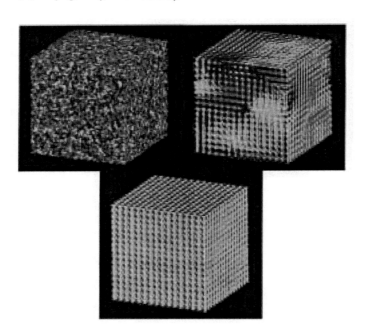

Plate 4 - *Formation of the bicontinuous cubic "P" mesophase from an initial random mixture of oil and water plus surfactant added, as simulated with a lattice-Boltzmann model. Time steps shown are (a) 0, (b) 400, and (c) 5600, respectively, and the surface displayed is where the surfactant mainly resides. The lattice size is 64^3. (Reproduced with permission from reference 20, copyright 2001, American Chemical Society).*
(See page 3 of color insert.)

Plate 5 - Formation of the lamellar mesophase from an initial random mixture of oil and water plus surfactant, as simulated with a lattice-Boltzmann model. Although the initial condition is identical to that of figure 9, the surfactant-surfactant coupling has been switched, leading to a different evolution. Time steps shown are (a) 0, (b) 800, and (c) 23200, respectively, and the surface displayed is where the surfactant mainly resides. The lattice size is 64^3. (Reproduced with permission from reference 20, copyright 2001, American Chemical Society).
(See page 4 of color insert.)

viscosity to assume vanishingly low values and hence many prove of value in the study of turbulence (24), one of the last outstanding challenges in classical physics.

(iii) Dissipative particle dynamics

As I said earlier, evolving the full microscopic description of molecular dynamics (MD) for complex, self-organizing systems remains today far too great a computational task. Moreover, by dint of their intrinsically simple structure, the mesoscopic hydrodynamic lattice-gas and lattice-Boltzmann methods just discussed generally lack the appropriate statistical mechanical basis to establish a straightforward connection between the simple but fictitious lattice particles and the 'real' molecules of a fluid which move in continuous space.

The dissipative particle dynamics (DPD) method is a continuous-time Langevin scheme with momentum conservation, including conservative, dissipative and fluctuating forces (26,27). The important statistical-mechanical properties of detailed balance, Gibbsian equilibrium distributions, and H-theorems all hold in this model. Here, the basic units are dissipative particles, which undergo momentum conserving collisions with one another through soft conservative potentials, but they are also subject to dissipative and fluctuating forces. These mesoscopic dissipative particles are not afforded any lower level interpretation. Indeed, the algorithm is foundationless in the sense that there is still no connection to the molecular level of DPD.

We have, however, managed to derive a new version of dissipative particle dynamics starting from the atomistic level of molecular dynamics (MD). We have achieved this through a systematic coarse-graining procedure which introduces new dissipative particles (defined as dynamical collections of MD particles), in such a way that mass, momentum, and energy are all conserved (28,29). The coarse-graining procedure naturally defines the dissipative particles as individual cells within a space-filling Voronoi lattice (see Figure 6); these cells evolve through exchange of mass, momentum and energy according to equations of motion which formally resemble the original equations of DPD (27).

The extent of the coarse-graining can be chosen at will (subject to the requirement that statistical mechanical approximations should be applicable within each separate dissipative particle, which means in practice that each such particle should typically contain several hundred molecules). This coarse-graining procedure is a form of renormalization procedure which filters out more and more unimportant details from the finer length and time scales as it is repeatedly applied (28,29)

The scheme provides an intrinsically multi-scale method for the description of complex fluid hydrodynamics, in which many different length scales are simultaneously present, as arises for example in colloidal and polymeric fluids, as well as fluid flows in confined geometries, such as porous media *(28)*.

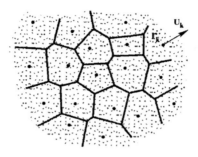

Figure 6 - Two-dimensional Voronoi tessellation of a field of massive, point particles. The mesoscopic scale is introduced by the partitioning into cells, where each cell is characterized by its centre of mass, located at position r_k, and with velocity U_k. Reproduced from reference (1)

Renormalization methods in nucleation and growth processes

Historically, the study of nucleation processes has not attracted much attention, largely on the grounds that such experiments are notoriously unreproducible: the nucleation and growth of crystals, for example, is extremely sensitive to numerous experimental variables, with impurities and imperfections in laboratory equipment often contributing to the observed effect. Moreover, the intrinsically non-equilibrium nature of the process makes it impossible to analyze by traditional methods based on equilibrium thermodynamics.

In the past few years, however, we have revisited the long-standing 'classical nucleation theory' from the perspective of modern statistical mechanics and non-linear dynamics, two tools appropriate for analyzing the behavior of complex systems. In this way, we have uncovered some remarkable structure underpinning these very complex processes. To illustrate our new approach, we describe how these methods can be applied to a description of the kinetics of surfactant self-assembly, and in particular to the self-reproduction of spherical micelles. The experimental work that originally motivated us is due to Luisi and his co-workers *(30)*, who have subsequently gone on to consider the analogous self-reproduction of vesicles (also known as liposomes), which are closed

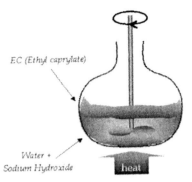

Figure 7 - Experimental setup of Bachmann et al. (30) for the production of self-reproducing micelles. Adapted from reference (1)

surfactant bilayers (and sometimes also multilayers) containing water, while surrounded by a continuous aqueous phase(31).

In the original experiments of Bachmann *et al.* (30), shown in Figure 7, immiscible oily ethyl caprylate ester is stirred in a vessel above an aqueous sodium hydroxide phase. The alkali hydrolyses the ester and produces the caprylate anion, which is a surfactant. Initially, this reaction proceeds slowly, and the surfactant formed largely adheres to the ester-water interface. However, when the surfactant concentration exceeds a certain threshold, related to the critical micelle concentration at which, under equilibrium conditions, spherical micelles form, then the hydrolysis reaction proceeds at a dramatically enhanced rate. The micellar clusters accumulating within the aqueous phase are able to solubilize unreacted ethyl caprylate, thereby dispersing this reactant and greatly enhancing the rate of reaction. The process is autocatalytic, and the micelles self-reproducing, as more surfactant leads to more micelles and increasing rate of reaction. The kinetics of the reaction are shown in Figure 8.

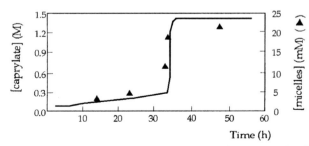

Figure 8 - Experimental measurement of the reaction kinetics in Bachmann et al.'s experiment (30).

Schematically, the rate processes involved can be summarized as follows,

$$EC \longrightarrow C$$

$$nC \longrightarrow C_n$$

$$EC + C_n \longrightarrow C_n + C$$

Eqs. (1), (2), and (3)

Here, EC represents ethyl caprylate, C the caprylate surfactant and C_n the caprylate micelles, all assumed to contain the same (mean) aggregation number, n, of monomers. Reaction (1) is the sodium hydroxide-induced hydrolysis of ethyl caprylate at the water-ester interface; rate process (2) summarizes the formation of spherical micelle clusters from monomers, while (3) describes the micellar-catalyzed hydrolysis of ethyl caprylate within the aqueous phase.

We describe the formation of micellar clusters on the basis of the Becker-Döring model of aggregation and fragmentation process *(32)*. In the Becker-Döring theory, which was originally introduced to describe the kinetics of first-order phase transitions*(33)*, such as that between liquid and vapor, the basic growth and fragmentation processes occur in a stepwise fashion:

$$C_r + C_1 \, \boxminus \, \boxplus \, C_{r+1}$$

Eq. (4)

In the formation of spherical micelles, it is generally thought that their growth is well described by the single, stepwise addition of monomer c_1 to clusters of size r, denoted c_r.

For the Becker-Döring processes (Equation 4), we can proceed to write down the kinetic equations for cluster growth and fragmentation based on the familiar law of mass action. Given that, for equation (4), the forward rate coefficient is a_r, and the backward rate coefficient is b_{r+1}, the kinetic equations may be written in a compact way as follows:

$$\dot{c}_r = J_{r-1} - J_r$$

$$\dot{c}_1 = -J_1 - \sum_{r=1}^{\infty} J_r$$

$$J_r = a_r a_r c_1 - b_{r+1} c_{r+1}$$

Eqs. (5), (6), and (7)

where $r=2,3,...$. The dots on c_1 and c_r denote the time derivatives of the concentrations of the monomer (c_1) and cluster (c_r). The quantity J_r represents the flux of matter flowing from cluster C_r to cluster C_{r+1} . Note that the rate equation for c_1, equation (6), has a different form to that of all other cluster sizes; this is because of the special status of C_1 monomers in the Becker-Döring theory.

Equations (5), (6), and (7) represent an infinite set of coupled, non-linear ordinary differential equations including two times an infinite number of generally unknown parameters $\{a_r, b_r\}$. This system evidently entails a very great deal of complexity, and in their full form, these equations are impossible to solve.

To make progress with the description of the extremely complex behavior of such nucleation problems, we have constructed a systematic coarse-graining contraction procedure which enables us to produce mesoscopic and macroscopic descriptions in a mathematically well-defined way from the full microscopic model, in a manner reminiscent of that described above in the derivation of dissipative particle dynamics from molecular dynamics (34). These reduced schemes can be selected to coincide with the level at which experimental observations are made on such systems - this is often at the maximally coarse-grained macroscopic level. The advantage of these reduced models is that they can be analyzed using the theoretical methods of nonlinear dynamics referred to in Section 1: we can then calculate many properties of these systems, including for example, the dependence of induction time on rate coefficients and initial reactant concentrations. We are also able to include additional rate processes in these schemes (other reactions, inhibitors, etc.), such as those involved in Luisi's experiments, summarized in Eqs. (1), (2), and (3).

Our coarse-graining procedure (34) essentially groups together clusters of differing aggregation number at the fine-grained, microscopic level, implied by Eqs. (5), (6), and (7). The scheme is shown diagrammatically in Figure 9. We define new, coarse-grained cluster concentration variables x_r as weighted sums over the fine-grained clusters, while leaving $x_1 = c_1$ unchanged. The resulting coarse-grained Becker-Döring equations have the form:

$$\dot{x}_r = L_{r-1} - L_1$$

$$\dot{x}_1 = -L_1 - \sum_{r=1}^{\infty} rL_r$$

$$L_r = \alpha_r x_r x_1^{\lambda_r} - \beta_{r+1} x_{r+1}$$

Eqs. (8), (9), and (10)

where r = 2,3,... .

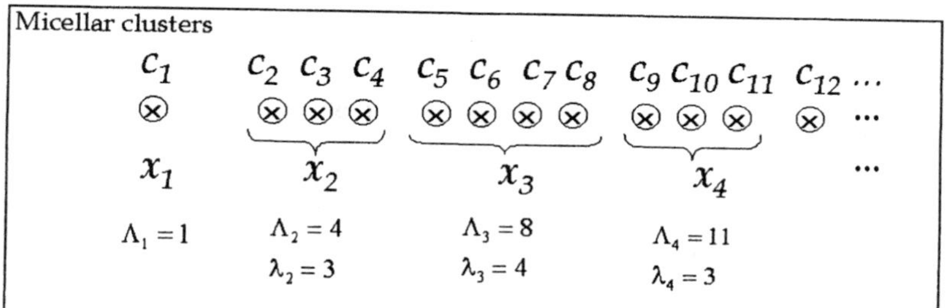

Micellar clusters

$$c_1 \quad c_2 \ c_3 \ c_4 \quad c_5 \ c_6 \ c_7 c_8 \quad c_9 \ c_{10} \ c_{11} \quad c_{12} \cdots$$

$$x_1 \qquad x_2 \qquad x_3 \qquad x_4$$

$$\Lambda_1 = 1 \qquad \Lambda_2 = 4 \qquad \Lambda_3 = 8 \qquad \Lambda_4 = 11$$

$$\lambda_2 = 3 \qquad \lambda_3 = 4 \qquad \lambda_4 = 3$$

$$x_r := \frac{1}{\lambda_r} \sum_{j=1}^{\lambda_r} c_{\Lambda_r + j} \quad \text{with } \Lambda_1 = 1, \ \Lambda_r = \Lambda_{r-1} + \lambda_r$$

Figure 9- The general coarse-graining scheme for the Becker-Döring equations. Reproduced from reference (1)

Comparison of Eqs. (8), (9), and (10) with Eqs. (5), (6), (7) shows that the basic structure of the Becker-Döring equations is unchanged by this coarse-graining scheme, other than the presence in equation (10) of x_1 raised to the power λ_r in place of simply c_1 in the fine-grained description. This is because in the coarse-grained description, λ_r monomers must be added to go from cluster size x_r to x_{r+1}. In fact, this rescaling of the variables amounts to a renormalization transformation. In the coarse-grained description, equations (8) to (10), the renormalized aggregation and fragmentation rate coefficients have the form

$$\alpha_r = T a_{\Lambda_r} a_{\Lambda_{r+1}} a_{\Lambda_{r-1}}$$

$$\beta_{r+1} = T b_{\Lambda_{r+1}} b_{\Lambda_{r+2}} b_{\Lambda_{r+1}}$$

Eqs. (11) and (12)

being expressed in terms of products of the fine-grained rate coefficients, where T is a change of time scale. A contraction of mesh size p followed by one of size q is also a contraction, of size pq, which is the fundamental renormalization property. All the basic mathematical properties of the Becker-Döring equations remain unchanged by this procedure.

In the case where the rate coefficients are of power-law form in the cluster size, a situation of particular relevance to nucleation and growth processes, repeated application of the renormalization procedure to the asymptotic limit of large cluster sizes reveals a dramatic simplification in the macroscopic dynamics.

Figure 10 shows the fixed points of this mapping in the two-dimensional space of the variables $\theta = \left(a_p / b_p \right)$ and cluster size p; systems with $\theta > 1$ are dominated by aggregation, those with $\theta < 1$ are dominated by fragmentation, while $\theta = 1$ is the marginal case where aggregation and fragmentation are precisely balanced.

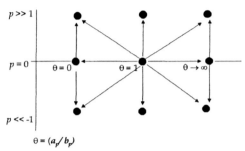

Figure 10, Fixed points of the renormalization mapping in the two-dimensional space of the variable $\theta = (a_p / b_p)$ and cluster size p for the case of the Becker-Döring equations with power-law rate coefficients

It is clear from Figure 10 that there are only nine *universality classes* into which all possible long time, macroscopic kinetic behaviors can fall, despite the fact that the fully microscopic description contained an infinite number of variables and rate parameters(35-37).

One important point concerns the algebraic form of the renormalized rate coefficients, Eqs. (11) and (12). In the physically relevant case where a_r and b_r depend on powers of cluster size r, the renormalized rate coefficients only have this form in the asymptotic limit of large cluster sizes, so that in this case the renormalization is only asymptotically valid.

Returning to Luisi's self-reproducing micelle experiments, we can now propose a detailed theoretical formulation of the microscopic rate processes involved. This is:

$$EC \longrightarrow C_1$$

$$C_r + C_1 \rightleftharpoons C_{r+1}$$

$$EC + C_r \longrightarrow C_r + C_1$$

Eqs. (13), (14), and (15)

where $r = 1, 2, ...$, corresponding to all possible micelle cluster sizes. Writing the concentration of ethyl-caprylate as e, and applying the coarse-graining scheme of

Figure 9 to micelle clusters, we find that, in the macroscopic limit, the rate equations for self-replicating micelle have the form *(34,38)*

$$\dot{e} = -\lambda e \left(k_0 + k_m x_m \right)$$
$$\dot{x}_1 = -e \left(k_0 + k_m x_m \right) - \left(\lambda + 1 \right) \left(\alpha x_1^{\lambda+1} - \beta x_m \right)$$
$$\dot{x}_m = \alpha x_1^{\lambda+1} - \beta x_m$$

Eqs. (16), (17), and (18)

together with

$$\rho = e + x_1 - \left(\lambda - 1 \right) x_m$$

Eq. (19)

which is the condition for conservation of total caprylate in the system, ρ. x_1 and x_m are the monomer and mean micelle concentrations, respectively, k_0 and k_m are the rate coefficients of hydrolysis and micelle-catalyzed hydrolysis of ethyl caprylate, while α and β are the coarse-grained aggregation and fragmentation rate coefficients for micelle formation. The micelles are taken to have cluster size λ (≈ 63). Because of equation (19), the three equations (16), (17), and (18) reduce to two, which may be solved by standard phase-plane methods. The temporal evolution of the solutions are shown in Figure 11 *(34)* and should be compared with the experimental data shown in Figure 8.

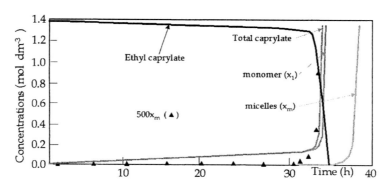

Figure 11 - Predictions of the coarse-grained generalized Becker-Döring model of self-reproducing micelles (compare with Figure 8).

(Adapted with permission from reference 34. Copyright 1996 Royal Society of (London.)

In conclusion, therefore, we see that the renormalizability of the Becker-Döring equations implies that simple universal behavior can be identified in the asymptotic time evolution of what is microscopically a very complex system involving nucleation and growth processes. Hence it is not necessary to solve the full kinetic equations in order to determine the large-scale macroscopic properties of interest in these situations. This is encouraging, since it dispenses with the need to specify many generally unknown fine-grained rate coefficients —the kinetic description in fact becomes very low dimensional.

As noted above, our theory has been established as an asymptotic property, valid only in the macroscopic limit of large aggregation numbers. Our latest results show that there are hierarchies of sets of aggregation and fragmentation rate coefficients for which the renormalization procedure is exact *(38)*. These are models for which the rate coefficients depend exponentially on cluster size. The exact renormalizability of these schemes means that all mesoscopic and macroscopic contractions of the full microscopic system are given *exactly,* not only asymptotically in the limit of large cluster sizes. The analysis of such complex self-organizing systems is thus rendered analytically (and numerically) tractable for the first time. The equations themselves describe numerous complex self-assembly processes and await detailed investigation.

In passing, it is worth pointing out that we have applied a similar approach to describe the formation and selection of self-reproducing sequences during RNA polymerization from prebiotic soups containing initially only the four nucleotide monomer units comprising RNA. We found that these kinetic schemes lead to the emergence of self-replicating RNA polymers from within a plausible pre-biotic environment *(39),* a key finding in support of the RNA world view of the origin of life on Earth *(40)*.

Acknowledgements

This chapter is partially adapted from "Self-Organization and Complexity: A New Age for Theory, Computation and Experiment" presented by the author at the the Nobel Symposium on Self Organization, Karolinska Institutet, August 25-27, 2002 *(1)*. I am grateful to Nélido González-Segredo for his help in the preparation of this manuscript. Much of the work described here has benefited from funding by the UK EPSRC under grants GR/M56234 and RealityGrid GR/R67699.

References

1. Coveney, P. V. "Self-Organization and Complexity: A New Age for Theory, Computation and Experiment" presented at the the Nobel Symposium on Self Organization, Karolinska Institutet, August 25-27, **2002**
2. Israelachvili, J. N., Mitchell, D. J. & Ninham, B. W. *J. Chem. Soc., Faraday Trans. II* **1976**, *72*, 1525-1568.
3. Israelachvili, J. N. *Intermolecular and surface forces*; Academic Press. London, **1992**
4. Kawakatsu, T., Kawasaki, K., Furusaka, M., Okabayashi, H. & Kanaya, T. *J. Chem. Phys.* **1993**, *99*, 8200.
5. Gompper, G.; Bulk and interfacial properties of amphiphilic systems: A Ginzburg-Landau approach. In *Structure and dynamics of strongly interacting colloids and supramolecular aggregates in solution;* Chen, . S.-H. Huang, J. S. and Tartaglia P., Eds.; Kluwer Academic Publishers: Dordrecht, **1992**
6. Gompper, G. Schick, M.; Microscopic models of microemulsions. In *Handbook of microemulsion science and technology*. Kumar P. & Mittal K. L., Eds.; Marcel Dekker, New York, **1999**
7. Rivet, J. P. & Boon, J. P. *Lattice gas hydrodynamics*; Cambridge University Press: Cambridge, **2001**
8. Frisch, U., Hasslacher, B. and Pomeau, Y. *Phys. Rev. Lett.* **1986**, *56*, 1505.
9. Wolfram, S.; *Journal of Statistical Physics*, **1986**, *45*, 471-526
10. Rothman, D. H. & Keller, J. M. *J. Stat. Phys.* **1988**, *52*, 1119-1127.
11. Boghosian, B. M., Coveney, P. V., & Emerton, A. N. *Proc. R. Soc. Lond.* **1996**, *A452*, 1221.
12. Love, P. J., Maillet, J.-B., & Coveney, P. V. *Phys. Rev. E* **2001**, *64*, 61302.
13. Love, P. J. & Coveney, P. V. *Phil. Trans. R. Soc. Lond.* **2002**, *A360*, 357.
14. Boghosian, B. M., Coveney, P. V. & Love, P. J., *Proc. R. Soc. Lond.* **2000**, *A456*, 1431.
15. Succi, S. *The lattice-Boltzmann equation for fluid dynamics and beyond* Oxford University Press: Oxford, **2001**
16. Shan, X. & Chen H. *Phys. Rev. E* **1993**, *41*, 1815.
17. Shan, X. & Chen H. *Phys. Rev. E* **1994**, *49*, 2941
18. Chin, J. & Coveney, P. V. *Phys. Rev. E* **2002**, *66*, 016303.; González-Segredo, N.; Coveney, P. V. *Phys. Rev. E* **2003**, *67*, 046304

19. Chen, H., Boghosian, B. M., Coveney, P. V., and Nekovee, M. *Proc. R. Soc. Lond.* **2000**, *A456*, 2043.
20. Nekovee, M. & Coveney, P. V. *J. Am. Chem. Soc.* **2001**, *123*, 12380.
21. González-Segredo, N. & Coveney, P. V. *The self-assembly kinetics of the cubic double gyroid mesophase: lattice-Boltzmann simulations.* **2003** (Unpublished.)
22. Coveney, P. V. & Highfield, R. R. *The Arrow of Time*; W. H. Allen: London, **1990**
23. Boltzmann, L. *Vorlesungen über Gastheorie;* 1st. edn.; J. A. Barth.: Leipzig, **1896**
24. Boghosian, B. M., Yepez, J., Coveney, P. V. & Wagner, A. *Proc. Roy. Soc. Lond.* **2001**, *A457*, 717.
25.. Boghosian, B. M., Love, P. J., Coveney, P. V., Karlin, I., Succi, S., and Yepez, J., *Galilean-invariant lattice-Boltzmann models with H-theorem* **2002** (Submitted for publication; see also http://arxiv.org/abs/cond-mat/0211093)
26. Hoegerbrugge, P. J. & Koelman, J. M. V. A. *Europhys. Lett.* **1992**, *19*, 155.
27. Español, P. & Warren, P. *Europhys. Lett.* **1995**, *30*, 191.
28. Flekkøy, E. G. & Coveney, P. V. *Phys. Rev. Lett.* **1999**, *83*,1775.
29. Flekkøy, E. G., Coveney, P. V. & De Fabritiis, G. *Phys. Rev. E* **2000**, *62*, 2140.
30. Bachmann, P. A., Luisi, P. L. & Lang, J. *Nature* **1992**, *357*, 57.
31. Walde, P., Wick, R., Fresta, M. Mangone, A. & Luisi, P. L. *J. Am. Chem. Soc.* **1994**, *116*, 11649
32. Becker, R. & Döring, W. *Ann. Phys.* **1935**, *24*, 719.
33. Gunton, J. D., San Miguel, M. & Sahni, P. S. The dynamics of first order phase transitions. In *Phase transitions and critical phenomena*, Domb C. and Lebowitz J. L. Eds.; vol. 8; Academic Press: London, **1983**
34. Coveney, P. V. & Wattis, J. A. D. *Proc. Roy. Soc. Lond.* **1996**, *A452*, 2079.
35. Coveney, P. V. & Wattis, J. A. D *J. Phys. A: Math. Gen.* **1999**, *32*,7145.
36. Wattis, J. A. D. & Coveney, P. V. *J. Phys. A: Math. Gen.* **2001**, *34*, 8679.
37. Wattis, J. A. D. & Coveney, P. V. *J. Phys. A: Math. Gen.* **2001**, *34*,8697.
38. Wattis, J. A. D. & Coveney, P. V. *Exact renormalisation of the Becker-Döring equations for exponential size-dependent rate coefficients.* **2002** (Unpublished.)
39. Wattis, J. A. D. & Coveney, P. V. *J. Phys. Chem. B* **1999**, *103*, 4231.
40. Zubay, G. *The origins of life on Earth and in the cosmos*, 2nd edn. Academic Press: London, **2000**

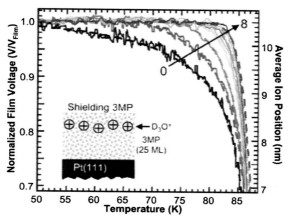

Plate 1. Shielding ions from the interfacial fluidity perturbations: Comparison of the experimental (solid) and simulated (dashed) normalized film voltage versus T for the top 3.1 nm of the 25 ML 3MP films with 0 to 8 ML of additional 3MP shielding the ions from the free interface. As the ions are displaced more than 6 ML from the interface, the mobility of the ions resembles that predicted by the position-independent bulk film viscosity. Additional shielding layers (not shown) have no further effect. Adapted from ref. 21.

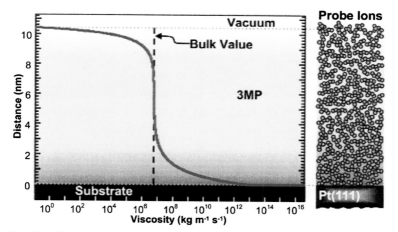

Plate 2. Viscosity change for a 25 ML 3MP film at 85 K using D_3O^+ as the probe ions. Notice the uniform change in viscosity from either interface to the bulk value in the middle of film. Adapted from ref. 21.

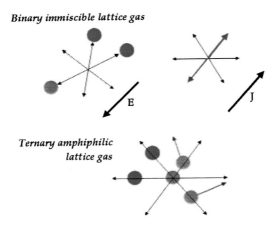

Binary immiscible lattice gas

Ternary amphiphilic lattice gas

Plate 1 - Directions of color flux and color field in a lattice-gas automaton model for immiscible fluids

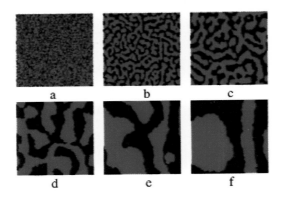

a   b   c

d   e   f

Plate 2 -Phase separation kinetics in a binary immiscible fluid mixture (oil and water) as simulated with a two-dimensional lattice-gas automaton model. Time steps shown are (a) 0, (b) 50, (c) 150, (d) 500, (e) 1000, (f) 6000. The lattice size is 128^2.

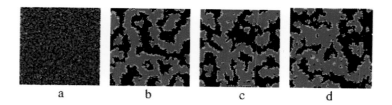

Plate 3 - *Arrest of domain growth in a binary immiscible mixture (black and red for water and oil, respectively) when surfactant (green) is added, as simulated with a two-dimensional lattice-gas automaton model. Time steps shown are (a) 0, (b) 4000, (c) 16000, and (d) 40000. The lattice size is 128^2.*

(Reproduced with permission from reference 11. Copyright 1996 Royal Society of London.)

Plate 4 - *Formation of the bicontinuous cubic "P" mesophase from an initial random mixture of oil and water plus surfactant added, as simulated with a lattice-Boltzmann model. Time steps shown are (a) 0, (b) 400, and (c) 5600, respectively, and the surface displayed is where the surfactant mainly resides. The lattice size is 64^3. (Reproduced with permission from reference 20, copyright 2001, American Chemical Society).*

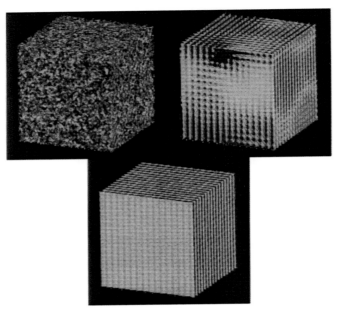

Plate 5 - Formation of the lamellar mesophase from an initial random mixture of oil and water plus surfactant, as simulated with a lattice-Boltzmann model. Although the initial condition is identical to that of figure 9, the surfactant-surfactant coupling has been switched, leading to a different evolution. Time steps shown are (a) 0, (b) 800, and (c) 23200, respectively, and the surface displayed is where the surfactant mainly resides. The lattice size is 64^3. (Reproduced with permission from reference 20, copyright 2001, American Chemical Society).

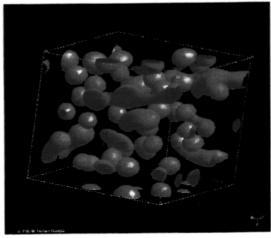

Plate 1a. Pluronic 85 in water at 15 °C from MesoDyn simulation. A predominantly spherical micellar emulsion is seen.

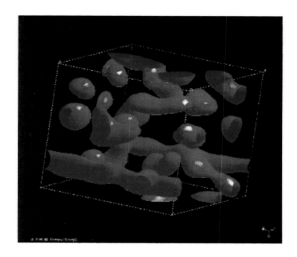

Plate 1b. Pluronic 85 in water at 70 °C from MesoDyn simulation. The system appears to be in a coexistence region between spherical and rod-like micelles.

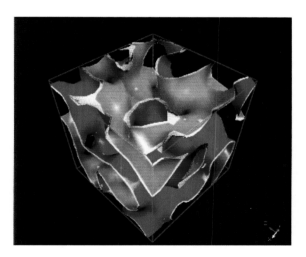

Plate 2a. Snapshot from MesoDyn simulation of polystyrene and polybutadiene equi-molar mixture. The surfaces indicate where each component falls to half its maximum density (unsurprisingly these surfaces coincide).

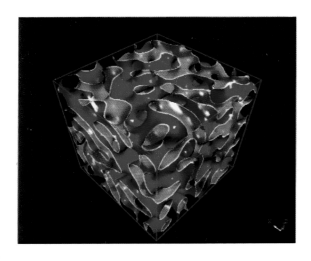

Plate 2b. Snapshot from a MesoDyn simulation of an equi-molar mixture of polystyrene and polybutadiene with 10% of the system comprising a symmetric diblock copolymer of styrene and butadiene. This copolymer acts as a compatibilizer, lowering the interfacial tension and leading to a more complex morphology

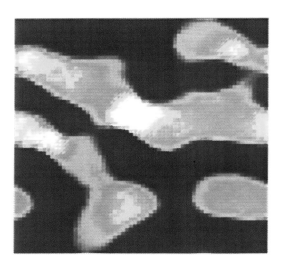

Plate 3. Flux rates for a 2d slice through the simulation cell, obtained using Palmyra. Blue implies zero flux, white corresponds to maximum flux. Oxygen does not permeate the styrene rich regions

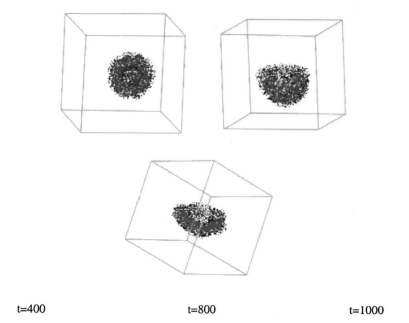

t=400   t=800   t=1000

Plate 1. Three snapshots of the DPD simulation starting from a patch of crystalline bilayer. After some time the bilayer expands, and starts oscillating, until the symmetry is broken and it bends preferentially onto one side into a cup-shape. The cup closes up into an almost perfect spherical vesicle.

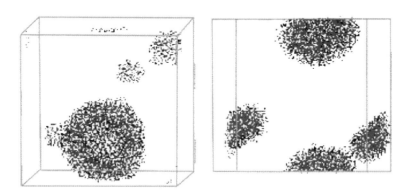

Plate 2. (a) A flat aggregate coexisting with a larger vesicle, and (b) two small micelles coexisting with a larger vesicle.

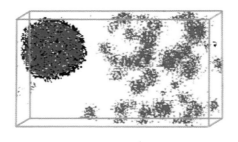

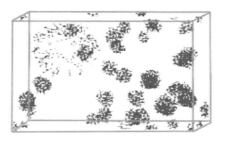

t=1000   t=3000

Plate 3. Two simulation boxes have been prepared independently to make a fully equilibrated mono-lamellar vesicle and a homogeneous solution of micelles. When the two simulation boxes are merged, we observe vesicle growth via micellar deposition. Only the distribution of the surfactant tails is shown in the second image.

Plate 4. Snapshot of the DPD simulation of a free-standing bilayer

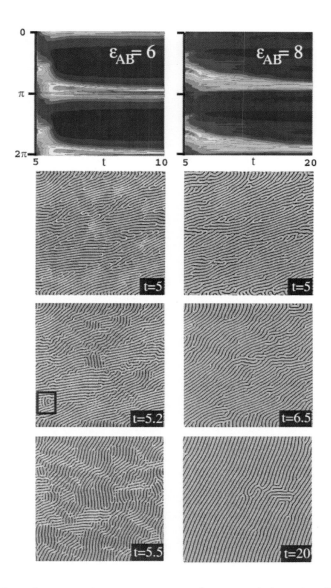

Plate 1: Top: the scattering intensity as a function of the azimuthal angle and dimensionless time ($t = \tau/\Delta\tau \times 10^3$) for a A_4B_4-melt in 2D. Below: simulation snapshots. Left: closer to ODT ($\chi N = 19.2$). Right: further from ODT ($\chi N = 25.6$). The axis in which the electric field is applied is vertical. The electric field strength parameter $B = 0.2$. (Reprinted with permission from Zvelindovsky et al. Phys. Rev. Lett. 2003, 90, 049601. Copyright 2003, American Physical Society).

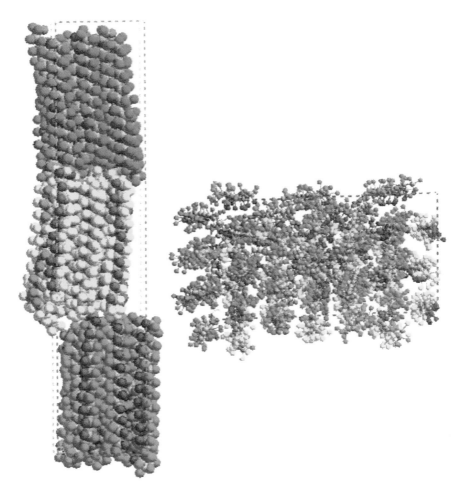

Plate 1. High-temperature structure of the SAM.

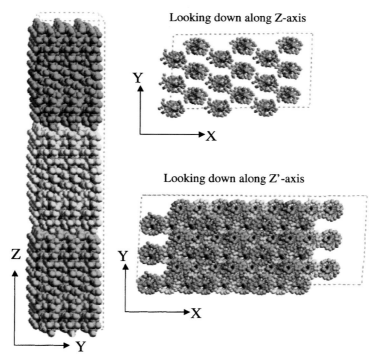

Plate 2. Low-temperature structure of the SAM.

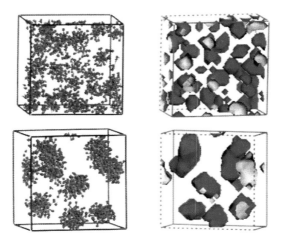

Plate 1 – mesoscale structure predicted for a 12% v/v solution of a surfactant with a strongly repulsive head group. Top: snapshot of the surfactants (head groups red, tails green) and the tail density distribution for a HT surfactant model. Bottom: results for a HTT (longer tail length). Water is not displayed

Plate 2 –DPD simulation of a system with less repulsive head group interactions. Predicted evolution of structure over 10,000 DPD time steps starting with a random distribution of surfactant molecules.

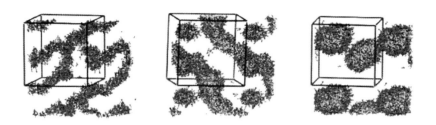

Plate 3 – Predicted structure for 12% HT surfactant (parameters as in Figure 3) with 0%, 5% and 10% oil.

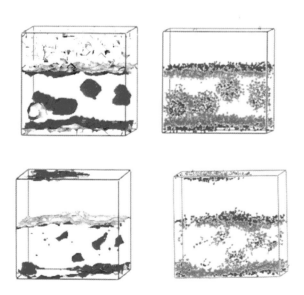

Plate 4 – DPD simulation of top: 45% A, 45% B, 10% C (ABB amphiphile), bottom: 47% A, 47% B, 6% C (ABB amphiphile. Predicted distribution of A (isodensity surface averaged over 500 time steps), and a snapshot of the amphiphile (surfactant) distribution (A beads = red, B beads = green)

Chapter 14

Mesoscale Simulations: Industrial Applications

Simon McGrother[1], Lam Yeng Ming[2,*], and Gerhard Goldbeck-Wood[1]

[1]Accelrys Inc., 9685 Scranton Road, San Diego, CA 92121
[2]Nanyang Technological University, School of Materials Engineering, Nanyang Avenue, Singapore 639798

The industrial importance of mesoscale phenomena is increasingly appreciated. Techniques for modeling such length and time scales are increasingly validated and available in standard software packages. Accelrys offers DPD (dissipative particle dynamics) and MesoDyn. Both tools coarse-grain the familiar atomistic representation of the molecule to gain orders of magnitude in both length and time scale relative to traditional atomistic scale simulation. The chemistry is captured through effective potentials based on the energy of mixing of the binary pairs in the system. The methods yield structural and dynamic information on phase morphology and can be used as input to finite element analysis. In this chapter the accuracy of such tools is assessed with particular emphasis on the communication between various length scales. Case studies from the fields of polymer science and surfactant science are discussed, recent advances are presented and ideas for future applications and directions are suggested.

Introduction

The need for mesoscale modeling is clear. Structures which form at length-, and over time-scales that are large compared to atomistic motion critically effect the properties and hence end0uses of a material. There are several schemes available to study fluids where the underlying materials are faithfully represented, but the dimensions of the system are coarse-grained in order that large-scale phenomena may be interpreted or predicted. This paper examines the current state of the art of mesoscale simulation and the applications that such modeling has found in industry.

Materials' modeling is an increasingly integrated tool in the R&D portfolio. The unique insights available through simulation of materials at the quantum, molecular, mesoscale and finite element levels can provide discontinuous scientific advances. The most significant challenge presently facing the development of materials modeling is the communication between the length and time-scales spanned by the various methods. The envisaged analogy is with a microscope: ideally one would like to be able to zoom the modeling, seamlessly between the levels of detail.

This paper will address recent efforts to integrate the scales. The principal focus will be on the derivation of accurate input parameters for mesoscale simulation, and the subsequent use of finite element modeling to provide quantitative information regarding the properties of the simulated mesoscale morphologies.

In polymer modelling, traditional molecular dynamics may be used to obtain thermodynamic information about a pure or mixed system. Properties obtained using these microscopic simulations assume that the system is homogeneous in composition, structure and density: this is a limitation. The slow dynamics of polymers also dictates that smart methods are needed to probe phase space.

A mesoscale modelling approach can overcome many of these limitations. These tools achieve longer length scales by uniting many atoms into a single bead, and longer time scales by integrating out the fast motions of the underlying particles leaving only soft, effective interactions. The familiar atomistic description of the molecules is coarse-grained leading to beads of fluid (representing the collective degrees of freedom of many atoms). These beads interact through pair-potentials which crucially, if meaningful data are to be obtained, capture the underlying interactions of the constituent atoms. Complex self-assembling fluids that have long-range order can be studied. The use of atomistic modeling to parameterise mesoscale simulation by suggesting sensible coarse-graining and effective interactions between species will be discussed. The major hole in the technology remains the reverse mapping from the mesoscale to the atomistic, where no adequate method has been developed

The primary output of mesoscale modeling is phase morphologies with sizes up to the micron level. These morphologies are of interest, but it is non-trivial to relate such morphologies to material properties. Finite element modeling can be used to predict physical and mechanical properties of arbitrary structures. Details of the link that has been established between Accelrys' MesoDyn and MatSim's Palmyra are given and highlighted with some recent validation work on polymer blends. These results suggest that the combination of simulations at multiple scales can yield important insights.Whilst much of a materials' properties can be determined or inferred from chemistry of the underlying molecules, a wide-range of polymeric systems have functionality that is dependent purely on the structures that the material adopts at the mesoscale. At this scale atomistic information is not necessary but bulk (e.g., fluid dynamics) simulation is too coarse. A perfect example is that variety of applications for HIPS, all of which are made from polystyrene, yet each has distinct application because of the structure that the material gains on processing.

MesoDyn and DPD are coarse-grained dynamics methods that capture such structural information whilst retaining a link to the chemistry of the underlying materials. MesoDyn is a dynamic variant on mean-field Langevin density functional theory. The system evolves because of local gradients in the chemical potentials of the various species that make up the system. These gradients arise from a gaussian chain Hamiltonian decorated with an effective external potential to represent the non-ideal interchain interactions. DPD is a coarse-grained dynamics algorithm that accurately captures the hydrodynamic behavior of the fluid and the underlying interactions of the species. DPD is ideal for surfactant solutions and MesoDyn for polymer melts and blends.

The question that needs to be answered is: can we pose a question purely in terms of chemical formulation and accurately predict "nanoscale" morphology and, consequently, realistic material properties?

The strategy that we have employed to address this issue comprises three stages. Simulation at the fully atomistic level, employing force-fields and molecular dynamics; mesoscale simulation using parameters obtained at the previous level to define interactions and mesoscale-topology, and finally finite element study of the properties of a nano-scaled structure obtained from mesoscale modeling.

Method description

Many modeling methods are used individually and combined in industrial problem solving. These span the gamut of length and times from the sub-atomic to the macroscopic. With molecular modeling, the focus is on chemical defined species, with various levels of coarse-graining. A comparison of various levels of "molecular" modeling: quantum, classical atomistic and mesoscale is shown in Table I

Table I. Comparison of modeling techniques

	Quantum	Atomistic	Mesoscale
Length	Angstroms	nm	100s of nm
Fundamental Unit	Electrons/nuclei	atoms	Beads representing group of atoms
Time scale	fs	ns	ms
Dynamics	Not appropriate	F=ma	Hydrodynamics

The primary techniques we used for mesoscale modeling are MesoDyn and DPD. In this study we also use finite element methods to obtain material properties of mesoscopically-structured materials

MesoDyn

MesoDyn is a dynamic mean-field density functional theory for complex fluids *(1)*. The free energy comprises an ideal term based on a Gaussian Chain Hamiltonian representation of the polymeric materials; a Gibbs entropy contribution favoring mixing and a non-ideal term accounted for using a mean-field approximation. The key approximation is that in the time regime under consideration the distribution functions are optimized (i.e., the free energy is minimal). Applying appropriate constraints the optimal distribution can be obtained and related back to the free energy. We are left with a simple expression for the non-ideal term (obtained by invoking the random phase approximation-RPA):

$$F_{RPA}^{nid}[\rho] = \frac{1}{2}\sum_{IJ}\int_V \varepsilon_{IJ}\rho_I(r)\rho_J(r)dr .$$

which assumes a local mean-field. However, because the mean-field must account for the inter-chain interactions a non-local mean-field is preferred. A suitable choice leads to:

$$F^{nid}[\rho] = \frac{1}{2}\sum_{IJ}\int_V \int_V \varepsilon_{IJ}(|r-r'|)\rho_I(r)\rho_J(r')drdr'$$

where $\varepsilon_{ij}(|r-r'|)$ is a cohesive interaction defined by the same Gaussian kernel as in the ideal chain Hamiltonian. This parameter is then directly related to a calculable property, namely the Flory-Huggins interaction parameter χ. Further details are given in the chapter by Sevink, Zvelindovsky and Fraaije

DPD

Dissipative particle dynamics (DPD) uses soft-spheres to represent groups of atoms, and incorporates hydrodynamic behavior via a random noise which is coupled to a pair-wise dissipation. These terms are coupled so as to obey the fluctuation-dissipation theorem.

The force acting on one of these soft-spheres comprises three pair-wise terms, the first is a soft conservative repulsion related to the energy of mixing of

$$\Omega_{ij} = w(\mathbf{r}_{ij})\left[\alpha_{ij} + \sigma\,\theta_{ij} - \frac{\sigma^2}{2kT}w(\mathbf{r}_{ij})\,\hat{\mathbf{r}}_{ij} \cdot \mathbf{v}_{ij}\right]$$

the bead with its neighbor, the second is a random noise term and the last is a dissipative or drag term. The noise and dissipation are coupled (by the fluctuation-dissipation theorem) and act as a thermostat. Further details are given in the chapter by Noro and Warren and the chapter by Broze and Case.

The two methods overlap, but DPD is preferred where concentrations are low, and MesoDyn is ideal for systems that comprise polymer melts and blends.

Finite Element methods

Standard solvers interpolate pure component data to yield very accurate results for complex, mixed systems. In this study we use GridMorph and Palmyra (2,3) from MatSim GmbH.

Parameterization.

Both DPD and MesoDyn rely on links with the underlying chemistry for them to be useful in industrial settings. With MesoDyn the use of a non-local mean field for the non-ideal contribution to the free energy leads to an expression involving a cohesive energy interaction between neighbor species. The choice of interaction is not unique, but cohesive energy is a readily calculable property (4,5) and therefore makes a sensible choice.

For DPD Groot and Warren (6) established very clearly a link between the conservative repulsion and the Flory-Huggins parameter of the interacting species. In addition they make sensible arguments for the value of repulsion between beads of like kind (relating this value to the compressibility of the pure fluid). Finally they established bounds on the noise (and dissipation) at which the system behaves sensibly.

The Flory-Huggins parameter χ, which captures the enthalpic contributions to the mixing free energy for two materials, can fitted from experimental data or calculated using atomistic scale simulation. Methods are described in the paper by Case and Honeycutt (7). One particularly promising approach is to fit χ to the enthalpy of mixing obtained from atomistic simulations of the bulk pure and mixed phases. The Compass forcefield has been shown to provide accurate predictions of cohesive properties such as energy of mixing (4,5).

For both methods the molecules must be described, at a coarse-grained level, by connected beads. Ideally we would represent the real chain by a bead-spring model with the same response functions. In practice, it is more convenient to use the ideas of Kuhn and represent a statistical unit of the chain by a single bead (8,9).

Validation Studies

1) Predicting phase diagrams for block copolymers in water

Both mesoscale methods have been validated by comparison with experimental phase diagrams of block copolymers and their aqueous solutions (10-12). Pluronic molecules (ethylene oxide, propylene oxide, ethylene oxide triblock copolymers) are used extensively in lubrication and drug delivery systems. They are benign, stable, cheap to manufacture and self assemble into complex phases depending upon the relative block lengths and the concentration of solution. Consequently they have enormous industrial value.

When a typical pluronic is added to water the relative hydrophobicity of the core propylene oxide leads to meso-phase separation. Depending upon the concentration of pluronic in the solution various phases are observed. With Gaussian chain representations obtained by including several monomers into each bead, and interaction parameters obtained from experimental vapor-pressure data MesoDyn predicts the correct phase at the correct point in phase space (11).

2. Calculation of critical micellar concentration (CMC)

The CMC is a characteristic concentration of a surface-active agent (surfactant) in solution defined by the appearance of aggregate structures such as micelles. Experimentally, CMC can be obtained from a discontinuity in the interfacial tension at an oil/water or water/air interface as a function of surfactant concentration. DPD can be used to measure the interfacial tension at a planar interface from the difference in transverse and planar pressures, and predict CMC (12). At concentrations above CMC, micelles can be observed forming and detaching from the interface. To relate interfacial tension (DPD units) to real surface tension it is necessary to stipulate the size and mass of the bead, this fixes a time frame for the simulation and permits conversion to real units.

3. Effect of Temperature

MesoDyn has been used to study the temperature dependence of a pluronic (P85) in water at a concentration of 27%. Experimentally such systems are seen to evolve from a spherical, micellar solution at 15°C to a rod-like dispersion at 70°C. The parameters for the pluronic plus water are known at 25°C, but little validation work exists to suggest whether such parameters can be used at other temperatures. From vapor pressure data (13) the temperature dependence of the Flory-Huggins parameter can be inferred. The usual expression is:

$$\chi = \alpha + \beta\!\big/\!_T$$

For ethylene oxide with water, $\alpha=2.85$ and $\beta=-439K$, for propylene oxide in water $\alpha=2.023$ and $\beta=-97.9K$. These parameters are used to obtain χ at both temperatures. The phases observed using MesoDyn at these temperatures can be seen in color Plates 1a and 1b. These phases agree well with experiment (14), and the dependence on T of the χ parameter proves to be important.

Case Studies

1. Mesoscale modeling of drug delivery

Pluronic solutions can be used to deliver hydrophobic drugs. The self-assembly of pluronics leads to hydrophobic regions where the drug resides. In a

234

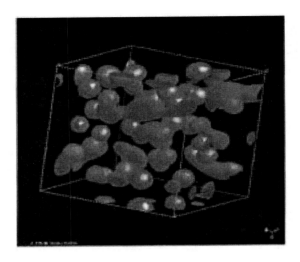

Plate 1a. Pluronic 85 in water at 15 °C from MesoDyn simulation. A predominantly spherical micellar emulsion is seen.
(See page 4 of color insert.)

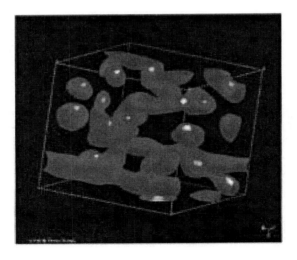

Plate 1b. Pluronic 85 in water at 70 °C from MesoDyn simulation. The system appears to be in a coexistence region between spherical and rod-like micelles.
(See page 5 of color insert.)

typical example haloperidol (an anti-psychotic) is dissolved in a solution of pluronic in water. From electron micrographs (Figure 1) the drug is seen to significantly distort the size and shape of the micelles formed, from a uniformly spherical distribution when no drug is present.

MesoDyn simulations are show in Figure 2. When 1% drug is present significant distortion is seen. The behavior can be understood by comparing the contents of each micelle: some contain no drug, while others are quite drug rich. This distribution of concentrations contributes to the non-sphericity and size-distribution obtained.

The core size increases with concentration of drug, with good agreement between experimental and simulated values (See Figure 3). Mechanistically, it

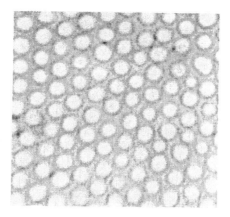

Figure 1a. Experimental TEM image of a 10% solution of pluronic F127 in water.

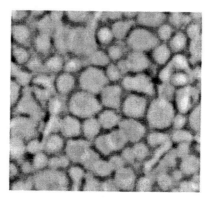

Figure 1b. Experimental TEM image of a 10% solution of pluronic F127 in water with a small amount of haloperidol present.

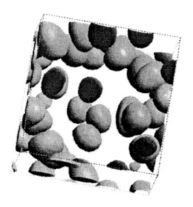

Figure 2a. MesoDyn simulation snapshot of pluronic F127 in water at 24% concentration. Uniform, spherical micelles form with propylene oxide in the center

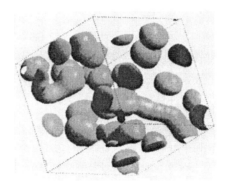

Figure 2b. MesoDyn simulation snapshot of pluronic F127 in water at 24% concentration with 1% haloperidol present. Significant size and shape distribution of the hydrophobic region is observed.

can be envisaged that the haloperidol aggregates quickly into small units and the slower moving pluronic self-assembles around these units. This leads to a range of drug concentrations in the micelles. Control of size distribution might be achieved by addition of a dispersant (some component which prevents or delays the aggregation of drug). A more homogeneous solution of drug would be better distributed between the various micelles. Since the drug must cross the blood-brain boundary, it is imperative that aggregate size distribution is adequately controlled.

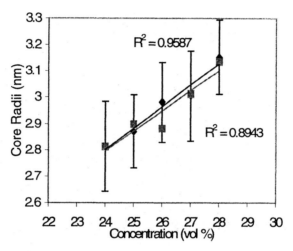

Figure 3. Core radii of self-assembled micelles in a solution of pluronic in water as a function of concentration of pluronic. Experimental values are shown as diamonds, simulated values as squares.

2. Mesoscale and Finite Element Simulation of Polymer Blends

The structures predicted by mesoscale modeling give rise to diverse and interesting material properties. Such properties can be calculated with Finite Element Methods such as Palmyra from MatSim GmbH. These methods require as input the structure of the material, and the property of the pure components that comprise the mixture. Using standard solvers the finite element code can then predict the property for the structured material. As a test case we studied the oxygen diffusion through a material designed to act as a gas separation membrane. A binary blend of polystyrene and polybutadiene was simulated with MesoDyn using parameters obtained from atomistic level modeling. These polymers tend to phase segregate and large domains form with little interface. Upon addition of a diblock copolymer of both species (styrene and butadiene), the blend is compatibilized and the interfacial tension is lowered. The resulting morphology is far more complex with much smaller domains, more interfacial zones and frustrated regions. Snapshots of the uncompatibilized and compatibilized blends are show in color Plate 2. Both of these structures were analyzed for oxygen diffusion using MatSim's Palmyra. The pure component oxygen diffusions for polystyrene and polybutadiene were obtained using classical dynamics with fully atomistic interactions (and again the Compass force-field (4)). It is also possible to extract such information from structure property relationships such as those proposed by Bicerano (15). The results of the FEM calculation are given in Table II.

Table II – predicted oxygen diffusion rates

System	Oxygen diffusion (Dow Units)
Without Compatibilizer	970
With Compatibilizer	1040

The compatibilized blend shows increased diffusion of oxygen, which can be attributed to an increase in the number of channels that the oxygen can choose to diffuse through. The finite element method also indicates the diffusion rates for various spatial regions. Color Plate 3 shows flux rates predicted by Palmyra for the blend. All the flux is seen to be in the polybutadiene rich regions, with maximum rate in the "necking" regions. The finite element calculation also indicates that there is no preferred direction for flux (diffusion in each Cartesian coordinate is equal), indicating that our system size is probably sufficiently large to represent the bulk system.

This study therefore has used interaction energies and diffusivities obtained from atomistic scale simulations to parameterize mesoscale methods and inform finite element tools, in order that mesoscopically calculated structures can be analyzed for diffusion rates of the true material. This is an exciting development that we intend to pursue.

Conclusion

Mesoscale modeling methods are increasingly being applied to predict materials properties and behavior, as shown by the validation and case studies in this chapter and in other chapters in this book.

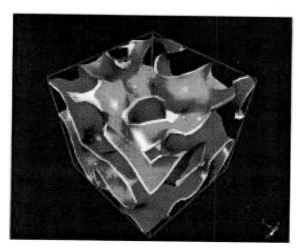

Plate 2a. Snapshot from MesoDyn simulation of polystyrene and polybutadiene equi-molar mixture. The surfaces indicate where each component falls to half its maximum density (unsurprisingly these surfaces coincide).

(See page 5 of color insert.)

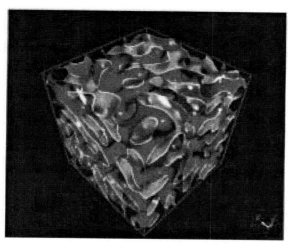

Plate 2b. Snapshot from a MesoDyn simulation of an equi-molar mixture of polystyrene and polybutadiene with 10% of the system comprising a symmetric diblock copolymer of styrene and butadiene. This copolymer acts as a compatibilizer, lowering the interfacial tension and leading to a more complex morphology
(See page 6 of color insert.)

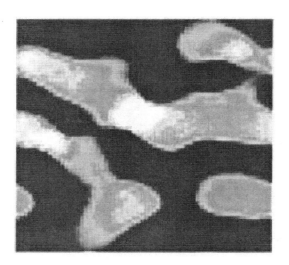

Plate 3. Flux rates for a 2d slice through the simulation cell, obtained using Palmyra. Blue implies zero flux, white corresponds to maximum flux. Oxygen does not permeate the styrene rich regions
(See page 6 of color insert.)

The use of accurate classical force-fields to generate material parameters, coupled with the novel ability to simulate system on a longer length and time-scale using mesoscale techniques has been demonstrated. The novel approach herein is to take the simulated mesoscale morphology and to use finite element methods to predict realistic material properties based on the morphology obtained. This now gives the modeler a direct route from the atomistic description of the system to a trust-worthy estimate of the properties of a material. The material properties of course depending on processing and formulations, but these effects are not beyond the scope of multilevel simulation.

Acknowledgements

We are most grateful to Albert Widmann-Shupack and Andrei Gusev of MatSim GmbH for their collaboration and support.

References

1. Fraaije, J. G. E. M.; van Vlimmeren, B. A. C.; Maurits, N. M.; Postma, M.; Evers, O. A.; Hoffmann, C.; Altevogt, P.; Goldbeck-Wood, G. *J. Chem Phys*, **1997**, *106*, 4260.
2. Gusev, A. A. *J. Mech. Phys. Solids* **1997**, *45*, 1449.
3. Gusev, A.A. *Macromolecules* **2001**, *34*, 3081.
4. Rigby, D.; Roe, R. J. *J Chem. Phys.* **1988**, *89*, 5280,
5. Eichinger, B. E.; Rigby, D.; Stein, J; *Polymer*, **2002**, *43*, 599.
6. Groot R. D.; Warren, P. B.; *J. Chem. Phys.* **1997**, *107*, 4423.
7. Case F. H.; Honeycutt, J. D.; *Trends in Polymer Science*, **1994**, *2*, 259.
8. Kuhn, W.; *Kolloid Z.*, **1936** *76*, 258
9. Kuhn, W.; *Kolloid Z.* **1939** *87*, 3.
10. Groot, R. D.; Madden, T. J.; *J. Chem. Phys.*, **1998**, *108*, 8713
11. van Vlimmeren, B.A.C.; Maurits, N.M.; Zvelindovsky, A.V.; Sevink, G.J.A.; Fraaije, J.G.E.M.; *Macromolecules*, **1999**, *32*, 646.
12. Nicolaides, D., *Molecular Simulation* **2001**, *26* , 51.
13. Malcolm G. N.; Rowlinson, J. S. *Trans. Faraday Soc.*, **1957**, *53*, 921.
14. Mortensen, K.; Pedersen, J.S., *Macromolecules*, **1993**, *26* , 805-812.
15. Bicerano, J. *Prediction of Polymer Properties*, **1993**, 1st edition, Marcel Dekker: New York.

Chapter 15

Application of Dissipative Particle Dynamics to Materials Physics Problems in Polymer and Surfactant Science

Massimo G. Noro, Frederico Meneghini, and Patrick B. Warren

Unilever R&D Port Sunlight, Bebington, Wirral CH63 3JW,
United Kingdom

The basic dissipative particle dynamics (DPD) method is described, and a useful extension to allow many-body interactions is detailed. The application of DPD to problems in materials physics is discussed for polymers, surfactants and self-assembling systems.

Dissipative particle dynamics (DPD) is a mesoscale simulation technique introduced initially by Hoogerbrugge and Koelman for applications to hydrodynamics problems (1). Here we describe some of the recent developments for DPD and its application to problems in materials science. But first we say a few words about the basic philosophy.

For every mesoscale problem, a central challenge is to determine the appropriate mesoscale parameters in the simulation, so that we obtain reasonable results. One strategy, championed for example by Coveney and co-workers in the previous chapter, is a *bottom-up* approach in which one calculates mesoscale parameters rigorously from the molecular properties of the material. A complementary strategy which we have pursued for DPD is a *top-down* approach, in which one adjusts the mesoscale parameters to make contact with macroscopic properties such as compressibility, solubility, surface tension, viscosity, bending rigidity, etc. Just as in density functional theory, the top-down approach seeks to reproduce the important properties of homogeneous systems at equilibrium, with the intention of applying the models to inhomogeneous systems or non-equilibrium problems. Unlike density functional theory though, the emergent properties from DPD are still most easily determined by simulation, therefore calibration of the method is very important. Once one accepts that calibration by computer simulation cannot be avoided, it immediately follows that the method can be and should be *as simple as possible*, whilst retaining compatibility with the basic underlying physics. This idea is present in the very first DPD paper by Hoogerbrugge and Koelman, and continues to motivate virtually all our choices in the development of DPD.

DPD is a particle-based simulation method. The particles move according to Newton's equations of motion under the influence of their mutual interactions and possibly external forces. Following the above philosophy, the interactions are chosen for the convenience of the computer simulator, rather than for any underlying physical reasons. They are short ranged, so that neighbour-list methods can be used efficiently, and soft, so that large time steps can be used in the integration of the equations of motion. Furthermore we wish to have a well-defined temperature and to preserve hydrodynamic modes, so a momentum-conserving thermostat was incorporated into the original algorithm. We recognise now that the original (Espanol-Warren) thermostat proposed for DPD (2), and described below, is perhaps not the best choice. Rather we would suggest to anyone starting in this field that the Lowe-Anderson thermostat is simpler and more efficient (3), although it means that some basic calibration simulations may have to be re-done.

It turns out that the use of soft interaction potentials, as proposed originally, severely limits the thermodynamic properties obtained from DPD. This restricts the possibility of adjusting the parameters in DPD to fit real systems, such as the non-ideal mixing behaviour for water-soluble polymers for example. This restriction can be lifted however by introducing an additional density dependence in the interaction laws, at a very modest additional computational cost. This idea, marketed as *many-body DPD* (or sometimes multi-body DPD) has been explored by several workers (4, 5) including ourselves. However, many-body DPD provokes a fundamental shift in the interpretation of the soft repulsions.

We will now discuss in more detail the basic ideas of DPD and many-body DPD, before giving some examples of applications to polymer, surfactant and self-assembling systems.

Basic DPD algorithm

The particles in DPD have positions $\mathbf{r}_i$ and velocities $\mathbf{v}_i$, where $i = 1$ to N runs over the set of particles, moving in a simulation box of volume V. They move according to the kinematic condition $d\mathbf{r}_i/dt = \mathbf{v}_i$, and Newton's second law $m_i\, d\mathbf{v}_i/dt = \mathbf{F}_i$ where m_i is the mass of the ith particle. Here $\mathbf{F}_i = \mathbf{F}_{i,\text{ext}} + \Sigma_j\, \mathbf{F}_{ij}$ is the total force acting on the ith particle, comprising a possible external force $\mathbf{F}_{i,\text{ext}}$ and the forces $\mathbf{F}_{ij}$ due to the interaction between the ith and jth particles. The interaction forces are decomposed into conservative, dissipative and random contributions, $\mathbf{F}_{ij} = \mathbf{F}_{ij}^{C} + \mathbf{F}_{ij}^{D} + \mathbf{F}_{ij}^{R}$. The individual contributions all vanish for particle separations larger than some cutoff interaction range r_c, and all obey Newton's third law so that $\mathbf{F}_{ij} + \mathbf{F}_{ji} = 0$.

The conservative force is $\mathbf{F}_{ij}^{C} = A_{ij}\, w_C(r_{ij})\, \mathbf{e}_{ij}$ where $\mathbf{r}_{ij} = \mathbf{r}_j - \mathbf{r}_i$, $r_{ij} = |\mathbf{r}_{ij}|$, and $\mathbf{e}_{ij} = \mathbf{r}_{ij}/r_{ij}$. The weight function $w_C(r)$ vanishes for $r > r_c$, and for simplicity is taken to decreases linearly with particle separation, thus $w_C(r) = (1-r/r_c)$. Different species of particles are differentiated by their repulsion amplitudes A_{ij}. This force corresponds to a pair potential $U(r) = (A/2)(1-r/r_c)^2$, in other words the particles are soft and can sit on top of each other without paying an infinite potential energy price. The dissipative and random forces are $\mathbf{F}_{ij}^{D} = -\gamma\, w_C(r_{ij})$ $(\mathbf{v}_{ij} \cdot \mathbf{e}_{ij})\, \mathbf{e}_{ij}$ and $\mathbf{F}_{ij}^{R} = \sigma\, w_R(r_{ij})\, \xi_{ij}\, \mathbf{e}_{ij}$. In these $w_D(r)$ and $w_R(r)$ are additional weight functions, also vanishing for $r > r_c$, γ and σ are amplitudes, $\mathbf{v}_{ij} = \mathbf{v}_j - \mathbf{v}_i$, and $\xi_{ij} = \xi_{ji}$ is *pairwise* continuous white noise with $\langle \xi_{ij}(t) \rangle = 0$ and $\langle \xi_{ij}(t)\xi_{kl}(t') \rangle = (\delta_{ik}\delta_{jl} + \delta_{il}\delta_{jk})\delta(t - t')$. The combination of the dissipative and random forces acts as a thermostat, provided the weight functions and amplitudes are chosen to obey a fluctuation-dissipation theorem, $\sigma^2 = 2\gamma k_B T$ and $w_D = (w_R)^2$, where T is the desired temperature and k_B is the Boltzmann's constant (2). We use the same weight function as for the conservative forces, basically for historical reasons, setting $w_R = w_C$ and $w_D = w_C^2$.

Usually all the particles are assumed to have the same mass, and to fix units of mass and length a convenient choice is to set $m_i = r_c = 1$. Often we also fix the units of energy (and hence time) by setting $k_B T = 1$, but for equilibrium problems it can be convenient to keep $k_B T$ as a free parameter.

The integration of the equations of motion is a non-trivial matter since one has to manage the random forces somehow. For an integration algorithm, Groot and Warren (6) investigated a version of the velocity-Verlet scheme used in molecular dynamics simulations, but it was later shown by den Otter and Clarke (7) that this is not a real improvement over a simple Euler type integration scheme. Much has been written on this subject and for a newcomer to the field we recommend the article by den Otter and Clarke for a particularly clear discussion. But all the problems (and some other problems too) are obviated if the Lowe-Anderson thermostat is used, which is based on distinctly different physical ideas (3). However, all our simulations described below were carried out with the simple velocity-Verlet like algorithm described by Groot and Warren. Care has been taken to test for possible artefacts due to a finite time step.

For a single-component DPD fluid, the equation of state (EOS) gives the pressure p as a function of the density $\rho = N/V$. For the soft potential given above, the EOS is now well established to be $p = \rho k_B T + \alpha A \rho^2$ where A is the amplitude of the soft repulsive force, and $\alpha = 0.101 \pm 0.001$ is very close the mean field prediction $\alpha = \pi/30 = 0.1047$ (6). The first term is an ideal gas term, and the second term is the excess pressure, which is found to be almost perfectly quadratic in the density (there is a very small correction of order ρ^3). Note though that αA is *not* the second virial coefficient (9), so the above EOS breaks down as $\rho \to 0$. It seems that a quadratic EOS like this is unavoidable for soft potentials (9). This represents the fundamental limitation to basic DPD mentioned in the introduction. Moreover we have to take $A > 0$ otherwise the pressure diverges negatively at high densities, so we are restricted to a strictly positive compressibility $\partial p/\partial \rho > 0$. In fact, making $A < 0$ throws the DPD pair potential into a formal class of catastrophic potentials for which it can be rigorously proved there is no thermodynamic limit (9, 10). The situation is not as grim as it might seem though, since considerable progress can be made for mixtures of particles, using the above soft interaction potential and allowing the interaction amplitudes to vary.

Many-body DPD

An obvious way to get around the problem of a quadratic EOS is to make the amplitude A in the force law somehow dependent on density. Such a scheme has been examined by several workers including ourselves (4, 5), and proves to be a simple extension to DPD. This *many-body DPD* requires only a modest additional computational cost, but throws open the possibility of simulating systems with an arbitrarily complicated EOS. Care must be taken though, since density-dependent interactions are strewn with pitfalls (11). Our approach is to use a local density in the force law. It thus falls into the class II of potentials with an active density dependence considered by Stillinger *et al* (12), and satisfies one of Louis' safety criteria (11).

In many-body DPD we write $\mathbf{F}_{ij}^{\,C} = [A(\boldsymbol{\rho}_i) + A(\boldsymbol{\rho}_j)]/2 \; w_C(r_{ij}) \, \mathbf{e}_{ij}$, for a one-component fluid; for a multi-component generalization see Trofimov *et al* (5). A *partial amplitude* $A(\boldsymbol{\rho})$ is introduced, depending on a *weighted local density*, which we define by $\boldsymbol{\rho}_i = \Sigma_j \, w_\rho(r_{ij})$. The weight function $w_\rho(r)$ vanishes for $r > r_c$ and for convenience is normalized so that $\int d^3\mathbf{r} \, w_\rho(r) = 1$, although in principle the normalization could be absorbed into the definition of $A(\boldsymbol{\rho})$. The weighted local density is readily computed by an additional sweep through the neighbor list; hence there is only a modest additional computational overhead. If $A(\boldsymbol{\rho}) = A$, the method reduces exactly to the standard DPD model. In mean field theory, it is easy to show that the modified force law should give an EOS $p = \rho k_B T + (\pi/30)A(\rho)\rho^2$, and thus an arbitrary dependence on density can be recovered *(13)*.

Applications

We discuss first the application of DPD to polymers, but briefly since this is a subject that has been well studied, then move on to surfactant dissolution simulations. Finally we discuss in more detail applications to vesicle formation and growth, and to membrane self assembly.

Polymers

To make polymers, we string DPD particles together with spring forces which can be as simple as Hookean springs, $U(r) = kr^2/2$, with spring constant k. One can use a single species to model a polymer melt, or multiple species to model polymer solutions or polymer blends. In the latter case, a connection to Flory-Huggins theory for polymer mixtures was established by Groot and Warren *(6)*. We consider a blend of two species of homopolymers labeled A and B, with lengths N_A and N_B, for which Flory-Huggins theory predicts a demixing transition into A-rich and B-rich phases, as the so-called Flory χ-parameter increases. As is well known, the theory captures many of the aspects of polymer incompatibility in real polymer blends. Flory χ-parameters have been tabulated for many pairs of polymers.

To approach this problem in DPD, we use the standard DPD model and make the interaction amplitudes obey $A_{AA} = A_{BB} = A$, and $A_{AB} = A + \Delta A$. As ΔA increases, it is observed that the two polymers phase separate into A-rich and B-rich phases exactly as in Flory-Huggins theory. By mapping the observed phase behavior onto Flory-Huggins theory, a connection can be made between ΔA and χ, for a given set of parameters. For example, at baseline repulsion $A = 25$, overall density $\rho = 3$, spring constant $k = 2$, and for $2 < N_A = N_B < 10$, it was found that χ is proportional to ΔA, with $\chi / \Delta A = 0.306 \pm 0.003$. In principle, the density chosen for the simulation is a free parameter, and it is determined in order to mimic the macroscopic behavior of the real system to be simulated. This is achieved by mapping the isothermal compressibility κ of the real fluid onto the product, properly scaled, of the interaction parameter A and the number density ρ *(6)*.

The simple proportionality between χ and ΔA appears to have its origin in the use of soft potentials, and suggests that the excess free energy of mixing in

DPD is predominantly quadratic in the composition, similar to the quadratic EOS observed for a single component system. The fact that the χ-ΔA proportionality rule holds down to dimers ($N_A = N_B = 2$) should not, however, be taken as indication that the DPD model is already in the long chain limit at this point!

The χ-parameter mapping described above is appropriate to the *strong segregation limit* where phase separation occurs. One might also want to consider the *weak segregation limit* in which species are miscible in all compositions. In such a case, the χ-parameter mapping has to proceed via some other thermodynamic measure. The one we most favor is the low wave-vector limit of the static partial structure factors, or the partial compressibilities, for which there are also well established theoretical connections to χ-parameters (*14*).

The enormous advantage that a particle-based simulation method like DPD has over other methods becomes apparent when one realizes that it is possible, with virtually no additional effort, to simulate melts and blends of *copolymers with arbitrary architectures*. In particular diblock copolymer melts have been studied in detail by Groot *et al* (*15, 16*). The microphase separated states seen for real diblocks can be recovered in the DPD simulations; even some of the more exotic phases with cubic symmetries, provided one is careful to correct for fluctuations due to the small values of the DPD polymer length.

The method also gives insights into the kinetics of microphase separation. For instance it suggests that the hexagonal phase found for asymmetric diblocks forms first by developing interconnected tubes, then by aligning the tubes, and finally by annealing out the connections (*16*). It is also found that the inclusion of hydrodynamics in the DPD algorithm is essential to reach the final equilibrium state. If one replaces the momentum-conserving thermostat in DPD by a simple free-draining Brownian dynamics algorithm, then the hexagonal phase gets stuck in the state of interconnected tubes. Presumably the hydrodynamic modes allow for the relaxation of the subtle pressure differences in the improperly aligned hexagonal phase.

Dissolution of surfactants

We now turn our attention to the phase behavior of surfactants. Our interest was driven by open questions on the early-time kinetics of surfactant dissolution. We started by constructing a simple model that reproduces the important elements of the phase behavior observed in real water/surfactant mixtures (*17*). Our model has three species, A, B and C, of DPD particles, and is run according to the standard DPD algorithm. The A and B particles are bound together into AB dimers, representing the surfactant, and the C particles form a solvent representing water. The dimers are rigid rod dimers of length l (it is very easy in a simulation to constrain the dimer length). After some experimentation, we chose $A_{AA} = A_{BB} = A_{CC} = 25$, $A_{AB} = 30$, $A_{AC} = 0$, $A_{BC} = 50$, and $l/r_c = 1/2$. With this choice, at an overall density $\rho = 6$, we found that the phase behavior of the AB/C mixture shows hexagonal and lamellar phases very similar to those seen for many surfactant/water systems (we did not find any cubic phases). We were also careful to construct the repulsion amplitudes to recover a re-entrant disordered fluid phase for the pure AB dimers, a feature which is also seen in water/surfactant mixtures (ie surfactants require some moisture before they realize that they are amphiphilic).

The model allowed us to set up dissolution experiments in the computer as follows (18). We fill one simulation box with an equilibrated AB dimer fluid, and another with equilibrated solvent C. We bring the two boxes together and allow the two components to interdiffuse. Mesophases appear at the interface and are tracked as a function of time. Note the importance of the re-entrant fluid phase for the AB dimers in setting up the initial conditions. By matching diffusion coefficients in the DPD model with real systems, we estimate that the simulation resolves the dissolution process on time scales of 10 µs to 0.1 ms. Experiments, on the other hand, are restricted to time scales of minutes to several hours. But mesophase growth in the simulation was observed to follow exactly the same laws as seen in the experiments, thus we have some confidence in extrapolating the growth laws observed experimentally at late times to the early stages of surfactant dissolution.

Vesicle formation and growth

Amphiphilic molecules, composed of a hydrophilic head bonded to one or more hydrophobic tails, self-assemble spontaneously in aqueous solutions into mesoscopic structures such as micelles, membranes or vesicles, depending on the concentration and the molecular architecture. Here we focus on vesicles, which are spherical lipid bilayer structures. The bilayer is a membrane enclosing a portion of the solution thereby allowing different equilibrium conditions inside and outside this natural barrier. As such, it is a simple model of a biological cell. Many experimental and theoretical studies have been proposed to investigate the static and dynamic properties of vesicles, such as shape transformation, bending rigidity, fusion of more vesicles into one, or, conversely, fission of one vesicle into more. Coarse-grained surface models have been used to study membranes treated as thin continuous surfaces (3). These models don't take into account the molecular detail, but they are useful to study shape transformations based on the minimization of the bending energy. Atomistic simulation studies, on the other hand, try to investigate the fine structure of the bilayers and the dynamics of the single molecules, but are restricted to very small systems. The shortcoming of this approach, however is that events on much longer time scales, or at much larger length scales, cannot be handled due to limitations of computer power.

For example, a burning question which could be addressed at the mesoscale level is the process of self-assembly of amphiphilic molecules into micelles or vesicles. Various approaches have been proposed, such as coarse grained amphiphilic molecules (20, 21, 22, 23), hard sphere MD simulations (24), Monte Carlo methods (25, 26), Brownian dynamics simulations (27). We believe that DPD could be beneficial in this field, especially when addressing a dynamic question, because its simulation scheme preserves collective hydrodynamics modes, which are essential to reproduce the path towards the equilibrium structure.

In this study we have used a simple model of an amphiphile, in the presence of an explicit solvent (water), to follow the pathway of formation of vesicles from a patch of bilayer. This has the additional benefit of following quite closely the industrial procedure of milling a concentrated lamellar phase. If the starting configuration is a patch of crystalline bilayer, after some time the bilayer expands, and starts fluctuating, until the symmetry is broken and it bends preferentially on one side into a cup-shape. At this point the cup closes up into an almost perfect spherical vesicle. The water is caught in the middle of the bilayer. Snapshots from such a simulation are shown in Plate 1.

A recent report describes DPD studies of the self-assembly of molecules into vesicles taking into account the solvent explicitly (28). Different types of models for the molecules and different initial conditions were used. The authors observed the formation of vesicles after relatively few simulation steps. The first objective of our study was to reproduce these results, using the same conditions.

Three different models were studied for the amphiphilic molecules: (1) a head formed by one particle connected to a single tail formed by three particles, A_1B_3, (2) a head formed by two particles connected to a single tail formed by five particles, A_2B_5, and (3) two connected heads of one particle, each connected to a tail formed by three particles, the triblock $B_3A_2B_3$. The connections between the particles were strain hardening springs with a potential $U(r) = k(r^2/2 + \beta r^4/4)$, with $k = 1$ and $\beta = 2$. The amphiphiles were immersed in a solvent formed from particles of a third species C.

For these studies we set $k_BT = 1$ to fix the energy scale. We set $A_{AB} = A_{BC} = 40$ and all other interaction amplitudes to a default value $A_{ij} = 25$. Simulations were performed using the standard DPD algorithm. Two initial configurations were taken: a random distribution 4032 amphiphile particles and 37440 solvent particles in a cubic box of size 24 units^3 and a bilayer distribution of 4032 amphiphile particles and 76968 water particles in a cubic box of size 30 units^3. The mole fraction of amphiphile particles is 0.097 for the random distribution and 0.05 for the bilayer distribution. Since the number of particles in

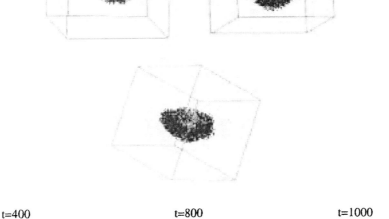

t=400   t=800   t=1000

Plate 1. Three snapshots of the DPD simulation starting from a patch of crystalline bilayer. After some time the bilayer expands, and starts oscillating, until the symmetry is broken and it bends preferentially onto one side into a cup-shape. The cup closes up into an almost perfect spherical vesicle.

(See page 7 of color insert.)

amphiphiles is fixed, the number of amphiphiles is changed for the three models: 1008 for A_1B_3, 576 for A_2B_5, and 504 for $B_3A_2B_3$. The overall density is $\rho = 3$ beads/unit^3 for all systems.

The choice of the size and density of the bilayer initial configuration is crucial. If the patch is too small or too big, it explodes to form a ring that fuses at the boundary of the box. On the other hand one could use the same simple model system, and start the simulation from a perfectly random mixture, as in a quench from a high temperature isotropic phase. The pathway is completely different. We observe first the formation of small micelles, which slowly coalesce to form larger disk-shaped flat aggregates. If the flat aggregates are large enough, they tend to fluctuate, and sometimes they initiate the bending on one side only, leading to the closure of a vesicle. In the example in Plate 2a one can clearly see a flat aggregate coexisting with a larger vesicle, and in Plate 2b a vesicle coexisting with two small micelles, still excluded from the big vesicle.

One can immediately recognize a threshold value for the disk radius. If the disk is larger than the threshold, the system is willing to pay the free energy price of bending, in order to expose more surface area to the water contact. There are situations where a big vesicle coexists with small micellar aggregates. One natural question we wanted to address at this point was to study the micellar fusion with the big vesicle. In order to test this, we have introduced a second amphiphilic component into the mixture, which preferentially self assembles into micelles at the same conditions of temperature. Thus the simulation has three components (two amphiphiles and one solvent), and five species of particles. For the micelle-forming amphiphile, we use D_1E_3, with interaction parameters $A_{DC} = 0$, $A_{EC} = 50$, and $A_{DE} = 30$, where D is the head, E is the tail, and C is again the solvent species. All unspecified interaction parameters in the mixture simulation are set to the default $A_{ij} = 25$.

In order to test the sensitivity to the parameters, we tried increasing the head-tail and tail-water interaction amplitudes to $A_{DE} = A_{EC} = 40$. Once again we observed the formation of micelles: this indicates that the key issue is that the head-water interaction should not be repulsive, or very low compared to the tail-water one. The higher the difference in the repulsion parameters, stronger the segregation of the tails from water, and the smaller the structures that are formed (micelles instead of vesicles).

One possibility is to start from a mixed completely random configuration. In this case we have observed the formation of mixed vesicles. In some cases this isotropic state is even stabilized by occasional bridges provided by the di-chain vesicle-forming surfactant (blue).

In a second study, depicted in Plate 3, we have prepared the two systems separately. In one simulation box, the vesicle forming amphiphile was allowed to reach the equilibrium single vesicle. In the other box, the curvature-liking amphiphile was allowed to self-assemble into an isotropic micellar phase. Then

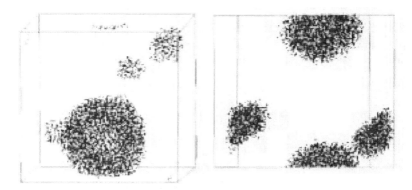

Plate 2. (a) A flat aggregate coexisting with a larger vesicle, and (b) two small micelles coexisting with a larger vesicle.

(See page 7 of color insert.)

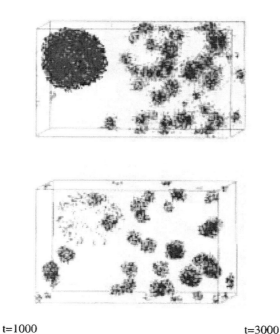

t=1000   t=3000

Plate 3. Two simulation boxes have been prepared independently to make a fully equilibrated mono-lamellar vesicle and a homogeneous solution of micelles. When the two simulation boxes are merged, we observe vesicle growth via micellar deposition. Only the distribution of the surfactant tails is shown in the second image.

(See page 8 of color insert.)

the partition was removed, and the micelles started diffusing towards the vesicle. We can follow the diameter of the vesicle increase over time, but even more dramatic is the picture of the second amphiphile species perfectly mixed in the bilayer of the vesicle.

Membranes

In the second part of our work we focused our attention on the bilayer structure that defines the vesicle, aiming at understanding the static and dynamic properties of this membrane. Many studies have been reported at the micrometer and at the nanometer scale. In a continuum model (3), membranes can be seen as liquid sheets described by differential geometry: the most important parameter is the bending rigidity, which describes the resistance of the membrane against bending. At a molecular level the individual motion of the molecules can be seen and protrusions roughen the surface (29, 30). The membrane was modeled at the molecular level, but the system was large enough to observe both length scales of the motions. The connection between the simulation results and the macroscopic relevant quantities, such as the surface tension and the bending rigidity is achieved by studying the fluctuation spectrum of a large patch of membrane.

It is clear at this point that many research groups have successfully self assembled a lipid bilayer in a computer model, taking into account the solvent explicitly. But if one is not particularly interested in the solvent, it seems computationally wasteful to have a vast excess of solvent particles in the simulation just to stabilize a membrane. The challenge we set ourselves was to develop a simple model that exhibits self-assembly of amphiphiles into bilayers, but *without the explicit solvent*. The computational gain would be significant and allow us to study at least the equilibrium properties of much larger areas of membranes. We were motivated to do this by the many-body DPD model discussed earlier in which one can find a liquid phase coexisting, for all practical purposes, with empty space (the vapor phase has an extremely small density). Can we dress up the model in some way such that the liquid phase becomes a bilayer?

The obvious thing to do is to add the many-body interactions discussed earlier to the pure amphiphile models. We start with the many-body DPD model described earlier, omit the solvent, and try to tune the standard DPD interaction amplitudes A_{ij} (which may be positive or negative) to get a structured liquid which wants to turn into a bilayer. Despite much experimentation, we have not been successful with this approach, although we emphasize the parameter space and space of possible amphiphile architectures is very large and nowhere near exhausted. It seems that making a self-assembling bilayer by this route is very hard, and we appear to have thrown away some essential piece of physics.

A second line of attack was necessary. We start with the above model but focus exclusively on AB dimers. We add new elements to the basic DPD interactions to encourage formation of a freestanding bilayer structure. After some experimentation with different ideas, we discovered that we could stabilize a monolayer with an extra four-body interaction of the form $U_4 = -D (\mathbf{r}_A - \mathbf{r}_{A'}) \cdot (\mathbf{r}_B - \mathbf{r}_{B'}) \, w_4(r_{AA'}) \, w_4(r_{BB'})$ where the head groups are located at $\mathbf{r}_A$ and $\mathbf{r}_{A'}$, with $r_{AA'} = |\mathbf{r}_{A'} - \mathbf{r}_A|$, the tail groups are located at $\mathbf{r}_B$ and $\mathbf{r}_{B'}$, with $r_{BB'} = |\mathbf{r}_{B'} - \mathbf{r}_B|$, and the amplitude $D > 0$. The two weight functions are taken to be $w_4(r) = (1 - r^2/r_c^2)$; this simplifies the computation of the forces which are determined by differentiation and added to the standard DPD and many-body DPD interactions.

This functional form is very successful in promoting monolayer configurations. However there is nothing that gives the system the tendency to aggregate two monolayers into a bilayer. One solution to this problem is a potential that promotes the formation of the bilayer by differentiating between parallel and antiparallel interaction between two amphiphiles. Otherwise, following a simpler approach, one can promote the formation of the bilayer by adding an isotropic attraction term between the tails. The rational for doing this is to compensate with some kind of lipophilic attraction for the 'missing' hydrophobic effect. Both these approaches seem to work well and they also show self-assembly of the molecules starting from a random configuration (see plate 4).

In order to secure this study onto solid scientific ground one needs, once again, to give significance to the mesoscale parameters used in the model. Using the same top-down strategy, we study the fluctuation spectrum of a freestanding bilayer, to extract quantities such as elastic coefficients.

We first determine a reference plane by least-squares fitting the parameters a_i in $z = a_0 + a_1 x + a_2 y$ to the positions of the tails (or heads). The closest distance of the tails (or heads) to this supporting plane is then used to construct a the height function $h(x, y)$. If $\langle h^2 \rangle$ remains small and constant, the membrane is considered stable. We consistently find $\langle h^2 \rangle \approx 0.2$ for all stable membranes. Another test for stability can be devised by checking the orientation of the molecules with respect to the normal to the average membrane plane.

We calculate $h_\mathbf{q} = (1/N) \sum_i \exp[-i(q_x x_i + q_y y_i)]$ and $S(q) = \langle h_\mathbf{q}^2 \rangle$. We expect to see three regimes: undulations of the whole membrane dominated by surface tension σ at small q, dominated by bending rigidity κ at intermediate q, and fluctuations of individual molecules (protrusion modes) at large q. In Figure 1 we plot $\log S(q)$ against $\log q^2$. We can clearly see the two collective modes. By fitting to $S(q) = k_B T / \sigma q^2 + k_B T / \kappa q^4$, we can extract the surface tension and bending rigidity. This is work in progress, and we still have some problems in the normalization, so the numbers in the fit are not yet clearly significant.

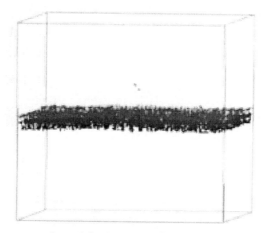

Plate 4. Snapshot of the DPD simulation of a free-standing bilayer
(See page 8 of color insert.)

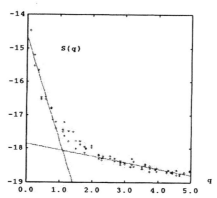

Figure 1 - a plot of the fluctuation spectrum (log s(q)) as a function of the wavevector (log q^2).

Conclusions

Dissipative particle dynamics (DPD) was first introduced for microhydrodynamics problems, but has proved to be very successful across a range of problems from polymers to surfactants. The development of many-body DPD promises to open up new possibilities, for problems containing implicit solvents, or for mixtures with non-trivial thermodynamics of mixing. As our ongoing work on self-assembled membranes indicates, the range of possible applications of DPD continues to expand.

References

1. Hoogerbrugge, P. J.; Koelman, J. M. V. A. *Europhys. Lett.* **1992,** *19,* 155.
2. Español, P.; Warren, P. B. *Europhys. Lett.* **1995,** *30,* 191.
3. Lowe, C. P. *Europhys. ett.* **1999,** *47,* 145.
4. Pagonabarraga, I.; Frenkel, D. *J. Chem. Phys.* **2001,** *115,* 5015.
5. Trofimov, S. Y.; Nies, E. L. F.; Michels, M. A. J. *J. Chem. Phys.* **2002,** *117,* 9383.
6. Groot, R. D.; Warren, P. B. *J. Chem. Phys.* **1997,** *107,* 4423.
7. den Otter, W. K.; Clarke, J. H. R. *Europhys. Lett.* **2001,** *53,* 426.
8. Vattulainen, I.; Karttunen, M.; Besold, G.; Polson, J. M. *J. Chem. Phys.* **2002,** *116,* 3967.
9. Louis A. A.; Bolhuis, P. G.; Hansen, J.-P. *Phys. Rev. E* **2000,** *62,* 7961.
10. Ruelle, D. *Statistical mechanics: rigorous results,* World Scientific: Singapore, 1999.
11. Louis, A. A. *J. Phys. Cond. Mat.* **2002,** *14,* 9187.
12. Stillinger, F. H.; Sakai, H.; Torquato, S. *J. Chem. Phys.* **2002,** *117,* 288.
13. Warren, P. B.; *Phys. Rev. Lett.* **2001,** *87,* 225702.
14. P. G. de Gennes, *Scaling concepts in polymer physics,* Cornell University Press: Ithaca, NY,1979.
15. Groot, R. D.; Madden, T. J. *J. Chem. Phys.* **1998,** *108,* 8713.
16. Groot, R. D.; Madden, T. J.; Tildesley, D. J. *J. Chem. Phys.* **1999,** *110,* 9739.
17. Jury, S.; Bladon, P.; Cates, M. E.; Krishna, S.; Hagen, M.; Ruddock, J. N.; Warren, P. B. *Phys. Chem. Chem. Phys.* **1999,** *1,* 2051.
18. Prinsen, P.; Warren, P. B.; Michels, M. A. J. *Phys. Rev. Lett.* **2002,** *89,* 148302.
19. Lipowsky, R.; Sackmann, E. *Structure and Dynamics of Membranes;* Handbook of Biological Physics Vol. I; Elsevier: Amsterdam, 1995.
20. von Gottberg, F. K.; Smith, K. A.; Hatton, A. *J. Chem. Phys.* **1997,** *106,* 9850.
21. Widom, B. *J. Chem. Phys.* **1984,** *81,* 1030.

22. Karaborni, S.; Esselink, K.; Hilbers, P. A. J.; Smith, B.; Karthauser, J.; van Os, N. M.; Zana, R. *Science* **1994,** *266,* 254.
23. Goetz, R.; Lipowsky, R. *J. Chem. Phys.* **1998,** *108,* 7397.
24. Drouffe, J. M.; Maggs, A. C.; Leibler, S. *Science* **1991,** *254,* 1353.
25. Bernardes, A. T. *J. Phys. II (France)* **1996,** *6,* 169.
26. Bernardes, A. T. *Langmuir* **1996,** *12,* 5763.
27. Noguchi, H.; Takasu, M. *Phys. Rev. E* **2001,** *64,* 041913.
28. Yamamoto, S.; Maruyama, Y.; Hyodo, S. *J. Chem. Phys.* **2002,** *116,* 5842.
29. Pastor, R. W.; *Curr. Opin. Struct. Biol.* **1994,** *4,* 486.
30. Tobias, D.J.; Tu, K.; Klein, M.L. *Curr. Opin. Coll. Int. Sci.* **1997,** *2,* 15.

Chapter 16

Dynamics of Phase Separation in Polymeric Systems

G. J. A. Sevink, A. V. Zvelindovsky, and J. G. E. M. Fraaije

Leiden University, Soft Condensed Matter Group, P.O. Box 9502, 2300 RA Leiden, The Netherlands

We review the dynamic density functional (DDFT) method for 3D pattern formation in complex amphiphilic systems. The focus is on the supra-molecular or mesoscopic level. The building blocks consist of sequences of dissimilar monomers, connected in copolymer chain molecules. Internal factors such as composition and architecture of the polymers, but also external factors such as applied electric fields or confinement control the self-organisation phenomena. The specific examples presented here focus on the validation of our method by experimental data and its predictive value. The ambitious goal is the invention of methods for the rational design of truly complex biomimicing materials, in which we combine principles of chemical engineering, physics, chemistry and biology. The keyword is self-organisation. However, autonomous self-organisation leads to trouble, modulated self-organisation leads to beauty.

Introduction

For a long time, chemical engineers have analyzed macroscale properties using a variety of continuum mechanics models. In the last decade molecular modeling has grown to an essential tool for research and development in the chemical and pharmaceutical industry. Despite the considerable success of both molecular and macroscale modeling, in the past few years it has become increasingly apparent that in many materials the structures on the *mesoscale* level determine the material properties to a very large extend. Mesoscale structures are typically of the size of 10 to 1000 nm. Even more, recent developments in *nanotechnology* deal with tailored structures exactly in this range of scales. The industrial relevance of mesoscale modeling is therefore obvious. Nevertheless the development of the necessary general purpose computational tools is still at its infancy.

We are developing general purpose methods that are aimed at combining molecular detail and the computational efficiency of continuum approaches. The resulting molecular field method is now widely used for mesoscale soft condensed matter computer simulations. It is based on a functional Langevin approach for mesoscopic phase separation dynamics of complex polymeric liquids. The project aims to develop tools and facilitate answers for a wide range of factors of utmost importance in nano-technological applications, such as chemical reactions, convection and flow effects, surfaces and boundaries, applied electric field, special sample preparations *etc.*

The morphology formation in complex liquids has been studied by many authors using time-dependent Landau-Ginzburg models (1-5). These models are based on traditional free energy expansion methods (Cahn-Hilliard (6), Oono-Puri (7), Flory-Huggins-de Gennes (8)) which contains only the basic principles of phase separation (9) and are not well suited for specific chemical and biological applications. In contrast to these phenomenological theories we use dynamic density functional theory (10-13) where we do not truncate the free energy at a certain level, but rather retain the full polymer path integral by a numerical procedure. Other groups, such as Kawakatsu (14), Doi (15) and Müller (16), consider similar approaches. Although the calculation of the polymer path integral is computationally very demanding, it allows us to describe the mesoscopic dynamics of specific polymer liquids (17).

Mesoscale morphologies of complex polymer systems are often the result of a processing history. Our DDFT method facilitates a general tool to predict the morphology as a result of this history, for a wide range of molecular topologies, such as (multi-block) linear and (multiple) branced. The application area includes emulsion copolymerization, copolymer melts and softened polymer melts, polymer blends, polymer surfactant stabilized emulsions. In all of these external fields of choice can be included.

In this paper we illustrate the potential of our DDFT method with three cases of modulation. First we consider the dynamics of reorientation of a symmetric diblock

copolymer melt subjected to an electric field (where the initial anisotropic orientation was obtained by shearing). In this case, our simulations allowed us to identify
the underlying microscopic dynamics of the complex experimental observations. Following, we consider surface directed phase separation in a triblock copolymer melt,
and identify so-called surface reconstructions. This project was very successful; a
comparison with experiments showed a perfect match. Finally, we consider phase
separation inside a droplet of asymmetric diblock copolymer, where the droplet was
dispersed into weakly selective solvent. Especially this example offers a glimpse
at a research direction where the computational chemist designs in silico dispersed
morphologies with desired properties, using realistic molecular field models. Such
a design strategy can be used for the design of more stable morphologies, or morphologies which respond in a desired way to external stimuli.

The free energy functional

We give a short outline of the theory used in the simulations. For more details
see Ref (10–13). We consider a system of n Gaussian chains of N statistical units
(' beads') of several different species (for example, $A_{N_A}B_{N_B}$, for a general diblock
copolymer, $A_{N_A/2}B_{N_B}A_{N_A/2}$, $N = N_A + N_B$ for a symmetric triblock copolymer).
In case of solvent, it presence can be easily taken into account (17). The volume of
the system is V. In this volume the following fields are used: concentration fields
$\rho_I(r)$, external potentials $U_I(r)$ and intrinsic chemical potentials $\mu_I(r)$, where I
denotes the bead type (say, A or B).

Imagine that on a course grained time scale, there is a certain collective concentration field $\rho_I(r)$. Given this concentration field a free energy functional $F[\rho]$ can
be defined as follows (10–13)

$$\beta F[\rho] = -\ln \frac{\Psi^n}{n!} - \beta \sum_I \int_V U_I(r)\rho_I(r)dr + \beta F^{nid}[\rho] . \qquad (1)$$

Here Ψ is the partition function for the ideal Gaussian chains in the external field
U_I, and $F^{nid}[\rho]$ is the mean-field contribution due to the non-ideal interactions.
The external potentials and the concentration fields are bijectively related via a
density functional for ideal Gaussian chains:

$$\rho_I[U](r) = n \sum_{s'=1}^{N} \delta^K_{Is'} Tr_c \psi \delta(r - R'_s) , \qquad (2)$$

where $\delta^K_{Is'}$ is a Knonecker delta with value 1 if bead s' along the Gaussian chain
is of type I and 0 otherwise. The trace Tr_c is limited to an integration over the

coordinates of one chain

$$Tr_c(\cdot) = C \int_{V^N} (\cdot) \prod_{s=1}^{N} dR_s \tag{3}$$

with C a normalization constant. The single chain distribution function

$$\psi = \frac{1}{\Psi} e^{-\beta[H^G + \sum_{s=1}^{N} U_s(R_s)]} , \tag{4}$$

with $H^G = \frac{3}{2\beta a^2} \sum_{s=2}^{N} (R_s - R_{s-1})^2$ the Gaussian chain Hamiltonian of an unperturbed chain, and a the Gaussian bond length parameter, is such that the distribution ψ is optimal at each coarse-grained stage in time. There is no known closed analytical expression for the inverse density functional, but for our purpose it is sufficient that the inverse functional can be iteratively determined by an efficient numerical procedure.

The mean field part $F^{nid}[\rho]$ of the total free energy contains several terms, including a part $F^c[\rho]$ for the cohesive interactions and excluded volume interactions by $F^e[\rho]$, taking care of the incompressibility of the system. As external factors, we can have a contribution due to the energetic interactions with (filler) particles, giving rise to $F^{cM}[\rho]$, or an applied electric field denoted by $F^E[\rho]$. We can then write the non-ideal part as

$$F^{nid}[\rho] = F^c[\rho] + F^{cM}[\rho] + F^e[\rho] + F^E[\rho] \tag{5}$$

with

$$F^c[\rho] = \frac{1}{2} \sum_{I,J} \int_V \int_V \varepsilon_{IJ}(|r - r'|) \rho_I(r) \rho_J(r') dr dr' \tag{6}$$

which is a two-body mean field potential, where $\varepsilon_{IJ}(|r - r'|)$ are the reciprocal cohesive interactions between beads of type I at position r and J at r', often taken as a Gaussian kernel

$$\varepsilon_{IJ}(|r - r'|) \equiv \varepsilon_{IJ}^0 (\frac{3}{2\pi a^2})^{\frac{3}{2}} e^{-\frac{3}{2a^2}(r - r')^2} . \tag{7}$$

For a diblock system $A_{N_A} B_{N_B}$, the values of the ε^0 parameters are easily related to the more familiar Flory-Huggins χ parameter by the relation $\chi = 1/2\beta\nu^{-1}[\varepsilon_{AB}^0 + \varepsilon_{BA}^0 - \varepsilon_{AA}^0 - \varepsilon_{BB}^0] = \beta\nu^{-1}\varepsilon_{AB}^0$ (with ν the bead volume), since $\varepsilon_{AA}^0 = \varepsilon_{BB}^0 = 0$ in most cases. In some case, rigid (filler) particles that are stationary in time can be part of the system, like electrodes, substrates and long rods. We have then an extra contribution to the free energy for the energetic interaction between beads of different types and the rigid particle material. The extra term in the free energy due to this contribution can be written as (18)

$$F^{cM}[\rho] = \frac{1}{2} \sum_{I,M} \int_V \int_V \varepsilon_{IM}(|r - r'|)\rho_I(r)\rho_M(r')dr dr' \, . \tag{8}$$

Inside the filler particles, occupying a total volume $V^0 \subset V$, the concentration fields $\rho_I(r)$ of the different bead types is equal zero. The constant density field ρ_M (where M represents 'auxiliary' beads of the filler particle type, and we allow different interaction values ε_{IM} for different filler particles) that appears in (8), is defined as $\rho_M(r) = 1$ for $r \in V^0$ and $\rho_M(r) = 0$ elsewhere $(r \in V/V^0)$. The interaction kernels are chosen equal to the ones used in (7). Although more rigorous approaches have been considered (19), the excluded volume interaction is included by a phenomenological Helfand penalty function

$$F^e[\rho] = \frac{\kappa_H}{2} \int_V \left(\sum_I \nu_I(\rho_I(r) - \rho_I^0) \right)^2 dr \tag{9}$$

with κ_H the Helfand compressibility parameter and ρ_I^0 is the reference concentration field of bead type I in complete mixing. This term provides a mathematically simple and numerically stable way to account for compressibility effects in the system (19). In the case that an electric field is acting on the volume V, there is an extra contribution to the free energy. Under the assumption of a fully incompressible diblock copolymer $(A - B)$ system and expanding the dielectric constants up to second order in the order parameter field $\psi(r)$ (and neglecting higher orders) $(20, 21)$, this term can be written in Fourier space as function of this order parameter field as

$$F^E[\psi] = \frac{1}{16\pi^3}\beta^{-1}\nu B \int_V \psi(k)\psi(-k)(\frac{|k|}{k} \cdot e_i)^2 dk \tag{10}$$

where e_i is the unit vector in the direction of the electric field, and B is a dimensionless parameter

$$B = \beta\epsilon_0 \nu \frac{2(\epsilon_A - \epsilon_B)}{(\epsilon_A + \epsilon_B)}|E_0|^2 \tag{11}$$

where $|E_0|$ is the strength of the electric field and ϵ_I is the dielectric constant of the pure component I (A or B).

The intrinsic chemical potentials μ_I are defined by the functional derivatives of the free energy

$$\mu_I(r) \equiv \frac{\delta F}{\delta \rho_I} = -U_I(r) + \frac{\delta F^c}{\delta \rho_I(r)} + \frac{\delta F^{cM}}{\delta \rho_I(r)} + \frac{\delta F^e}{\delta \rho_I(r)} + \frac{\delta F^E}{\delta \rho_I(r)}$$

$$= -U_I(r) + \mu_I^c(r) + \mu_I^{cM}(r) + \mu_I^e(r) + \mu_I^E(r) \tag{12}$$

with

$$\mu_I^c(r) = \sum_J \int_V \varepsilon_{IJ}(|r - r'|)\rho_J(r')dr'$$

$$\mu_I^{cM}(r) = \frac{1}{2}\int_{V^0} \varepsilon_{IM}(|r - r'|)\rho_M(r')dr'$$

$$\mu_I^e(r) = \nu_I\kappa_H \int_V \sum_J \nu_J(\rho_J(r) - \rho_J^0)dr$$

$$\mu_I^E(k) = \beta^{-1}\nu B(\frac{|k|}{k} \cdot e_i)^2(\rho_I(-k) - (2\pi)^3\rho_I^0\delta(-k)) \tag{13}$$

Finding the minimum of free energy

Several methods can be employed to find the minimum of free energy (1) and equilibrium concentration fields $\rho_I^{eq}(r)$. They can roughly be divided into *static* and *dynamic* methods, although a number of hybrids exist which are generally referred to as quasi-dynamic methods (for instance (*22*)). A rather complete and recent review is given in Ref (*23*). In this overview article, we focus on a number of different dynamic schemes that have been developed within our group for the minimization of the free energy (1). An advantage of the dynamic schemes is that that they consider a dynamic pathways towards a free energy minimum, including visits to long-living metastable states. In this sense, the model can be seen to mimic the experimental reality when compared to static schemes, which are optimizations, based upon mathematical arguments.

In equilibrium $\mu_I(r)$ is constant; this yields the familiar self-consistent field equations for Gaussian chains, given a proper choice of $F^{nid}[\rho]$. When the system is not in equilibrium, the gradient of the intrinsic chemical potential $-\nabla\mu_I$ acts as a thermodynamic force which drives collective relaxation processes. In the standard method of so-called Mixed Dynamics employed in our group, we write stochastic partial differential equations that describe the collective diffusion of the concentration fields (*24*)

$$\frac{\partial\rho_I}{\partial t} = M_I\nabla \cdot \rho_I\nabla\mu_I + \eta_I \tag{14}$$

where we have used Onsager or local transport coefficients that are linear in concentration, M_I is a constant mobility for bead I and $\eta_I(r)$ is a noise field, distributed according to the fluctuation-dissipation theorem. In some cases, it is sufficient to consider constant transport coefficients and the diffusion equation becomes

$$\frac{\partial \rho_I}{\partial t} = M_I \Delta \mu_I + \eta_I \; ; \tag{15}$$

A numerically efficient scheme including non-local kinetic coupling can be arrived at by considering External Potential Dynamics instead of Mixed Dynamics (25). In this case, the governing equations for the dynamics read

$$\frac{\partial U_I}{\partial t} = -M_I \Delta \mu_I + \eta_I \tag{16}$$

A full discussion about the validity of the approximations used to arrive at equation (16) can be found in Ref (25). Recently, an alternative scheme was developed that has been derived from the External Potential Dynamics (16). In the notation of (25), the External Potential Dynamics algorithm is obtained by left-commutation of mobility and gradient operators $\nabla \cdot P_{IJ} \nabla \mu_J \to P_{IJ} \nabla^2 \mu_J$. An alternative algorithm is obtained by right-commutation $\nabla \cdot P_{IJ} \nabla \mu_J \to \nabla^2 P_{IJ} \mu_J$, which is valid provided $\mu_J \ll 1$. This gives rise to the dynamic equations

$$\frac{\partial \rho_I}{\partial t} = M_I \nabla^2 (\rho_I - \rho_I[\mu_{nid}]) \; , \tag{17}$$

where $\rho_I[\mu_{nid}](r)$ is the concentration functional of the polymer (or solvent) and can be calculated from equation (2) by replacing the external potential fields U_I by the intrinsic chemical potentials due to the non-ideal interactions. As the algorithm is fully explicit in the concentration variables, and thereby avoids the calculation of the external potentials, it is a computationally efficient scheme compared to the other ones. This algorithm is well suited for avoiding the compressibility problems that normally arise when concentration fields are zero or close to zero in a certain part of the simulation volume.

External fields

Recently, polymeric systems received increased interest because of their use in many applications in nanotechnology. One important aspect in nanotechnology is the ability of tailoring micro-phases for industrial needs by various external and internal conditions. The application of external fields, such as applied flow fields, applied electric fields or confinement in thin films, is known to have a strong effect on the structuring of micro-phases on a macro-scale level; it gives rise to a global orientation. In some case, experiments as well as simulations have shown that even the micro-phase type can be altered by the external conditions. For this reason, the behavior of systems under an external field is an extensively studied topic both experimentally and theoretically. On the other hand, also the preparation pathway can lead to altered phases. An example is polymeric vesicles, which can only be obtained following a specific preparation technique (26–28).

Earlier we considered the extra terms in the free energy due to the electric fields and confinement. In the dynamic scheme, application of an electric field results in

$$\frac{\partial \rho_I}{\partial t} = M_I \Delta \mu_I + M_I \beta^{-1} \nu B \nabla_i^2 \rho_I + \eta_I \; . \tag{18}$$

where an electric field applied in the e_i ($i = x, y$ or z) direction. In case of geometry constraints (the inclusion of (filler) particles), the dynamic equations are not altered by extra terms. However, for the particles we consider boundary conditions, in accordance with the conservation laws, that allow no flux through the solid particle surface (18), i.e.

$$\nabla \mu_I \cdot \mathbf{n} = 0 \; , \tag{19}$$

where $\mathbf{n}$ is the normal pointing towards the particle. The same boundary conditions apply for the calculation of the noise, that should satisfy the fluctuation-dissipation theorem. At all boundaries where no filler particles are present, periodic boundary conditions apply.

In the presence of a steady shear flow, with $v_x = \dot{\gamma} y$ and $v_y = v_z = 0$, the microphase separation is described by (31, 32)

$$\frac{\partial \rho_I}{\partial t} = M_I \nabla^2 \mu_I - \dot{\gamma} y \nabla_x \rho_I + \eta_I \tag{20}$$

which is equal to equation (15) apart from an extra term that accounts for the applied shear (where $\dot{\gamma}$ is the shear rate (the time derivative of the shear strain γ) and sheared boundary conditions apply (29–32)).

Examples

In this section we consider three systems, two of which are directly related to experiments (33, 34) carried out in the group of Georg Kraus and Robert Magerle (Bayreuth University, Germany). For the third system of polymeric particles in highly selective solvents, extensive experimental knowledge is present in the group of Adi Eisenberg (McGill University, Canada) (35). A comparison to their experiments is the focus of current collaboration.

It is worth mentioning that even in the calculation of the phase diagram of sections B, the dynamics plays a role. In our method, there are no means to distinguish between equilibrium states and states that are stable within the time range considered in our simulations. The same holds for the experimental data, as the STM data is all acquired at some fixed time.

Dynamics of electric field alignment

In recent experiments Böker *et al.* (*34*) identified two distinct microscopic mechanisms of electric field induced alignment of diblock copolymer microdomains. The system they considered was a dissolved (in toluene) polystyrene-*block*-polyisoprene (SI) diblock copolymer of nearly symmetric compositions, forming lamellae under normal conditions. Initially (before switching on the electric field) the lamellae align parallel to the electrodes (with a small distribution in orientation) due to the preferential attraction of the PS block to the gold surfaces. Böker *et al.* observed that after the electric field is switched on, the lamellae start to orient in the direction of the electric field. Interestingly, two different underlying mechanisms were identified. Close to order-disorder transition (ODT) the scattering peak from the parallel to the electrodes lamellae quickly disappears and a new peak from the perpendicular lamellae emerges. Further from ODT the peak merely rotates from the initial position towards the final one. We used two-dimensional DDFT simulations (with free energy (1), dynamic equations (15) and the appropriate term for the electric field contribution) to see whether we are able to reproduce the observed scattering behavior. It should be mentioned that our dynamics is purely diffusive, and does not include any hydrodynamic contributions.

In the simulations, the initial orientation parallel to the electrodes can be obtained by preferentially attracting surfaces (*18*). An alternative route is the application of shear for a certain amount of time. This seems reasonable, since, in the experiments, the scattering data was obtained from the center part of the sample, far away from the confining electrodes. This route facilitates an easy way to obtain the roughly aligned lamellae of the experimental study. One structure after shear is shown in plate 1 (at $t = \tau/\Delta\tau = 5$). After cessation of shear, the electric field is applied in the direction orthogonal to shear direction.

In their article Böker *et al.* argue that the two mechanisms (grain boundary migration and grain rotation) are distinguished due to kinetic factors (different initial grain size, different mobility and viscosity). Our simulation shows that the behaviors are discriminated by energetics, namely by the balance of the ponderomotive force (*20*) and the surface tension (and higher terms like bending) of a single lamella, which is related to the block interaction ε_{AB}, the only varied parameter. The initial grain size, plate 1 (at $t = 5$) was nearly the same. Close to ODT the surface tension cannot withstand the electric field, and undulations appear in lamellae, plate 1 (waves at $t = 5.2$). In addition to existing grains new perpendicular grains form in the body of the parallel lamellae. They appear via a transition to an intermediate micellar or/and bicontinuous phase. As this phase is a structured one (not "selective melting" into disorder) there is no considerable broadening of the scattering peak, as was already observed in (*34*). The migration of grain boundaries follows the same transition (row of micelles in the square, plate 1). Further from ODT the undulation instability is suppressed and the lamellae collectively rotate.

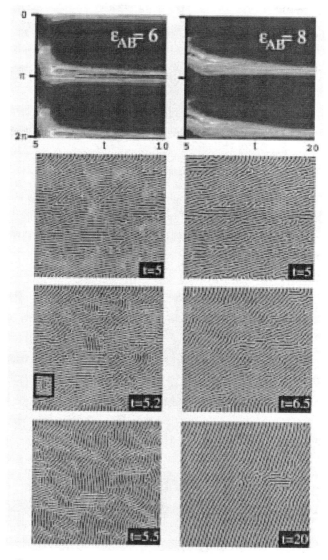

Plate 1: Top: the scattering intensity as a function of the azimuthal angle and dimensionless time $(t = \tau/\Delta\tau \times 10^3)$ for a A_4B_4-melt in 2D. Below: simulation snapshots. Left: closer to ODT $(\chi N = 19.2)$. Right: further from ODT $(\chi N = 25.6)$. The axis in which the electric field is applied is vertical. The electric field strength parameter $B = 0.2$ (Reprinted with permission from Zvelindovsky et al. Phys. Rev. Lett. 2003, 90, 049601. Copyright 2003, American Physical Society).

(See page 9 of color insert.)

The process is assisted by single defect migration and binary annihilation. The kinetic factors (34) will only modify the time scale of the processes. As to the initial misalignment (here created on purpose by shearing), it is indeed responsible for the choice of rotation direction (34).

Confined polymer melts

Most of the studies performed in confined polymer systems so far have focussed on the orientational phase transitions of the system of certain bulk symmetries (lamellae, cylinders and spheres) under some external field. It should be realized that the presence of external surfaces and the confinement of the material to certain film thicknesses considerably alters the microphase behavior and can stabilize novel, non-bulk, structures. It is important to study the phase behavior as the alteration increases the number of possible thin-film structures for a given polymeric system. This results in complex phase behavior that may be exploited in potential applications. We have used our model (the free energy (1) and dynamic equations (14)) previously ($37, 38$) to calculate the phase diagram for a confined cylinder-forming diblock copolymer (which was modeled as a $A_3 B_6$ chain and $\varepsilon_{AB} = 7.1$). We found that non-cylindrical structures are stabilized in thin films by the preferential attraction of one type of block to the surface (the *surface field*). In particular, we found a sequence of phase transitions, with increasing surface field, from cylinders oriented parallel to the surface via hexagonally perforated lamellae to lamellae. In a recent study, Knoll et al (33) have performed similar calculations for the triblock analog of the polymer used in Huinink et al ($37, 38$), a cylinder forming triblock copolymer modeled as a $A_3 B_{12} A_3$ chain, and revealed phase behavior quite similar to the diblock copolymer case ($37, 38$). These simulation results were supported by experiments on thin films of concentrated solutions of a polystyrene-*block*-polybutadiene-*block*-polystyrene (SBS) triblock copolymer in chloroform. Films of different thickness were exposed to a well-controlled partial pressure of chloroform vapor. As chloroform is a relatively neutral solvent for both chemical blocks, the effective strength of the surface interactions was varied via the solvent concentration in the polymer film, giving rise to an experimental thin-film phase diagram. The effective interaction was then determined using the experimental data as $\varepsilon_{AB} = 6.5$ (all other parameters are as in ($37, 38$)) and the phase behavior simulated as a function of slit width H and polymer-particle interaction difference $\varepsilon_M = \varepsilon_{AM} - \varepsilon_{BM}$. A stunning agreement between the calculated and the measured phase diagram was found (33). In particular, non-cylindrical structures were indeed observed in certain, well-defined regions of the experimental phase diagram.

As an example, figure 1 shows a comparison of the experimental results (top and middle pictures) and the simulations (bottom picture) for a given strength of the surface field ($\varepsilon_M = 6$). The simulation box contains two filler particles: a horizontal planar particle (one gridcell thick) at the bottom and a tilted planar particle (also one gridcell thick) at the top. This simulation volume was chosen to mimic the lateral increase in film thickness of the experimental observation (schematic middle picture in fig 1). In figure 1, a rich variety of structures is observed as the film thickness increases from left to right. With increasing film thickness, both experiment and simulation show the same sequence of thin-film phases: a featureless film surface for the smallest thickness, isolated domains of the A phase, parallel-oriented A cylinders, perforated A lamella, parallel-oriented A cylinders of ellipsoidal cross section, perpendicularly oriented A cylinders, and finally two layers of parallel-oriented cylinders. The phase transitions occur at well-defined film thickness, as can be seen from the white iso-thickness lines that have been calculated from the scanning force microscopy (SFM) height images. From a fundamental point of view, these study nicely demonstrates that block copolymer thin films are governed by the interplay of two parameters: i) the surface field can either orient the bulk structure or stabilizes non-bulk structures, so-called surface reconstructions (36) ii) the film thickness modulates the stability regions of the different phases via interference and confinement effects. The range of the surface reconstructions seems to be limited to a near-surface region comparable to a characteristic microdomain spacing (33, 37, 38). However, even for thicker films the surface regions can be of crucial importance for the function of the film, as they establish the contact between the film and its surrounding media, such as electrodes, electrolytes or etching media. The mere existence of surface reconstructions leads to the important general remark that it is not straightforward to draw conclusions about the volume structure of a thin film from the surface structure.

The above simulation study considers 'hard' films where both plates or particles are inpenetrable and completely incompressible. Here, we did not consider 'half-soft' films where one of the confining sides is highly compressible, for instance in the presence of air. The results for the 'hard' films can be extrapolated (37, 38) to films that are able to adjust their shape, as in the experiments, to the local tensions in the film. An alternative route, that is currently under investigation, is to introduce neutral solvent. Moreover, in experiments of supported films, dissimilar substrates often exist, i.e. at one particle surface the favored microdomain orientation or surface reconstruction is different than at the other particle surface (for a simulation example, see figure 2). In such a situation a hybrid structure may form and a grain boundary is stabilized in the middle of the film (39). Two experimental examples of such hybrid structures reported for cylinder-forming block copolymer are 'a perforated lamellae with spheres' (40) and 'cylinders with necks' (41). Both of them exhibit the same surface structure. However, due to the different underlying structures, the mechanical response to external forces will be quite different. The phase diagram for this type of system with dissimilar substrates has been calculated and a detailed article will be submitted soon. The method and concepts presented here, together with recent developments in experimental three-dimensional imaging

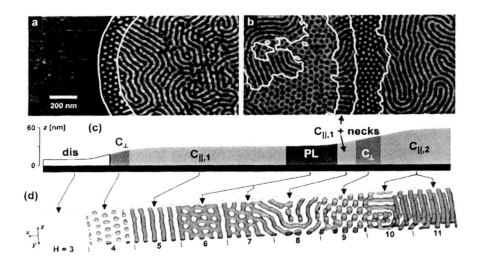

Figure 1: a,b) Tapping-mode SFM phase images of thin SBS films on silicon substrates after annealing in chloroform vapor. The surface is covered with a homogeneous ≈ 10 nm thick polybutadiene (PB) layer. Bright [dark] corresponds to polystyrene (PS) [PB] microdomains below this top PB layer. Contour lines calculated from the corresponding height images are superimposed. c) Schematic height profile of the phase images shown in (a,b). d) Simulation of an $A_3B_{12}A_3$ block copolymer film in one large simulation box with film thickness increasing from left to right. The iso-concentration surface $\rho_A = 0.5$ is shown. (Reprinted with permission from Böker et al. (34) Copyright 2002, American Physical Society).

a)   b)

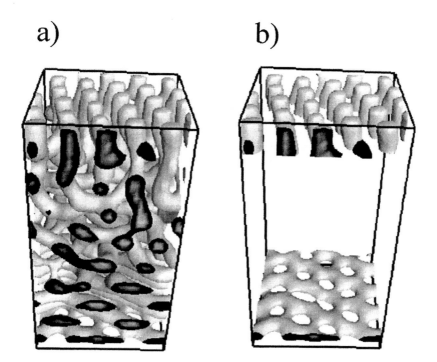

Figure 2: Representative simulation result in case of dissimilar substrates located at the top and the bottom of the simulation box: one substrate favors a perpendicular microstructrure, the other a parallel (perforated lamellae) one. Left: crop of the field in the vicinity of the hard particles for iso-concentration surface $\rho_A = 0.5$, right: total field (same iso-value).

techniques ($40, 42$) provide the means to control and understand block copolymer thin-film structures of various symmetry.

Smart nanoparticles

Through simulation we discovered remarkable bicontinuous structures in dispersed droplets of polymer surfactant (figures 3-4). The nanogels structures are soft and fragile: the molecular bonding energy is weak, derived from self-assembly of polymer molecules. Nanogels may have many applications in soft nanotechnology, in drug delivery, templates for heterogeneous catalysts, aerosols and personal care products. Patterned nanostructures are the key in all these applications, and a better understanding of their formation is of paramount interest. In these calculations, we generate the nanogels by quenching a homogenous droplet of polymer surfactant in an aqueous bath. The free energy model is given by (1) and the dynamic equations by (17).

The simulation parameters are for diblock polymer surfactant $A_{N-M}B_M$ with $N = 20$ in weakly selective solvent and mild segregation, $\chi_{AB}N = 40$, $\chi_{AS} - \chi_{BS} = 0.3$, so that A is solvo-phobic and B is solvo-phylic. These are essentially the parameters we verified before by comparison with experimental microphase diagrams of concentrated Poly Propylene Oxide – Poly Ethylene Oxide aqueous solutions in ambient conditions (17), in which case each bead or statistical unit corresponds to 3-4 monomers. Of course one should realize any polymer surfactant solution with the same properly scaled interaction parameters will behave in exactly the same way, according to the mean-field model.

The free energy model is that for an nVT ensemble, and accordingly we do not calculate the global equilibrium in an open system, but rather a local equilibrium morphology of an isolated droplet. The situation is analogous to that of the classic study of the shape of an isolated lipid vesicle, when interactions between vesicles are less relevant. With the selected values of the Flory-Huggins parameters, and N and M, the polymers are all insoluble, hence the deformations of the droplets are at constant mass of polymer.

The simulations proceed by a sudden quench of a homogeneous droplet in a solvent bath. Following the quench, the droplet takes up or releases additional solvent locally and globally, depending on the particular morphology being formed. The polymer concentration outside the droplet is zero, and this remains so during the morphology adaptation. The simulations are stopped when the order parameters do not change any more: the free energy is then steady in a minimum, and the pattern is a solution to the self-consistent field condition $\rho_I = \rho_I[\mu_I^{nid}]$. We have found it advantageous to add a small uncorrelated white noise field to the mean-field chemical potentials. The uncorrelated noise does not obey the fluctuation-dissipation theorem, but it helps small barrier crossings in the free energy landscape. The droplets (initial radius R^0 cell units, each cell has the size of the statistical unit) are placed in the centre of the box with sufficient space, $N/2$ cells, between the droplet surface

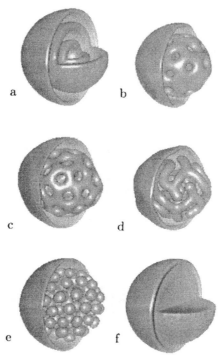

Figure 3: Morphologies of $A_{N-M}B_M$ polymer surfactant nanodroplets (isosurfaces partly removed for visualisation). Solvophylic A concentration field for different block ratios $f = M/N$. 0.35(a), 0.30(b), 0.25(c), 0.20(d), 0.15(e), 0.10(f). (Reprinted with permission from Fraaije et al. (accepted) Copyright 2003, American Chemical Society).

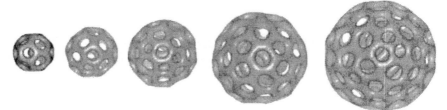

Figure 4: Morphologies of $A_{N-M}B_M$ polymer surfactant nanogels ($f = M/N = 0.25$) gels for different initial radii R^0. From left to right: $R^0 = 20, 23, 26, 30, 33$ (in units of polymer statistical bead size). Notice that in this viewing mode the outer fuzzy shell is not visible. (Reprinted with permission from Fraaije et al. (accepted) Copyright 2003, American Chemical Society).

and the boundaries of the computational box, thereby avoiding artifacts resulting from the periodic boundary conditions. In all cases the droplets develop an outer fuzzy layer of the solvo-phylic B block, but since the confinement of the polymers is not hard but soft, the droplet surface is not necessarily spherical. The surface topography is that of small valleys, ridges and bumps, reflecting the underlying gel morphology, very much like earths topography is an image of deeper events. The nanogel internal structures, shown in figure 3, depend on the size ratio $f = M/N$ similar to bulk block copolymer systems. More symmetric polymers $f = 0.35$ form into an onion structure of alternating A and B layers (figure 3a); slightly less symmetric polymers $f = 0.30 - 0.25$ form a bicontinuous phase (figure 3b-c), then at $f = 0.20$ a cylindrical phase (figure 3d) and an inverted micellar phase (figure 3e) at $f = 0.15$. Too asymmetric polymers $f = 0.1$ do not form any internal structure (figure 3f). In equilibrium the gels contain an appreciable amount of solvent (on the average ca. 15%), distributed inhomogeneously over the solvo-phobic A and solvo-phylic B-rich layers.

In the case of $f = 0.25$, figure 4, the layers are perforated with pores, and the entire structure is bicontinuous. The gel strikingly resembling a buckyball, with in each inner B layer a mixed pore pattern of pentagons, hexagons and septagons. In bulk solution or melt systems, perforated lamellae consist ideally of a perfectly hexagonal array of pores. In the curved nanogels, by rule of geometry, a perfect array of hexagons is impossible to form, and the perforation is mixed. In bulk, a mixture of 85% of this particular polymer surfactant and 15% solvent forms a gyroid bicontinuous structure (data not shown).

An interesting finding for the morphologies of figure 4 is that the equilibrium structures are more or less radial invariant. From figure 5, where the radial positions of internal maxima of B concentration are plotted as a function of initial droplet size R^0, it can be observed that, with increasing R^0, the position of the B rich shells shifts linearly. The system adds shells at the center if the confinement allows it. When an additional layer can be formed, the inner layers move outward.

Since the domain size is more or less constant, the pore density in each layer is constant too, and the number of pores in each shell is determined from its radius. There is no intrinsic magic '60' number, as in C_{60}, associated with additional stability of the nanogels. The resemblance with a carbon buckyball is purely coincidental, based on a geometrical rule for packing pores in a spherical shell.

The experimental investigation of these systems is at its infancy. Perforated lamellar or similar bicontinuous structures have been observed experimentally in thin polymer films (33), and in simulations of confined systems (33, 37, 38). In the area of polymersomes (26), closely related to the systems of interest here, results so far indicate hollow bilayer vesicles. At first glance the system of bicontinuous cubic phases or cubosomes (see also the article by Pat Spicer in this volume) resembles the buckyball structures in this article. Although this looks very promising, the cubosomes are cubic liquid crystalline phases that autonomously self-assemble from

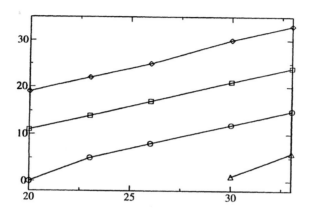

Figure 5: Shell position versus initial radius for different layers. (Reprinted with permission from Fraaije et al. (accepted) Copyright 2003, American Chemical Society).

solution, whereas our structures depend on the processing conditions (starting from dispersed nano droplets) to a high degree and are made of rather floppy polymers (due to the Gaussian chain model involved). Therefore, more work should be carried out to determine the leading factors. Very recently it was shown that polymersomes may fold in giant high-genus vesicles (*27*), which also superficially resemble the bicontinuous structures we have found. But in the perforated polymersome case the holes are very large, much larger than single coil domain-sizes we observe in our gels. In solutions of crew-cut amphiphilic polymers Eisenberg and co-workers have determined a entire wonderland of structures (*35*). Some exhibit onion phases, others are bicontinuous, and the raspberry droplets with the inverted micelles (figure 3e) resemble so-called Large Compound Vesicles. All agreements at this point are qualitative. A more detailed comparison, using realistic experimental parameters, is in progress.

Conclusions

In a recent review, it was remarked that field-theoretic computational methods are a powerful alternative to particle simulations in soft matter (*23*). The results shown in our examples section emphasize this remark and offer a glimpse of the wide range of research areas where this type of methods is valuable. We showed that our molecular field method is: 1) capable of capturing the diffusive dynamics of phase separation in experimental systems and provide a detailed microscopic picture of the underlying phenomena, 2) able to reproduce the static experimental phase behavior of thin films and allow to distinguish determining factors, 3) suitable for the prediction of new phenomena, based upon the proven reliability of the method. Future developments include an extension to hydrodynamic contributions: this will open a new era in soft condensed matter research.

Acknowledgment

The authors acknowledge NWO-DFG project DN 72-216. Part B of the examples was carried out in close collaboration with Armin Knoll (experiments), Andriana Horvat and Robert Magerle (simulations) at Bayreuth University: they are gratefully acknowledged. We acknowledge Katya Lyakhova (Leiden University) for providing figure 2 of this article. The supercomputer resources were provided by NCF (Nationale Computer Faciliteiten) at the national computer facilities at SARA, Amsterdam for the electric field alignment and the High-Performance Computing Centre, University of Groningen, for the smart particles. We thank Adi Eisenberg (McGill University), Alexander Böker and Georg Krausch (Bayreuth University) for stimulating discussions.

References

1. Qiwei He, D.; Naumann, E. B. *Chemical Engineering Science* **1997**, *52*, 481.
2. Gonella, G.; Orlandini, E.; Yeomans, J. M. *Phys. Rev. Lett.* **1997**, *78*, 1695.
3. Pätzold, G.; Dawson, K. J. *Chem. Phys.* **1996**, 104, 5932.
4. Kodama, H.; Komura, S. *J. Phys. II France* **1997**, *7*, 7.
5. Ohta, T.; Nozaki, H.; Doi, M. *Physics Letters A*, **1990**, *145*, 304.
6. Cahn, J. W.; Hilliard, J. E. *J. Chem. Phys.* **1958**, *28*, 258.
7. Oono, Y.; Puri, S. *Phys. Rev. Lett.* **1987**, *58*, 836.
8. de Gennes, P. G. *J. Chem. Phys.* **1980**, *72*, 4756.
9. Maurits, N. M.; Fraaije, J. G. E. M. *J. Chem. Phys.* **1997**, *106*, 6730.
10. Fraaije, J. G. E. M. *J. Chem. Phys.* **1993**, *99*, 9202.
11. Fraaije, J. G. E. M.; van Vlimmeren, B. A. C.; Maurits, N. M.; Postma, M.; Evers, O. A.; Hoffmann, C.; Altevogt, P.; Goldbeck-Wood, G. *J. Chem. Phys.* **1997**, *106*, 4260.
12. Maurits, N. M.; Fraaije, J. G. E. M. *J. Chem. Phys.* **1997**, *107*, 5879.
13. van Vlimmeren, B. A. C.; Maurits, N. M.; Zvelindovsky, A. V.; Sevink, G. J. A.; Fraaije, J. G. E. M. *Macromolecules* **1999**, *32*, 646.
14. Kawakatsu, T. *Phys. Rev. E* **1997**, *56*, 3240.
15. Hasegawa, R.; Doi, M. *Macromolecules* **1997**, *30*, 3086.
16. Reister, E.; Muller, M.; Binder, K. *Phys. Rev. E* **2001**, *64*, 041804.
17. van Vlimmeren, B. A. C.; Maurits, N. M.; Zvelindovsky, A. V.; Sevink G. J. A.; Fraaije, J. G. E. M. *Macromolecules* **1999**, *32*, 646.
18. Sevink G. J. A.; Zvelindovsky, A. V.; van Vlimmeren, B. A. C.; Maurits N. M.; Fraaije, J. G. E. M. *J. Chem. Phys.* **1999**, *110*, 2250.
19. Maurits, N. M.; van Vlimmeren, B. A. C.; Fraaije, J. G. E. M. *Phys. Rev. E* **1997**, *56*, 816.
20. Landau, L. D.; Lifshitz, E. M. *Electrodynamics of continuous media*; Pergamon: Oxford, 1960.
21. Kyrylyuk, A. V.; Zvelindovsky, A. V.; Sevink, G. J. A.; Fraaije, J. G. E. M. *Macromolecules* **2002**, *35*, 1473.
22. Drolet F.; Fredrickson, G. H. *Phys. Rev. Lett.* **1999**, *83*, 4317.
23. Fredrickson, G. H.; Ganesan, V.; Drolet, F. *Macromolecules* **2002**, *35*, 16.
24. van Vlimmeren, B. A. C.; Fraaije, J. G. E. M. *Comput. Phys. Commun.* **1996**, *99*, 21.
25. Maurits, N. M.; Fraaije, J. G. E. M. *J. Chem. Phys.* **1997**, *107*, 5879.
26. Discher, B.; Won, Y.; Ege, D.; Lee, J.; Kossuth, M.; Discher, D.; Bates, F.; Hammer, D. *Biophysical Journal* **1999**, *76*, A435.
27. Haluska, C. K.; Gozdz, W. T.; Döbereiner, H. G.; Forster, S.; Gompper, G. *Phys. Rev. Lett.* **2002**, *89*, 238302.
28. Schillen, K.; Bryskhe, K.; Melnikova, Y. S. *Macromolecules* **1999**, *32*, 6885.
29. Doi, M.; Chen, D. *J. Chem. Phys.* **1989**, *90*, 5271.
30. Ohta, T.; Enomoto, Y.; Harden, J. L.; Doi, M. *Macromolecules* **1993**, *26*, 4928.
31. Zvelindovsky, A. V.; Sevink, G. J. A.; van Vlimmeren, B. A. C.; Maurits, N. M.; Fraaije, J. G. E. M. *Phys. Rev. E* **1998**, *57*, R4879.
32. Zvelindovsky, A. V.; Sevink, G. J. A.; Fraaije, J. G. E. M. *Phys Rev E* **2000**, *62*, R3063.

278

33. Knoll, A.; Horvat, A.; Lyakhova, K. S.; Krausch, G.; Sevink, G. J. A.; Zvelin-dovsky, A. V.; Magerle, R. *Phys. Rev. Lett.* **2002**, *89*, 035501.
34. Böker, A.; Elbs, H.; Hansel, H.; Knoll, A.; Ludwigs, S.; Zettl, H.; Urban, V.; Abetz, V.; Muller, A. H. E.; Krausch, G. *Phys. Rev. Lett.* **2002**, *89*, 135502.
35. Cameron, N. S.; Corbierre, M. K.; Eisenberg, A. *Can. J. Chem.* **1999**, *77*, 1311.
36. Rehse, N.; Knoll, A.; Konrad, M.; Magerle, R.; Krausch, G. *Phys.Rev. Lett.* **2001**, *87*, 035505.
37. Huinink, H. P.; Brokken-Zijp, J. C. M.; van Dijk M. A.; Sevink, G. J. A. *J. Chem. Phys.* **2000**, *112*, 2452.
38. Huinink, H. P.; van Dijk, M. A.; Brokken-Zijp, J. C. M.; Sevink, G. J. A. *Macromolecules* **2001**, *34*, 5325.
39. Fasolka, M. J.; Banerjee, P.; Mayes, A. M.; Pickett, G.; Balazs, A. C. *Macromolecules* **2000**, *33*, 5702.
40. Harrison, C.; Park, M.; Chaikin, P. M.; Register, R. A.; Adamson, D. H.; Yao, N. *Polymer* **1998**, *13*, 2733.
41. Konrad, M.; Knoll, A.; Krausch, G.; Magerle, R. *Macromolecules* **2000**, *33*, 5518.
42. Magerle, R. *Phys. Rev. Lett.* **2000**, *85*, 2749.

Chapter 17

Simulations of Polymer Solutions: A Field-Theoretic Approach

Alfredo Alexander-Katz[1], André G. Moreira[2], and Glenn H. Fredrickson[2]

[1]Physics Department and [2]Materials Research Laboratory, University of California, Santa Barbara, CA 93106

We used field-theoretic simulations to study the equilibrium behavior of polymers in a good solvent confined to a slit of width L. In particular, we obtained density profiles across the slit for different values of the monomer excluded volume over a wide range of concentrations C. We also obtained mean field results for the profiles. The effective correlation length ξ_{eff} was calculated from the density profiles and compared to the mean field result (valid in the limit of high concentrations). For small excluded volume parameters we found that ξ_{eff} is well described by the mean field result, while for larger excluded volume interaction the correlation length shows a $C^{-3/4}$ scaling behavior, which is compatible with the behavior expected for this system in the semi-dilute regime.

Field theory can be a useful tool in the study of condensed matter systems and especially complex fluids. Examples of its application in the latter context include polymer solutions(*1,2,3*), polyelectrolytes(*4*), or ionic solutions(*5*). As usual in such approaches, one often has to invoke approximations that make the calculations amenable to analytical treatment, e.g. perturbation theory, mean field approximation, etc. Although this is some times enough to study the systems under consideration, it is highly desirable to have a numerical scheme that allows the exact solution of the field theory. On one hand, this would allow a direct comparison with the aforementioned approximations (i.e., one can directly test the validity of the approximations in different limits). On the other hand, it is reasonable to expect that under some conditions (e.g. dense systems), a field-theoretic simulation approach can be more effective than a particle-based one.

Recently, Fredrickson *et. al.*(*6,7*) proposed a new method to simulate the equilibrium properties of polymer solutions based on field theory, the so-called Field-Theoretic Polymer Simulation (FTPS). Although there has been previous work on the simulation of polymers using similar techniques(*8*) (see also the article by Sevink and Fraaije in this volume), the new technique is capable of yielding "exact" results (aside from numerical errors, as well as finite size and discretization effects) that go beyond the mean field approximation.

The FTPS method has been successfully applied(*6*) to the study of the order-disorder transition for an incompressible two-dimensional diblock copolymer melt. Although incompressible melts are interesting from the scientific (and industrial) point of view, there is a wide range of systems where polymers are in solution. It is then desirable to understand how the FTPS approach works in the latter case. We will focus our attention in this short communication on a homopolymer solution under good solvent conditions confined to a slit of width L. This problem has received some attention in the past(*9–16*), which provides a theoretical basis for comparison with our simulation results.

This article is organized as follows. In section II we provide a brief theoretical background on the field-theoretic formulation of a homopolymer solution. In section III we introduce the simulation method, where the sampling algorithm is discussed. In section IV we present the simulation results for the homopolymer solution confined to a slit. In particular we discuss the density profile across the slit for several concentrations, ranging from the dilute to the concentrated regime. We compare these results with mean field results obtained both analytically as well as numerically. Finally, section V contains some concluding remarks.

Theoretical Background

Polymers can be described in a coarse-grained fashion by the so called Gaussian thread model(*3*), which corresponds to the continuum limit of a sequence of beads (monomers) attached by harmonic springs. The chains are represented by continuous curves in space $\mathbf{R}_\alpha(s)$, where $\alpha = 1,...,n$ counts the different chains and s is a continuous variable between 0 and 1 along each chain contour ($\mathbf{R}(0)$

corresponds to the position of one end of the chain while $\mathbf{R}(1)$ corresponds to the position of the other end). The energy of a given configuration contains two parts, the first one being an elastic contribution which is given (in units of $k_B T$) by

$$H_O = \frac{1}{4} \sum_{\alpha=1}^{n} \int_0^1 ds \left| \frac{\partial \mathbf{R}_\alpha(s)}{\partial s} \right|^2. \tag{1}$$

In the latter expression, all lengths were scaled by R_{go}, the ideal radius of gyration, given by $R_{go}^2 = b^2 N/(2d)$. The parameter b represents the statistical segment length, d is the dimensionality of space, and N is the degree of polymerization.

The second contribution to the total energy comes from the interaction between non-adjacent monomers in the same polymer and between monomers in different polymers. This interaction can be attractive or repulsive depending on the solvent: in good solvents it is repulsive, in bad solvents it is attractive and in theta solvents it is absent (and the chain behaves accordingly as a Gaussian chain). A typical simple model used for a good solvent is a pairwise repulsive $v\delta(\mathbf{r})$ potential which prevents the monomers from overlapping. In the case of a poor solvent, $v < 0$, one would have to include a repulsive three-body interaction to stabilize the system. Here we will focus only on the case of a good solvent. In terms of the microscopic monomer density, $\hat{\rho}(\mathbf{r}) = \sum_\alpha \int_0^1 ds \delta(\mathbf{r} - \mathbf{R}_\alpha(s))$, the interaction energy H_I can be written as

$$H_I = \frac{B}{2} \int d\mathbf{r} |\hat{\rho}(\mathbf{r})|^2, \tag{2}$$

where $B = vN^2/R_{go}^3$ is a dimensionless parameter, and v is the excluded volume parameter that dictates the effective repulsion among monomers mediated by the solvent. As in Eq. (1), all lengths are measured in units of R_{go}.

The partition function is given by

$$Z = \frac{1}{n!} \int \mathcal{D}\mathbf{R} \exp(-H_O - H_I) \tag{3}$$

where the integration $\int \mathcal{D}\mathbf{R}$ is performed over all possible configurations of the chains. After decoupling the quadratic density interactions with a Hubbard-Stratonovich transformation, one can show that the configurational integral becomes

$$Z = \mathcal{N} \int \mathcal{D}\Omega \exp(-H[\Omega]), \tag{4}$$

where $\mathcal{N}$ is a normalization constant, and Ω is a real scalar field. The field Ω corresponds to a local fluctuating chemical potential. After some straightforward algebra, one can show that the action $H[\Omega]$ reads

$$H[\Omega] = \frac{1}{2B} \int d\mathbf{r} \Omega(\mathbf{r})^2 - CV \ln Q[i\Omega], \tag{5}$$

where $C = n/V$ is a dimensionless polymer concentration, V is the volume in units of R_{go}^d, and $Q[i\Omega]$ is the single-polymer partition function in the presence of an inhomogeneous field Ω, viz.

$$Q[i\Omega] = \frac{\int \mathcal{D}\mathbf{R}\exp[-H_O - i\int_0^1 ds\Omega(\mathbf{R}(s))]}{\int \mathcal{D}\mathbf{R}\exp[-H_O]}. \tag{6}$$

One can show(2) that the latter path integral can be rewritten as

$$Q[i\Omega] = \frac{1}{V}\int d\mathbf{r}q(\mathbf{r},1), \tag{7}$$

where the function $q(\mathbf{r},s)|_{s=1}$ is the probability of finding a polymer that ends at $\mathbf{r}$ (with the other end placed anywhere in the solution). This function satisfies the complex diffusion equation

$$\frac{\partial q(\mathbf{r},s)}{\partial s} = \nabla^2 q(\mathbf{r},s) - i\Omega(\mathbf{r})q(\mathbf{r},s) \tag{8}$$

with initial condition $q(\mathbf{r},0) = 1$.

It is important to notice that the single polymer partition function Q is in general complex. This implies that the action Eq.(5) is also complex, and that its exponential cannot be used as a probability distribution (since it is not positive definite). This is related to the so-called sign problem(17) that appears in lattice gauge theories. Note though that since all transformations involved in going from a particle-based into a field-theoretic description are exact, the partition function Z remains a real quantity.

The saddle point for this field theory follows from the solution of

$$\frac{\delta H[\Omega]}{\delta\Omega}\bigg|_{\Omega^*} = 0 \tag{9}$$

which leads to $\Omega^*(\mathbf{r}) = -iBC\phi(\mathbf{r};[\Omega^*])$, where $\phi(\mathbf{r}) = \rho(\mathbf{r})/\rho_o$ is the reduced density, and ρ_o is the average monomer density. The reduced density can be computed for some particular field configuration $\Omega(\mathbf{r})$ by the expression

$$\phi(\mathbf{r}) = \frac{1}{Q}\int_0^1 ds\, q(\mathbf{r},1-s)\, q(\mathbf{r},s) \tag{10}$$

In the case of an unconfined polymer solution, the saddle point is homogeneous, and $\Omega^*(\mathbf{r}) = -iBC$. For the slit geometry this is not true, and the saddle point solution has to be calculated numerically. We note that the saddle point solution can be associated with a mean field approximation to the field theory commonly referred to as "self-consistent field theory"(18).

Simulation Method

The sampling algorithm we used to simulate the field theory given by Eqs. (5-8) is a complex Langevin algorithm(*19*). This technique was designed to address the problem of sampling complex actions, and it basically consists of extending the field Ω to the complex plane, i.e. $\Omega = \Omega_R + i\Omega_I$. Although this method has not been proved to converge for an arbitrary action, it has been shown(*20*) that if the averages computed with the Complex Langevin become stationary and are independent of the initial conditions used to start the simulation, then the averages obtained are the correct (in our case thermodynamic) values. Formally, this means that the complex Langevin generates a probability distribution $P[\Omega]$ such that thermodynamic averages follow from

$$\langle A \rangle = \int \mathcal{D}\Omega P[\Omega] A[\Omega] =$$
$$= \frac{1}{Z}\frac{1}{n!} \int \mathcal{D}R \exp(-H_O - H_I)A \tag{11}$$

where A is some arbitrary observable. The complex Langevin dynamics is given by

$$\frac{\partial}{\partial t}\Omega_R(\mathbf{r},t) = -\Gamma Re\left[\frac{\delta H[\Omega]}{\delta\Omega(\mathbf{r},t)}\right] + \eta(\mathbf{r},t)$$
$$= -\Gamma\left[-C\phi_I(\mathbf{r},t;[\Omega]) + \frac{\Omega_R(\mathbf{r},t)}{B}\right] + \eta(\mathbf{r},t) \tag{12}$$

and

$$\frac{\partial}{\partial t}\Omega_I(\mathbf{r},t) = -\Gamma Im\left[\frac{\delta H[\Omega]}{\delta\Omega(\mathbf{r},t)}\right]$$
$$= -\Gamma\left[C\phi_R(\mathbf{r},t;[\Omega]) + \frac{\Omega_I(\mathbf{r},t)}{B}\right] \tag{13}$$

where Γ is a constant, and it is chosen such that the fluctuation dissipation theorem is satisfied. The field $\eta(\mathbf{r},t)$ is a Gaussian real noise with

$$\langle \eta(\mathbf{r},t) \rangle = 0 \tag{14}$$

and

$$\langle \eta(\mathbf{r},t)\eta(\mathbf{r}',t') \rangle = 2\Gamma\delta(\mathbf{r}-\mathbf{r}')\delta(t-t'). \tag{15}$$

It is important to notice that if the noise term in Eq.(12) is removed, then Eqs.(12, 13) relax the field Ω to the saddle-point solution (mean field). This can be seen immediately by noting that without the noise source, the solution will be attained once $\partial\Omega/\partial t = 0$, which corresponds the saddle point Eq. (9). This is

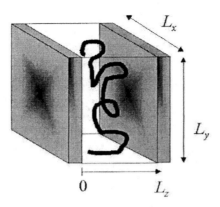

Figure 1: *Schematic representation of the simulation box (reproduced with permission from reference (21). Copyright 2003 American Institute of Physics).*

an important feature of this method, since mean field solutions are automatically obtained by removing the noise source.

Our simulations are carried out on a lattice with lattice spacing given by $\Delta x, \Delta y, \Delta z$ in the x, y, z directions respectively. We employ a simple explicit , forward time integration scheme for solving Eqs.(12-13):

$$(\Omega_R)_{\mathbf{j},t+\Delta t} = (\Omega_R)_{\mathbf{j},t}$$
$$- \frac{\Delta V \Delta t}{2}[-C(\phi_I)_{\mathbf{j},t} + \frac{(\Omega_R)_{\mathbf{j},t}}{B}] + \sqrt{\Delta t}\,\eta_{\mathbf{j},t} \qquad (16)$$

and

$$(\Omega_I)_{\mathbf{j},t+\Delta t} = (\Omega_I)_{\mathbf{j},t} - \frac{\Delta V \Delta t}{2}[C(\phi_R)_{\mathbf{j},t} + \frac{(\Omega_I)_{\mathbf{j},t}}{B}] \qquad (17)$$

where $\Delta V = \Delta x \Delta y \Delta z$, Δt is a dimensionless time step, $\mathbf{j} = (j,k,l)$ denotes the lattice site coordinates, and the Gaussian noise source $\eta_{\mathbf{j},t}$ has first and second moments given now by

$$\langle \eta_{\mathbf{j},t} \rangle = 0 \qquad (18)$$

$$\langle \eta_{\mathbf{j},t} \eta_{\mathbf{j}',t} \rangle = \delta_{\mathbf{j}\mathbf{j}'} \qquad (19)$$

The simulations were performed in a cubic lattice of side lengths $L_x = L_y = L_z = L = 1$ with periodic boundary conditions in the x-y directions, and having two impenetrable walls, one at $z = 0$ and the other one at $z = L = 1$ (Fig. 1). The lattice discretization was chosen to be $\Delta x = \Delta y = .05$, and $\Delta z = .025$. The choice of these parameters is based on preliminary tests with $\Delta z = 0.1, 0.0625, 0.05, 0.04$, where we observed that the average density profile across the slit was insensitive to the discretization for $\Delta z \lesssim 0.05$. We also performed simulations with $\Delta x = \Delta y = .1$, and $L_x = L_y = 2$ to explore finite-size effects in the lateral directions. The results obtained were not statistically different from those reported here with $L_x = L_y = 1$. The diffusion equation Eq.(8) was solved using an alternating direction implicit algorithm (ADI) with a step $\Delta s \leq 0.0009$. The ADI method is only condition- ally stable in three dimensions, and the parameter Δs has to be chosen such that $\Delta s / (\Delta z)^2 \leq 1.5$. The simulations were performed on a single AMD Athlon 1300 processor. The typical performance was about 1.2×10^3 Langevin steps per 24 hours of CPU time. After equilibrating the system, the reduced density profile across the slit was calculated for each configuration of the field Ω, and then aver- aged over all the configurations using the following expression

$$\langle \phi(l) \rangle_\Omega = \langle \frac{1}{N_x N_y} \sum_{j,k} \phi(j,k,l;[\Omega]) \rangle_\Omega \tag{20}$$

where N_x, N_y are the number of lattice points in the x, y directions respectively. The reduced density $\phi(j,k,l;[\Omega])$ is determined using Eq.(10).

Results

In Fig. 2 we compare the mean field results for the density profiles from our simulation (without noise) to an approximate analytical solution (ground state dominance approximation) for the density across the slit(*21*)

$$\phi(z) = \psi_m^2 \left\{ \tanh[\frac{\psi_m}{2\xi_b} z] + \tanh[\frac{\psi_m}{2\xi_b}(z-L)] - \tanh[\frac{\psi_m}{2\xi_b} L] \right\}^2, \tag{21}$$

where as before $\phi = \rho/\rho_o$, and $\xi_b = 1/\sqrt{2BC}$ is the Edwards correlation length. The reduced density in the middle of the slit ψ_m^2 is calculated using the normaliza- tion condition $\int_0^L \phi(z)dz = L = 1$. The agreement between the analytical solution and the numerical mean field solution is excellent, as can be seen from Fig. 2. This approximation works very well for $BC \gtrsim 1$.

Figure 3 shows the typical data for the reduced density profile from the sim- ulation with fluctuations (filled squares) compared to the mean field result (open squares). The analytical solution (solid line) given by Eq.(21) was applied in the mean field case with the parameter $BC = 12.5$. As expected, the effect of fluctua- tions is to increase the effective correlation length ξ_{eff} compared to the mean field result ξ_b, which can be seen directly from the plot since the data for $B = 25; C = .5$ exhibits a higher peak at the center of the slit when compared to the mean field re- sult, which indicates that the polymers in the former case are being more strongly

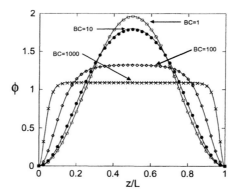

Figure 2: *Profile of the reduced density* ϕ *across the slit. The points are the data form the simulation (without noise) for* $BC = 1$ *(open circles),* $BC = 10$ *(filled circles),* $BC = 100$ *(open diamonds), and* $BC = 1000$ *(x's). The solid lines represent Eq.(21) for the same set of parameters respectively (reproduced with permission from reference (21). Copyright 2003 American Institute of Physics).*

pushed away from the wall. Furthermore, the effective correlation length ξ_{eff} can be extracted from a fit to the density profile using Eq.(21). In particular, we assume that the functional form of Eq.(21) still holds in the case of fluctuations (excluded volume correlations), but substitute the bulk correlation length ξ_b by its effective correlation length $\xi_{eff}(22)$. The parameter ψ_m is fixed by the normalization condition discussed above, which leaves us with a fitting function whose only dependence is ξ_{eff}, the effective correlation length. For this particular case, ξ_{eff} was calculated to be $\xi_{eff}(B = 25; C = .5) = 0.35 \pm 0.06$.

Figure 4 shows the effective correlation lengths ξ_{eff} plotted as a function of the parameter $BC = B \cdot C$ for various concentrations. The solid line represents ξ_b as defined above, and the dotted lines have a slope of -0.75, which is the same as the scaling behavior expected for the correlation length in the semi-dilute regime, $\xi_{sd} \propto C^{-3/4}$. As can be seen from the plot, the correlation lengths extracted for $B = 1$ (filled diamonds) lie on top of the self consistent mean field result $\xi_{eff} = 1/\sqrt{2BC}$. This behavior is expected since the parameter B dictates the effective repulsion among monomers, and in the limit $B \to 0$ one should recover the Gaussian result, ξ_b. As the parameter B increases, excluded volume correlations become increasingly important and the polymers tend to swell, implying that the correlation length (which is related to the blob size) increases as a function of B. This behavior is clearly seen in Fig. 4, where for any given concentration the effective correlation length $\xi_{eff}(B)$ increases with B up to a threshold $\xi_{max} = L = 1$.

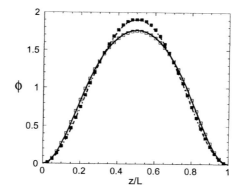

Figure 3: *Profiles of the reduced density* ϕ *across the slit for the case* $C = 0.5, B = 25$. *The points are the data from the simulation with noise (filled squares), and without noise (open squares). The lower solid line represents Eq.(21) with parameter* $BC = 12.5$. *The upper dashed line represents a best fit to the data using Eq.(21) and an effective correlation length (reproduced with permission from reference (21). Copyright 2003 American Institute of Physics).*

This threshold is imposed by the geometry of the slit, since the maximum blob size ξ_{max} allowed in the z direction is clearly $\xi_{max} = L$. For large concentrations, the correlation lengths become smaller for both $B = 10$ (open circles) and $B = 25$ (filled squares), and they tend toward the concentrated regime solution, in which the correlation length is given by $\xi_b = 1/\sqrt{2BC}$.

At intermediate concentrations, where $\xi_b < \xi_{eff} < L$, we expect to find the semi-dilute regime. Scaling arguments(9) predict this regime corresponds to concentrations within the range $1/B^{1/3} \lesssim C \lesssim B$. For the case of $B = 10$ (open circles), the semi-dilute regime should occur when $0.4 \lesssim C \lesssim 10$, while for the case $B = 25$ (filled squares), it should occur if $0.3 \lesssim C \lesssim 25$. As can be seen from Fig. 4, both systems with $B = 10$ and $B = 25$ have a ξ_{eff} that is compatible with the $C^{-3/4}$ scaling law for the semi-dilute regime (cf. dashed lines) in the above ranges.

Conclusions

Although the results presented above are quite encouraging, one should mention that the CPU time required to obtain sensible averages for the density profile for concentrations in the semi-dilute regime is still relatively long. In our simple explicit integration scheme, the Langevin time step has to be kept small to avoid numerical instabilities. It seems likely that semi-implicit schemes might alleviate

288

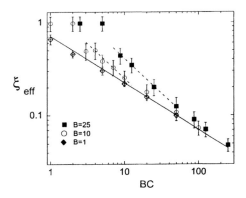

Figure 4: *Effective correlation length ξ_{eff} for $B = 1$ (filled diamonds), $B = 10$ (open circles), and $B = 25$ (filled squares) as a function of the parameter BC. The solid line denotes the Gaussian correlation length $\xi_b = 1/\sqrt{2BC}$. The dashed curves have a slope of -0.75 (reproduced with permission from reference (21). Copyright 2003 American Institute of Physics).*

this difficulty. Because the solution is not incompressible, the field has to generate local correlations that account for the monomer excluded volume, which tends to make the convergence process slower. On the other hand, as the overall density fluctuations were reduced by increasing C, the method presented here did converge quite rapidly, which is a promising feature since there exist a wide variety of important problems that involve concentrated solutions. In this respect, we expect that this method should prove to be very useful in the study of dense polymer phases and melts, which tend to be more difficult to treat with particle-based simulation methods.

In summary, we have presented a field theoretic simulation method to calculate in an essentially exact way the equilibrium properties of polymer solutions. As an example, we performed a simulation for a homopolymer solution under good solvent conditions confined to a slit. The density profiles and the correlation length were calculated for this particular case, and we showed that the simulations lead to the expected deviations from mean field solution. It is also possible to use this method to calculate other thermodynamic quantities like the chemical potential, the density-density correlations and the osmotic pressure.

This work was partially supported by the NSF through grant DMR-98-70785. Acknowledgment is also made to the donors of the Petroleum Research Fund, administered by the ACS. Extensive use of the UCSB-MRSEC Central Computing Facilities is also acknowledged. AAK would like to thank CONACYT for partial support through the UC Mexus program.

References

1 . Edwards, S. F. *Proc. Phys. Soc.* **1966**, *88*, 265.
2 . Freed, K. F. *Adv. Chem. Phys.* **1972**, *22*, 1.
3 . Doi, M.; Edwards, S. F. *The theory of polymer dynamics*; Clarendon Press: Oxford, 1986.
4 . Netz, R. R.; Andelman, D., to appear in *Phys. Reports.* **2003**.
5 . Moreira, A. G.; Netz, R. R. in *Electrostatic Effects in Soft Matter and Biophysics–Nato Science Series*, edited by Holm, C.; Kékicheff, P.; Podgornik R.; Kluwer: Dordrecht, 2001.
6 . Ganesan, V.; Fredrickson, G. H. *Europhys. Lett.* **2001**, *6*, 814.
7 . Fredrickson, G. H.; Ganesan, V.; Drolet, F. *Macromolecules* **2002**, *35*, 16.
8 . Fraaije, J. G. E. M. *J. Chem. Phys.* **1993**, *99*, 9202.
9 . Daoud M.; de Gennes P. G. *J. Physique (Paris)* **1977**, *38*, 85.
10 . Wang, Y.; Teraoka, I. *Macromolecules* **1997**, *30*, 8473.
11 . Wang, Y.; Teraoka, I. *Macromolecules* **2000**, *33*, 3478.
12 . Teraoka, I.; Cifra, P. *J. Chem. Phys.* **2001**, *115*, 11362.
13 . Yethiraj A.; Hall, C. K.*Macromolecules* **1990**, *23*, 1865.
14 . Milchev, A.; Binder, K. *J. Comput.-Aided Mater. Des.* **1995**, *2*, 167.
15 . Milchev, A.; Binder, K. *Macromolecules* **1996**, *29*, 343.
16 . Milchev, A.; Binder, K. *J. Phys. II (France)* **1996**, *6*, 21.
17 . Schoenmaker, W. *J. Phys. Rev. D* **1987**, *36*, 1859.
18 . Helfand, E. *J. Chem. Phys.* **1975**, *62*, 999.
19 . Gausterer, H. *Nucl. Phys. A* **1998**, *642*, 239c.
20 . Lee, S. *Nucl. Phys. B* **1994**, *413*, 827.
21 . Alexander-Katz, A.; Moreira, A. G.; Fredrickson, G. H. to appear in *J. Chem. Phys.* **2003**.
22 . de Gennes, P. G. *Scaling concepts in polymer physics*; Cornell University Press: Ithaca, 1979.

Chapter 18

Modeling Mechanical Properties of Resins Prepared by Sol–Gel Chemistry

Stelian Grigoras

Central R&D, Dow Corning Company, Midland, MI 48686
(email: s.grigoras@dowcorning.com)

This paper attempts to evaluate the elastic constants of nanoparticles, based on the analysis of thermal fluctuations from Molecular Dynamics atomistic simulations. These results are then used to evaluate mechanical properties, at mesoscale level, of the considered nanoparticles in a polymeric matrix using PALMYRA software. Particles with various shapes (spherical, disk-like, cigar-like) have been studied as nanocomposites in a polymeric matrix. The results indicate that these nanoparticles increase the modulus of the studied composites, within the limits of the predicted bounds (upper, lower) of the continuous models. The disk-like shape is the most efficient at raising the modulus. The same resins are evaluated as a matrix, containing pores, in low dielectric constant materials. The highest internal stress was shown in the case of the pores with disk-like shape.

Introduction

The electronic industry, which plans for ultra large scale integration, demands ILD materials prepared with standard methods developed by Sol-Gel chemistry; with lower dielectric constants, greater mechanical strength and greater chemical resistance to various processes involved in building the electronic chips. The development of metal-oxide-semiconductor technology was based, in the past, on the use of thermally grown silicon dioxide as the gate dielectric material. The current trend is to use spin-on resins that provide more versatility for building chips with smaller dimensions.

These products derived from Sol-Gel chemistry (called resins) can be characterized as ladders, cages, or highly branched macromolecules. After thermal cure these nanoparticles are stiffer than regular polymeric materials. One example, which gained high interest recently as an improved interlayer dielectric (ILD) material for electronic industry, is hydrogen silsesquioxane (HSQ) (*1-6*). The objective of this work is to:

1. Predict mechanical properties of various resin structures and indicate which type is most promising as the target for synthesis, or isolation.
2. Predict the properties of the resin nanoparticles as parts of a composite to indicate which type of shape and which special organization can provide the optimum performance for specific technological applications.

Computational Methods and Results

Atomistic simulations

The atomistic simulation method presented in this paper derives the elastic constants from the shape fluctuations, using Parrinello-Rahman (*7*) formula. The shape fluctuations are obtained from Molecular Dynamics (MD), or Monte Carlo atomistic simulations carried out on a single nanoparticle in vacuum at constant temperature. By selecting specific groups of atoms along a certain direction, the shape of the particle can be defined along a desired direction. This method was previously used (*8*) to estimate the elastic constants of carbon-based nanotubes. The results showed very good agreement with experimentally determined data.

A

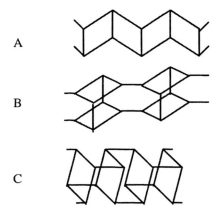

B

C

Figure 1: Ladder structures considered in this study

The types of ladder considered are shown in Figure 1. Each line represents a Si-O-Si moiety. MD calculations were carried out with a single structure consisting of 12 interconnected cyclic tetrasiloxane units, at 300°K using the Cerius2 software (9) with the Universal force field. The simulations were run for 5 ns, saving the trajectories at intervals of 2 ps. The trajectory files generated this way were then analyzed with software developed at ETH-Zurich (U.W. Suter group) to evaluate the fluctuations of distances between groups of atoms located at the ends of the principal axes of each type of ladder.

Table 1 shows the calculated values of the elastic constants at atomistic level for the ladders, as described in Figure 1, having hydrogen, or methyl as substituents at the Si atoms.

The results shown in Table I indicate that the type of ladder-structure plays an important role upon the stiffness of the resin nanoparticle. The type B ladder

Table I : Elastic constants along the main directions for ladder structures as described in Figure 1, at 300°K

Structure	E longitudinal (GPa) with H	E longitudinal (GPa) with Me	E transversal (GPa) with H
A	29.0	30.0	3.0
B	0.8	2.0	0.1
C	30.0	17.0	3.5

has the longitudinal elastic constant significantly lower than the other two types of ladder. The data indicate that structures of type A or C would be more suitable for stiff materials. The data also show that the transversal elastic constants are significantly lower that the longitudinal ones. Replacing the substituents at silicon from hydrogen to methyl would lower the longitudinal elastic constant for the structure assigned as type C. Overall the data indicate that the ladders type A or C would provide stiffer nanoparticles.

The calculated elastic constants were used to characterize materials consisting of the various ladder structures, which were then further studied as part of a composite material. The properties of the composite material, at a mesoscale level, have been evaluated with Palmyra (10).

Mesoscale calculations

The results obtained from atomistic simulations were used to evaluate mechanical properties of these nanostructures at mesoscale level in a polymeric matrix, using PALMYRA software (10). This method is based on Finite Element Analysis under periodic boundary conditions (11,12). In addition to the elastic constants, the program can evaluate the dielectric constant, thermal and electric conductivity, or diffusion coefficients. This paper will discuss two possible materials:

1. A typical polymeric matrix reinforced by one of the ladder nanostructures
2. A matrix consisting of a material derived from one of the ladder structure resins, which contains pores as inclusions.

In both cases the systems consists of matrix and inclusions. It is important to understand whether the shape of the inclusions has an impact upon the properties of the composite material.

For the first type of material (a typical polymeric matrix reinforced by the ladder nanostructures), three different inclusion shapes were considered, spherical, cigar-like and disk-like. The aspect ratios for the non-spherical shapes were 8 and 0.125 for the cigar like and disk-like shapes, respectively. In our models the orientation and the distribution of the inclusions was random as generated by a Monte Carlo algorithm. The size of the particles in each case was the same - monodiperse systems. Table II shows the results for the Young's moduli, shear and bulk moduli of systems containing 90% polymeric matrix and 10% by volume nanoparticles (inclusions).

Table II: Effect of the shape of the inclusions, as filler, in a polymeric matrix upon elastic properties

Shape	$E_{11}(GPa)$	$E_{33}(GPa)$	$G_{12}(GPa)$	$G_{13}(GPa)$	$K(GPa)$
Matrix-100%	0.20	0.20	0.091	0.091	0.167
Inclusion-100%	3.50	30.00	2.700	11.20	4.300
Sphere-10%	0.24	0.26	0.095	0.102	0.196
Disk-10%	0.26	0.26	0.098	0.102	0.201
Cigar-10%	0.24	0.25	0.096	0.096	0.197

Using the Hashin-Shtrikman model (13) upper and lower bounds for the Youngs modulii can be predicted for the continous system. The values for E_{11} are 0.23 and 0.32 GPa, and for E_{33} are 0.24 and 0.94 GPa.

The results indicate that the presence of 10% by volume inclusions can increase the modulus of the material by 20 to 30%. The disk shape is predicted to give the largest increase. A comparison of the simulated values with the intervals predicted by the continuum model shows that the simulations predict elastic properties towards the lower bound for the system.

The second type of material considered is one of the ladder-structure resins used as a thin-film interlayer dielectric material. In this type of application the resin is spin-coated on a silicon wafer, which is later heated at high temperatures. This process leads to a film that is described as a matrix in our approach. The exposure of the film to high temperature results in the formation of pores. Using Palmyra the matrix has been considered as an isotropic resin, with a value for the Young's modulus of 14.5 GPa, because we do not have knowledge about a preferential orientation of the resin particles in the film. The inclusions (pores) were assigned as vacuum. In this study, the volume loading of the pores was varied from 2 to 30%. The shape of the pores was considered to be either spherical, disk-like or cigar-like, with the same aspect ratios as in the previous case. They were either oriented, or distributed in regular manner.

The predicted dielectric constants (k) are shown in Figure 2. The model predicts that for low content of pores the values of k follow the upper bound of Voight-Reuss model, but at higher content of pores, the simulated value of k tends to have values below the upper limit. Our results suggest that the shape or the ordering of the pores does not significantly impact the dielectric constant.

Figure 3 shows the predicted values of the Young's modulus of the porous glass, where the content and the shape of the pores are varied. In several cases, the volume fraction of pores was maintained and the pores were ordered/oriented, in few cases the number of pores was varied. The plot shows that in general the simulated values follow the upper bound of the continuum model. A couple of results from the simulations are higher than the expected upper bound for Young modulus. Note that the comparisons are between calculated values, hence is not clear if this result should be attributed to modeling errors or not. In the case where the pores have disk-like shape the predicted modulus values are significantly lower.

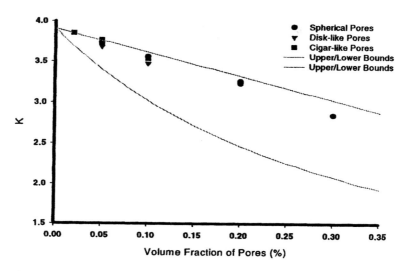

Figure 2: Simulated values for the dielectric constant of a porous resin

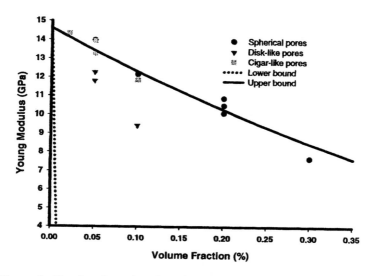

Figure 3: Simulated results of Modulus for porous material with different type of pores

We also predicted the highest values for von Mises stress in the system. While this is a local parameter, if comparisons are made between similar systems it may be considered as a rough estimation of the internal stress within the composite. The maximum values of the von Mises stress for the cases studied in this paper are plotted in Figure 4. The plot shows that the highest stress occurs for the pores that have disk-like shape. The stress becomes lower, by a factor of 3 to 5 in cigar-like pores and is very low for spherical pores.

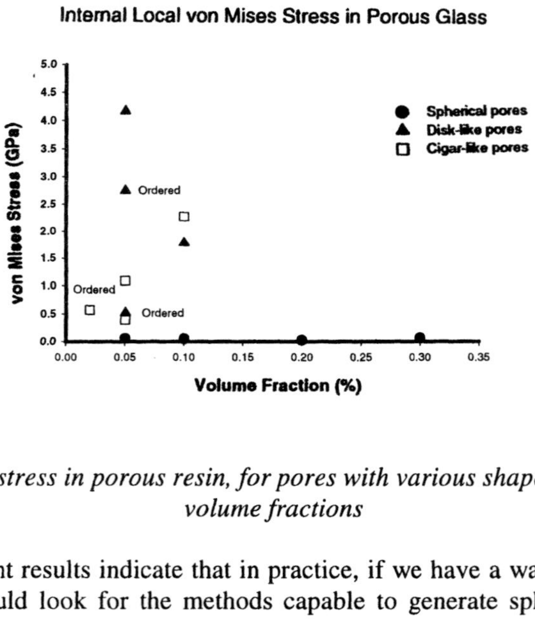

Figure 4: The stress in porous resin, for pores with various shapes, and different volume fractions

The current results indicate that in practice, if we have a way to control the pores, we should look for the methods capable to generate spherical pores in order to maintain low values for k, high modulus and low internal stress.

Conclusions

The paper shows that it is possible to calculate at atomistic level the elastic properties of nanoparticles which could lead the experiments to target the most suitable structure for synthesis, or isolation. The mesoscale approach was also used to evaluate the mechanical properties of porous glasses, as a function of the size of the pores and their fraction volume. Overall it is concluded that the chemical structure of the silica framework and the type of substituents, are the primary factors, in addition to the volume fraction of each phase, that control the properties of a heterogeneous system. It is envisioned that with current approaches we will be able to identify the architectures that would provide properties beyond the upper limits of the continuous models. Furthermore, complex models can be approached and developed by increasing the types of materials to be included in the matrix. The critical issue at this time is to gain the experimental capability to create the mesoscale architectures that can be predicted by the computational methods.

Acknowledgements

The author is very grateful to Professor U. W. Suter and his group at the Institute fur Polymere, in the Department of Materials at ETH-Zurich, for providing him the opportunity to use some of the most advanced methods available to study the properties of materials in his group.

References

1. Haluska, L.A.; Michael, K.W. ;Tarhay, L. US Patent 4756977, 1988.
2. Weiss, K.D.; Frye, C.L. US Patent 4999397, 1991
3. Bank, H.M.; Ciufentes, M.E. US Patent 5010159, 1991
4. Collins, W.T.; Frye, C.L. US Patent 3615272, 1972
5. Frye, C.L.; Collins, W.T. *J. Am. Chem. Soc.* **1970**, 92, 5586.
6. Agaskar, P.A.; Klemperer, W.G. *Inorg. Chimica Acta* **1995**, 119, 355
7. Parinello, M.; Rahman, A. *J. Chem. Phys.* **1982**, 76, 2662
8. Grigoras, S.; Gusev, A.A.; Santos, S., Suter, U.W. *Polymer* **2002**, 43, 489.
9. Cerius2, Accelrys, San Diego, CA.
10. PALMYRA, MatSim GmbH, Zurich, Switzerland, (www.matsim.ch)
11. Gusev, A.A. *J. Mech. Phys. Solids* **1997**, 45, 1449
12. Gusev, A.A. *Macromolecules* **2001**, 34, 3081.
13. Hashin, Z; Shtrikman, S. *J. Mech. Phys. Solids* **1962**, 10, 343.

Chapter 19

Grand Canonical Monte Carlo Simulations of Equilibrium Polymers and Networks

James T. Kindt

Department of Chemistry and Cherry L. Emerson Center for Scientific Computation, Emory University, Atlanta, GA 30322

Grand canonical Monte Carlo (GCMC) simulation is a useful technique for mesoscale modeling of systems of self-assembled aggregates – i.e., structures at equilibrium with each other and with a pool of constituent monomers. A new biased GCMC algorithm has been developed to efficiently handle the growth and equilibration of polydisperse structures in phenomenological simulations. Its efficiency is demonstrated in a preliminary study of the nematic ordering transition in equilibrium polymers. The adaptability and precision of the GCMC approach is further shown in simulations of self-assembled networks, in which loop formation is demonstrated to suppress the first-order condensation transition for flexible chains, an effect not anticipated by mean-field theory.

The Metropolis Monte Carlo algorithm (*1,2*) has been widely used for half a century to simulate systems at thermal equilibrium with an external fixed-temperature bath. The extension of the Metropolis algorithm to the grand canonical ensemble, in which the system's *composition* reflects an equilibrium with a reservoir of particles at fixed chemical potential μ (in addition to thermal equilibrium) was introduced in 1969 by Norman and Filinov (*3*) and also has become commonplace. Grand canonical Monte Carlo (GCMC) is a useful tool for the understanding of the thermodynamics and mesoscale structure of self-assembled systems, particularly those in which thermal fluctuations and aggregate polydispersity play an important role. It has been used extensively in lattice-based simulation, where it can be implemented very efficiently (*4-7*). Off-lattice models, however, allow a more realistic representation of curvature elasticity and symmetry-breaking transitions. GCMC, and the related Gibbs ensemble Monte Carlo method, are also effective in applications using atomically detailed force-fields (see chapters by Siepmann and McCormick in the present volume); nevertheless, in simulations of *phases* of self-assembled aggregates on the scale of 100's of nanometers, atomistic simulation is far from practical. This chapter will specifically address off-lattice coarse-grained simulation models.

An advantage of the grand ensemble is that the automatic generation of the density as a function of μ allows easy identification of first-order phase transitions and allows contact to be made with approximate analytic expressions for the free energy. In terms of efficiency, particle exchanges between the system and an infinite reservoir of particles get around the dynamic bottleneck of monomer diffusion between aggregates during equilibration of the aggregate size distribution. On the other hand, like other Monte Carlo moves designed to circumvent rather than reproduce true dynamics, use of this type of move renders essentially impossible the recovery of even qualitative dynamical information from the results.

One of the most significant advantage of Monte Carlo methods in phenomenological modeling is the degree of control they afford over the local structure and energetics of self-assembly. Most dynamics-based methods require smoothly varying potentials, which are difficult to develop for a particular morphological outcome (e.g., aggregation into chains, sheets, ribbons, helices, networks, etc.), and even more so for independently adjustable association constants and elastic constants of the structure. In contrast, non-differentiable (square-well or delta-function) potentials, which are conveniently and efficiently implemented in biased MC simulations, allow association constants and structural properties to be precisely engineered into the potential. This approach might be criticized for being a "top-down" method, in which the simulation results can only confirm presuppositions about the system behavior. It is perhaps better characterized as a "middle-up" approach: it does not address the question of how molecular structure leads to preferred local geometries, free energies of association, and elastic constants, but treats these as inputs to determine the next higher level of structure. (These inputs may come from experiment, from microscopic theory or simulation, or can be treated as a parameters to vary in order to determine whether and how they influence mesoscale structure.) As such, it can play a useful role in a multi-scale approach to understanding and predicting mesoscale phenomena.

The remainder of this chapter will describe a new and general "polydisperse insertion" (PDI) algorithm for the efficient simulation of systems with a polydisperse distribution of aggregates, followed by examples of the application of GCMC methods to two related problems in self-assembly: the isotropic-nematic transition in semiflexible equilibrium polymer systems, and the gas-liquid transition in self-assembled networks with 3-fold junctions.

A GCMC Method for Polydisperse, Self-Assembled Systems

GCMC on self-assembled systems can quite naturally be performed through directed addition or removal moves that change the size of an aggregate by one (coarse-grained) monomer at a time. There are several motivations, however, for attempting to improve on this approach by performing compound move in which multiple monomers or entire aggregates are added or removed at once. The first is that in a system with average number M of monomers per aggregate, the number of moves necessary to grow or remove an aggregate is of order $(M/a)^2$ with a the number of particles used in each step. Equilibration of the system composition – the number and size distribution of aggregates – becomes slow for systems with large M, but can be improved with an increased step size. Another motivation is that there may be barriers to aggregate growth, as observed in some surfactant systems where spherical micelles coexist with long wormlike micelles, but short cylindrical micelles are nearly absent. (8) For such systems, the ability to bypass unfavorable aggregate growth stages through direct addition or removal of large aggregates will improve equilibration efficiency. Multiple particle insertions are most likely to be useful for models that lack explicit solvent and for semidilute systems, where the influence of non-bonded interactions is important on the level of entire aggregates but weak on level of individual monomers.

The configuration-bias Monte Carlo (CBMC) approach of Siepmann and Frenkel (9) has been very successful for the insertion of chain molecules, and even branched structures and rings, (10) into crowded environments. A key step of the CBMC method is the generation of multiple possible configurations, from which one is chosen with a Boltzmann-weighted probability; the move's overall acceptance probability is determined by a Rosenbluth weight W (11) which incorporates the weights of all configurations considered. In the "polydisperse insertion" (PDI) method introduced here, a similar process of generation and selection is used; the configurations used are the sub-aggregates generated in the course of a directed sequence of aggregate growth or disassembly moves, and so contain different numbers of monomers.

Several algorithms have been developed for the reversible growth of clusters of associating particles through biased Monte Carlo moves. (12-14) For generality, we assume that the aggregate is defined by some connectivity criterion based on each monomer occupying a position within a volume V_{bond} defined with respect to one or more other monomer positions and orientations.

A given aggregate contains N_{free} unoccupied bonding regions to which monomer addition moves can be directed, and N_{rem} monomers that can be removed without affecting the integrity of the aggregate. (To preserve reversibility, no PDI move that splits or merges aggregates will be permitted.) For instance, if the aggregate is constrained to be an unbranched chain that can grow from either end, $N_{\text{free}} = N_{\text{rem}} = 2$ for any n-mer with $n>1$.

To show that the use of biased, multi-particle moves will satisfy detailed balance, it is sufficient to show that the resulting ratio of probabilities of adding and removing an n-mer equals the ratio of the probabilities of adding or removing the same structure through n unbiased single particle moves. The probability of building an n-mer in n successive unbiased additions depends on the product of the probability $\alpha_{\text{gen}}{}^{n}$ of attempting n such moves in succession, the probabilities $N_{\text{free}}V_{\text{bond}}/V$ that the n-1 moves after the first will be added to the growth sites of the aggregate, and the standard single-monomer GCMC acceptance probabilities, (2) $\min[1, V\Lambda^{-3}(N+1)^{-1}\exp(\beta\mu-\beta\Delta U)]$, of each move:

$$\pi(N_n \rightarrow N_n + 1, n \text{ steps}) = \alpha_{\text{gen}}^{n} \prod_{i=1}^{n-1} \frac{N_{\text{free},i}V_{\text{bond}}}{V} \prod_{i=1}^{n} \min\left[1, \frac{V\exp\left(\beta\mu - \beta(U_{i\text{-mer}} - U_{(i-1)\text{-mer}})\right)}{\Lambda^3(N+i)}\right] \quad (1)$$

The probability of removing the aggregate through a sequence of n monomer removal moves, preserving the aggregate by removing one of the N_{rem} removable monomers on the same one of N_n n-mers at each step until the last, is likewise:

$$\pi(N_n + 1 \rightarrow N_n, n \text{ steps}) = \alpha_{\text{gen}}^{n}(N_n + 1)\prod_{i=1}^{n} \frac{N_{\text{rem}}}{N+i} \min\left[1, \frac{\Lambda^3(N+i)}{V\exp\left(\beta\mu - \beta(U_{i\text{-mer}} - U_{(i-1)\text{-mer}})\right)}\right] \quad (2)$$

If we define a weighting function ω_n for an n-mer as,

$$\omega_n \equiv \frac{V\exp(n\beta\mu)\exp(-\beta U_{n\text{-mer}})}{\Lambda^{3n}} \prod_{i=1}^{N-1} \frac{N_{\text{free},i}V_{\text{bond}}}{N_{\text{rem},i+1}} \quad (3)$$

then the ratio of these transition probabilities may be written

$$\frac{\pi(N_n \rightarrow N_n + 1)}{\pi(N_n + 1 \rightarrow N_n)} = \frac{\omega_n}{N_n + 1} \quad (4)$$

In the PDI insertion move, which illustrated as a cartoon in Figure 1, a trial aggregate of n_{max} monomers is first generated by the insertion of a single monomer followed by biased single-monomer additions. Out of these n_{max} intermediates, an aggregate of size n is selected with probability proportional to ω_n, and the overall move is accepted with a probability that depends on the Rosenbluth-like weight W, the sum of the weights of all the intermediates:

$$acc\left(N_{ag}, N_n \to N_{ag}+1, N_n+1\right) = \min\left(1, \left(N_{ag}+1\right)^{-1}\sum_{i=1}^{n_{max}}\omega_i\right) = \min\left(1, \left(N_{ag}+1\right)^{-1}W\right) \tag{5}$$

The transition probability from a system containing N_{ag} aggregates, of which N_n are n-mers, to one containing an additional n-mer is given by the product of the probabilities of attempting a PDI insertion move ($\alpha_{PDI}/2$), of selecting the n-mer from a range of aggregate sizes for insertion, and the acceptance probability for the insertion:

$$\pi(N_{ag}, N_n \to N_{ag}+1, N_n+1) = \frac{\alpha_{PDI}}{2}\frac{\omega_n}{W}\min\left(1, \left(N_{ag}+1\right)^{-1}W\right) \tag{6}$$

If the acceptance probability for the PDI removal of an n-mer is set to $\min[1, (N_{ag}+1)W^{1}]$, then the transition probability for taking away an n-mer in the reverse move, in which one out of the N_{ag} aggregates in the system is randomly selected for a removal, becomes

$$\pi(N_{ag}+1, N_n+1 \to N_{ag}, N_n) = \frac{\alpha_{PDI}}{2}\frac{N_n+1}{N_{ag}+1}\min\left(1, \left(N_{ag}+1\right)W^{-1}\right) \tag{7}$$

and the ratio between eqs 6 and 7 is consistent with eq 4, demonstrating that the PDI moves yield the same probability distribution as the unbiased single-particle GCMC algorithm. Because the Rosenbluth weight W appears in the acceptance probability of the aggregate removal move, "dummy" addition moves to the aggregate must be performed solely to obtain values for ω_{n+1} through ω_{nmax}. By "super-detailed balance", (2) the values of W used in aggregate addition and removal moves will have the same statistical distribution and therefore converge to the proper probability distribution. For both aggregate addition and removal moves, the growth of the aggregate can be truncated at the first instance of an excluded-volume overlap before n_{max} is reached, as the weights ω_i beyond such an overlap will be zero.

The third type of move within PDI, resizing an existing n-mer, can be performed independently or as a fallback for a failed PDI aggregate removal. In the former case, an attempt is made to grow the aggregate to size n_{max} and the weights for all aggregates, from ω_1 through ω_{nmax}, must be calculated by eq 3; if resizing is coupled to removal moves, the necessary coordinates and weights are already available. Either way, a new size n' is selected for the aggregate with a probability given by $\omega_{n'}/(W-\omega_n)$ and accepted with a probability:

$$acc(N_n, N_{n'} \to N_n-1, N_{n'}+1) = \min\left(1, \frac{W-\omega_n}{W-\omega_{n'}}\right) \tag{8}$$

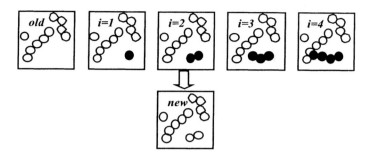

Figure 1: Schematic illustration of polydisperse insertion Monte Carlo step

to give the following ratio of transition probabilities for the $n \rightarrow n'$ resizing and its reverse, when the probabilities of choosing an n-mer or n'-mer for resizing are included:

$$\frac{\pi_{n \rightarrow n'}}{\pi_{n' \rightarrow n}} = \frac{N_n N_{ag}^{-1} \omega_{n'} (W - \omega_n)^{-1}}{(N_{n'} + 1) N_{ag}^{-1} \omega_n (W - \omega_{n'})^{-1}} \left(\frac{W - \omega_n}{W - \omega_{n'}} \right) = \left(\frac{N_n}{\omega_n} \right) \left(\frac{\omega_{n'}}{N_{n'} + 1} \right) \tag{9}$$

As the resizing move is equivalent to the coupled removal of an n-mer and addition of an n'-mer, eq 9 is consistent with eq 4. We note that attempting the removal move only upon the failure of the removal move does not break the symmetry of the Markov chain. Resizing from a given n-mer to a given n'-mer will yield the same distribution of W, as both cases will sample the same subset of "dummy" aggregates containing the larger of n' and n. The removal move success only depends on W and N_{ag}, so by super-detailed balance the probability of reaching the re-sizing move is the same for forward and reverse directions.

Compared with using fixed-n aggregate insertions and removals, the advantage of the PDI approach is that it is multiplexed: all size aggregates are considered in each insertion step with little additional cost. Comparing the efficiency of the method to biased addition an removal of single monomers, the creation or elimination of an n-mer (or a change in aggregate size by n particles) could be achieved in one step (with up to n_{max} substeps) as opposed to an average of order n^2 single-monomer steps in a random walk. With an appropriate choice of n_{max} this can represent an efficiency improvement by a factor on the order of the average aggregate size, in the most favorable cases where single-particle move acceptance rates would be very high. In general, the effective step length will be limited to mean number of monomers that can be added to an aggregate before overlap occurs. For very densely packed systems, where the success probability of success for compound insertion steps is very low, the algorithm's performance reduces to that of a single-particle method.

Application: Semiflexible Equilibrium Polymers

In this section the application of the PDI method to a simple system is described. The simplest mode of self-assembly is equilibrium polymerization, the reversible assembly of worm-like or chain-like structures. Many researchers have simulated systems of monomers that exhibit self-assembly into chains, with the earliest reports by Jackson et al. (15) dating back fifteen years. Recently, extensive MC simulations of flexible lattice and off-lattice equilibrium polymer models have been reported by Wittmer, Milchev and co-workers. (16) Lattice-based simulations with *semiflexible* bending potentials have been reported by Milchev and Landau (17) and by Rouault (18); restriction to the lattice appears to introduce significant artifacts in the nematic orientationally ordered phase formed due to the packing effects of stiff chains. A GCMC simulation study of a three-dimensional off-lattice semiflexible equilibrium polymer model was reported by Chatterji and Pandit (19); in that study, nematic ordering was obtained by reducing the temperature at fixed chemical potential, effectively changing K_{assoc}, persistence length, and density at once. While this may realistically represent a thermotropic transition, it complicates comparison with analytic theory. Using constant NVT molecular dynamics, Fodi and Hentschke (20) have studied the properties of isotropic, nematic, and hexagonal phases of an off-lattice semiflexible equilibrium polymer system. Their work yields many of the same insights as the results obtained in the present study, but in two respects is limited by the need to use smooth potentials. The first is that at a given temperature chains of different flexibility have different association constants; the second is that the intrinsic persistence length of the polymer was varied by less than a factor of five, presumably to avoid the high forces and torques that accompany a very stiff bending potential.

A phenomenological approach to equilibrium polymer self-assembly requires the definition of a minimum of three quantities: the polymer diameter, the free energy of breaking the polymer (or equivalently, the "end-cap" energy of a wormlike micelle) and the polymer persistence length. In the present work, we have attempted to construct a simulation model that incorporates these parameters without any complicating factors. All three quantities can be straightforwardly engineered into a hard-sphere association model, according to the following prescriptions:

- Diameter : hard sphere diameter σ
- End-cap free energy ε : $K_{assoc} = \exp(2\,\beta\varepsilon)$
- Intrinsic persistence length l_p : $l_p \approx 2\sigma/(1+\cos\theta_{min})$

Associating fluid models with angle- and bondlength-restricted square-well potentials, of the type used in early simulations (15) allow some freedom in choosing these properties. The model employed in the present studies represents, in a way, a limit of these square-well potentials: the bonding region for chain formation is defined to be a section of the spherical shell of infinitesimal thickness, defined by a diameter equal to the hard-sphere diameter (σ) consistent with the formation of a bond angle greater than some cutoff θ_{min}. When directed addition MC moves are used to form bonds, the restriction of the bond length distribution to a δ function is useful, as it removes a degree of freedom that is uninteresting (for chains not under tension or shear stress) and which would otherwise require some computational effort to sample. The depth U_{bond} of this potential well need not be explicitly defined; to calculate the acceptance probability for a biased addition to a chain end, one need only specify the product $V_{bond}\exp(+\beta U_{bond})$ which can be identified with K_{assoc}, the equilibrium association constant for the addition of a monomer to the end of a chain. (15) The cut-off bond angle θ_{min} effectively defines the bending potential of any three consecutive spheres in a chain to have an infinite square-well form (0 for $\theta \geq \theta_{min}$, otherwise infinite) from which the persistence length (by construction, independent of temperature) can be derived as shown above, approaching an exact result in the stiff-chain limit (as $\theta_{min} \to \pi$).

The mean-field prediction for the aggregate length distribution is independent of persistence length for a given value of K_{assoc}; this exact distribution will be recovered through this simulation model when excluded volume interactions are neglected. In the full simulation, with excluded volume, differences between results obtained for systems of equal K_{assoc} and different θ_{min} are entirely due to flexibility-dependent correlation effects.

Only two types of MC move were used: PDI insertion and PDI removal/resizing, as described in the previous section. For both types of move, n_{max} was chosen to be at least 2000, several times greater than the maximum chainlength observed in any of the simulations.

Results: semiflexible equilibrium polymers

Simulations were performed at a single value of association constant, K_{assoc} = 5000 σ^3, at four different intrinsic persistence lengths (4, 10, 100, and 1000 σ) and a range of chemical potentials. The cubic simulation box used was at least 40 σ to a side, holding up to ~20,000 monomers at the highest densities investigated. At low monomer concentration $\varphi < 0.1$, the concentration dependence of the mean chain-length M (Figure 2a) is very close to the ideal prediction (solid curve, $M = \frac{1}{2} + \sqrt{\frac{1}{4} + K_{assoc}\varphi}$). Deviation from the ideal distribution at higher concentrations for the two more flexible polymer systems is consistent with systematic simulation studies of flexible equilibrium polymers by Milchev et al. (16) Even more striking deviation is observed for the more rigid models. The jumps in mean chainlength M correspond to first-order phase transitions, indicated by the plateaus in the graph of chemical potential μ versus concentration φ in Figure 2b, between the low-concentration isotropic phase and a higher-concentration nematic (orientationally ordered) phase. The transition is coupled to a marked increase in mean chain length, as predicted by theory (21) and also observed in prior simulations (17-20). This increase has been explained by the decrease in rotational entropy loss associated with chain-end fusion; aligned chains are already oriented to favor attachment to another. Examples of coexisting isotropic and nematic phase configurations are shown in Figure 3.

The PDI algorithm yielded a significant performance improvement over the use of single-particle insertion and removal moves. As a simple test, the simulation was initiated with an empty box with an edge length of 60 σ, and a chemical potential chosen to reach a mean number of 17,000 particles in the box at equilibrium, assembled in an isotropic phase of chains of mean length 21.5 σ with an intrinsic persistence length of 100 σ. The CPU time required for the simulation to first reach the mean equilibrium system density was 1 minute with the PDI method versus 15 minutes with the single-particle insertion/removal method. In a more crowded system (N/V = 0.20), following a perturbation that caused the equilibrium mean chainlength to drop from 38 to 24 at roughly constant density, the relaxation time by the PDI method was approximately four times faster than by the single-particle method.

Self-Assembled Network

As mentioned above, the statistical thermodynamics of equilibrium polymerization – reversible self-assembly of rod-like, worm-like, or chain-like aggregates as exemplified by "giant micelles"– has been studied in depth by analytical and computational means. Recently, experimental observations, (22) that branchpoints or 3-fold junctions play an important role in structures of microemulsion and micellar systems have spurred theoretical study of networks of self-assembled chains. (Flory (23) first considered structural aspects of branched polymer networks, and structural and thermodynamic features of reversible gelation of *fixed-length* polymers (24) have been a very active topic for a number of years.) The "equilibrium network" problem was first addressed by Drye and Cates (25), who considered the limit of high endcap energy, in which all micelles end in junctions, with no dangling ends. They predicted a phase coexistence (in the case of flexible but not rigid micelles) between a network and excess solvent. Safran and co-workers (26-28) have predicted a re-entrant first-order phase transition, below a critical temperature, from a dilute phase rich in chain ends to a more concentrated network phase; or, under certain circumstances, a coexistence between networks of different density. With clear theoretical predictions, a computational model was devised to check the mean-field theory prediction of a re-entrant phase transition, and to determine the effects of chain persistence length on the system's behavior.

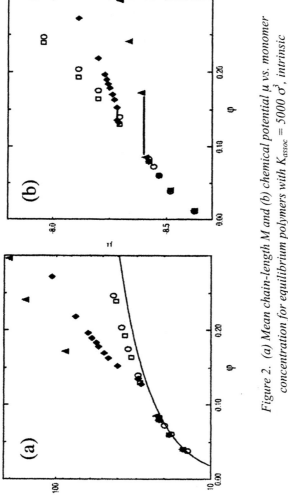

Figure 2. (a) Mean chain-length M and (b) chemical potential μ vs. monomer concentration for equilibrium polymers with $K_{assoc} = 5000 \, \sigma^3$, intrinsic persistence lengths $l_p = 4$ (circles), 10 (squares), 100 (diamonds), and 1000 (triangles) in units of hard-sphere diameter σ. Solid curve in (a) is mean-field result; horizontal lines in (b) are tie lines between isotropic and nematic phases.

308

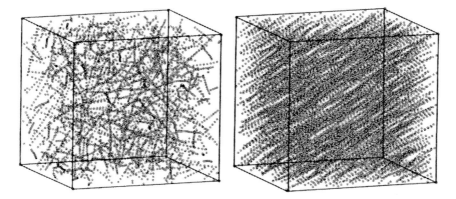

Figure 3. Simulation snapshots of isotropic (left) and nematic (right) phases of the model equilibrium polymer at coexistence, $l_p = 1000$ σ.

Self-Assembled Network: Methods

Two free energies define the properties of the mean-field equilibrium network model. One of these, the chain scission/extension energy, is the same as in the simple equilibrium polymer system. The other is the binding energy at the three-fold junction. A simulation study of the effect of chain flexibility on the system's structure and phase diagram requires a model in which these two energies can be set independently of one another and of the chain persistence length. For treatment of chain polymerization, we used the equilibrium polymer model described above. The additional feature of the network-forming system model is that a third particle can bind to any interior monomer of any chain, transforming the interior monomer into a 3-fold junction site. Junctions must be separated by at least one interior monomer. The third bond at a junction has a lower equilibrium constant for binding than the bonds within a single chain. It is also restricted to a length of 1.0 σ, but the third particle attached at a junction site can form any angle (consistent with non-overlapping hard spheres) with respect to the other particles attached to that same junction.

To sample the ensemble of configurations corresponding to a given temperature and chemical potential, the method employs five types of Monte Carlo move. Ten percent of move attempts are split between addition or removal of free monomers, using the standard GCMC algorithm for a hard-sphere system. Of the remainder, one-third are attempted chain-extension moves – directed insertions to an existing monomer or chain end. Whenever these moves result in an overlap with an existing particle, instead of being rejected automatically, they are converted into attempts to form a junction with the existing particle. The remaining moves are divided equally between attempted removal of particles from the free ends of chains, and the attempted removal of particles attached to junctions. Acceptance probabilities designed to satisfy the requirements of detailed balance for all of these moves have been published. (29) This work predated the development of the PDI method, which we predict would have improved the simulation efficiency by enabling rapid growth and regrowth of chain segments between junctions.

Self-Assembled Network: Results

A GCMC simulation of the model equilibrium network-forming fluid described above has been performed to test the mean-field predictions and to investigate effects of persistence length. A model with persistence length of 10 particle diameters gives a clear indication of a first-order transition through a plateau in the concentration/chemical potential isotherm, which decreased in width as temperature was lowered. (Figure 4, left) Qualitatively, the phase behavior matches the predictions of Tlusty and Safran. (26) In contrast, a model with a persistence length of 4 – in which the association constants for chain and junction formation were both matched to those of the $l_p=10$ system – showed a continuous transition to a network with no first-order phase transition (Figure 4, right). This is apparently due to the formation of intra-chain loops (as evident in the simulation snapshot shown in Figure 5) which reduce the driving force for the chains to join into a network by lowering the number of free chain ends. Interestingly, the observation that increasing flexibility suppresses the phase transition runs opposite the predictions of Drye and Cates (25), which were based on how flexibility impacts excluded volume considerations.

Summary

Monte Carlo simulation in the grand canonical ensemble can be a powerful method for studying mesoscale structure in self-assembled systems. The design of models and algorithms tailored to a particular mode of self-assembly facilitates the simulation of a system with a precise set of local aggregate properties (e.g., end-cap or edge free energy; persistence length or other elastic constants) to study the structural and thermodynamic properties of an ensemble of interacting aggregates. A general algorithm, based on the configuration-bias approach, is proposed for the insertion, removal, and re-sizing of aggregates through coupled multiple insertion and removal moves. This "polydisperse insertion" (PDI) approach is efficient because it allows growth or removal of an n-mer in a MC move with computational cost of order n rather than a random walk of order n^2 monomer insertion and removal moves. The PDI method was

310

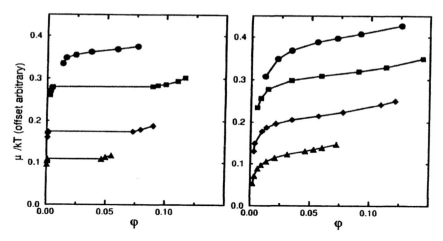

Figure 4: Phase behavior of equilibrium network model from GCMC simulation. Left graph, $l_p = 10$; right graph, $l_p = 4$. Curves represent isotherms of chemical potential vs. concentration, with temperatures increasing with higher placement on the graph. (Reproduced from reference 29. Copyright 2002 American Chemical Society.)

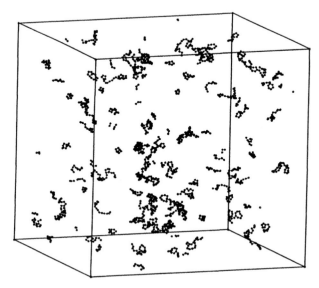

Figure 5. Simulation snapshot of self-assembled network-forming system at a concentration of $\varphi = 0.0028$, showing a high proportion of intrachain loops.

applied to a model of semiflexible equilibrium polymers, in which the coexistence region between isotropic and nematic phases is being mapped out. Further simulations, more detailed analysis, and close comparison to theoretical predictions and experimental results will be carried out for the simple equilibrium polymer case. The effect of chain flexibility on the phase diagram of self-assembled networks with threefold junctions has been addressed; when chains become sufficiently long and flexible, loop formation can suppress the first-order condensation predicted by mean-field theory.

Acknowledgments

Acknowledgment is made to the Cherry L. Emerson Center of Emory University, which is supported in part by a National Science Foundation grant (CHE-0079627) and IBM Shared University Research Award, for the use of its resources; and to the donors of The Petroleum Research Fund, administered by the ACS, for partial support of this research. The author is a Camille and Henry Dreyfus New Faculty Awardee.

References

1. Metropolis, N.; Ulam, S. *J. Am. Stat. Assoc.* **1949**, *44*, 335-41.
2 For further background on Monte Carlo algorithms, see Frenkel, D.; Smit, B. *Understanding Molecular Simulation*; Academic Press: San Diego, 1996.
3. Norman, G. E.; Filinov, V. S. *High Temp.* **1969**, *7*, 216-22.
4. Bohbot, Y.; Ben-Shaul, A.; Granek, R.; Gelbart, W. M. *J. Chem. Phys.* **1995**, *103*, 8764-82.
5. Jennings, D. E.; Kuznetsov, Y. A.; Timoshenko, E. G.; Dawson, K. A. **2000**, *112*, 7711-22.
6. Kim, S. H.; Jo, W. H. *Macromolecules* **2001**, *34*, 7210-8
7. Kim, S. Y.; Panagiotopoulos, A. Z.; Floriano, M. A. *Mol. Phys.* **2002**, *100*, 2213-20.
8. Bernheim-Groswasser, A.; Zana, R.; Talmon, Y. *J. Phys. Chem. B* **2000**, *104*, 4005-9.
9. Siepmann, J. I.; Frenkel, D. *Mol. Phys.* **1992**, *75*, 59-70.
10. Wick, C D.; Siepmann, J. I. *Macromolecules* **2000**, *33*, 7207-7218.
11. Rosenbluth, M. N.; Rosenbluth, A. W. *J. Chem. Phys.* **1955**, *23*, 356-359.
12. Busch, N. A.; Wertheim, M. S; Yarmush, M. L. *J. Chem. Phys.* **1996**, *104*, 3962-3975.
13. Wierzchowski, S.; Kofke, D. A. *J. Chem. Phys.* **2001**, *114*, 8752-8762.
14. Chen, B.; Siepmann, J. I. *J. Phys. Chem. B* **2000**, *104*, 8725-8734.
15. Jackson, G.; Chapman, W. G.; Gubbins, K. E. *Mol. Phys.* **1988**, *65*, 1-31.

16. Milchev, A.; Wittmer, J. P.; van der Schoot, P; Landau, D. *Europhys. Lett.* **2001**, *54*, 58-64.; Wittmer, J. P.; van der Schoot, P.; Milchev, A.; Barrat, J. L. *J. Chem. Phys.* **2000**, *113*, 6992-7005; Wittmer, J. P.; Milchev, A.; Cates, M. E. *J. Chem. Phys.* **1998**, *109*, 834-845.
17. Milchev, A; Landau, D. P. *Phys. Rev. E*, **1995**, *52*, 6431-6441.
18. Rouault, Y. *Eur. Phys. J. B* **1998**, *6*, 75-81.
19. Chatterji, A; Pandit, R. *Europhys. Lett.* **2001**, *54*, 213-219.
20. Fodi B.; Hentschke R. *J. Chem. Phys.* **2000**, *112*, 6917-6924.
21. van der Schoot, P.; Cates, M. E. *Europhys. Lett.* **1994**, *25*, 515-520.
22. Bernheim-Groswasser, A.; Wachtel, E.; Talmon, Y. *Langmuir* **2000**, *16*, 4131-40.
23. Flory, P. J.. *Principles of Polymer Chemistry*; Cornell University Press: Ithaca, 1953.
24. Rubinstein, M.; Dobrynin, A. V. *Curr. Opin. Colloid Interf. Sci.* **1999**, *4*, 83-87.
25. Drye, T. J.; Cates, M. E. *J. Chem. Phys.* **1992**, *96*, 1367-1375.
26. Tlusty, T.; Safran, S. A.; Strey, R. *Phys. Rev. Lett.* **2000**, *84*, 1244-1247.
27. Tlusty, T.; Safran, S. A. *Science* **2000**, *290*, 1328-1331.
28. Zilman, A.; Safran, S. A. *Phys. Rev. E* **2002**, *66*, 051107.
29. Kindt, J. T. *J. Phys. Chem. B* **2002**, *106*, 8223-8232.

Chapter 20

Study of the Effects of Added Salts on Micellization of Cetyltrimethylammonium Bromide Surfactant

B. Lin, S. Mohanty[1,2], A. V. McCormick[1,*], and H. T. Davis[1]

[1]Department of Chemical Engineering and Materials Science, University of Minnesota, Minnesota, Minneapolis, MN 55455
[2]Current address: 3M Corporate Headquarters, 3M Center, I-94 at McKnight Road, , St. Paul, MN 55144–1000

Sodium bromide (NaBr) or sodium salicylate (NaSal) can be added to aqueous solutions of cetyltrimethylammonium bromide (CTAB) to convert spherical micelles to rod-like (cylindrical) micelles that are believed to be necessary for the surfactant to be an effective drag-reducing agent. The effects of the added salt (inorganic or aromatic) on the micellization behavior have been tested by several available molecular-thermodynamic models. Models proposed both by Blankschtein and coworkers and by Nagarajan and coworkers show ability to predict the effects of nonadsorbing salts such as NaBr on the micellization of CTAB solutions. However, their application is limited when the additive is an organic salt (e.g., NaSal) providing that ions penetrate into the surfactant (e.g., CTAB) micelles. A model that combines the molecular thermodynamic approach with energetic and structural data obtained from atomistic scale Monte Carlo simulations of the micelle shell, has been developed (27) that predicts the micellization behavior of pure CTAB aqueous solutions and also performs well for the system with penetrating organic ions. The Monte Carlo simulation data from the complementary model show that intramicellar molecular ordering is crucial to the estimation of free energies and that ordering is affected by the curvature of the micelles. Finally, it was shown that the molecular ordering of the salicylate ions causes the deviation from the ideal mixing of the CTA^+ and Sal^- ions, especially for cylindrical micelles.

Introduction

Micellar solution phenomena (self-association of monomers, aggregate shape, growth of micellar aggregates, etc.) involve a delicate balance of various forces, including van der Waals, electrostatic, steric, hydration, and hydrophobic forces. The theoretical formulation of micellization generally incorporates the effect of surfactant molecular structures and solution conditions on the free energy associated with the hydrophobic attractions of the alkyl tails within the micellar core, the steric/electrostatic repulsions of the hydrophilic heads at the micellar core-water interface, and interfacial curvature effects and the conformational limitations due to alkyl-chain packing (1-5).

Solvent quality and other solution conditions strongly influence surfactant solution phenomena by changing the surfactant hydrophilicity, which can result in the variation of the amphiphilic characteristics and associated surface activity. Addition of certain electrolytes, particularly aromatic salts, into some cationic detergents can transform spherical micelles into lamellar or rod-like micelles. For example, the addition of sodium salicylate (NaSal) to a spherical micellar solution of cetyltrimethylammonium bromide (CTAB) causes a transformation to a rod-like (also known as wormlike or cylindrical) micellar solution, which can be used as an effective drag reducing agent (6-9). Quaternary ammonium surfactants coupled with several additives have been investigated experimentally for drag reduction (10-14). However, little understanding of the additive effects at a fundamental and molecular level has been achieved despite the commercial importance of these surfactant-additive mixtures.

There are several molecular thermodynamic theories available to predict the micellar properties and facilitate the molecular interpretation of these solution properties. The approach used by both Blankschtein (15-20) and Nagarajan (21-25) is based on detailed molecular models for the free energy of micellization with the various molecular contributions computed on the basis of surfactant architectures and solution conditions. For example, what we will refer to as the Blankschtein model (15-20) visualizes micellization as a series of reversible steps, each associated with a well-defined physico-chemical factor. The free energy of micellization is made up of the following components:

$$g_{mic} = g_{s/w} + g_{w/hc} + g_{hc/mic} + g_{int} + g_{st} + g_{elec},\qquad(1)$$

where $g_{s/w}$ is the free energy for transfer of the tail from salt solution to water (applicable only when the amphiphile is in the ionic electrolyte solution), $g_{w/hc}$ is the free energy for transfer of the tail from water to a hydrocarbon environment, $g_{hc/mic}$ is the free energy associated with packing the tail within the hydrocarbon core of the aggregate, g_{int} is the free energy associated with the surface energy, g_{st} is any steric interaction between headgroups (and accounts for the loss in free energy in transferring the headgroups from an electrolyte solution to a micelle shell (if applicable) and subsequent grafting at the micellar core interface), and g_{elec} accounts for the electrostatic contribution (Coulombic or dipolar) of the charged layer outside the micelle (applicable for ionic/zwitterionic surfactants).

Similarly, what we will refer to as the Nagarajan model (*21-25*) explicitly includes contributions reflecting the hydrophilic head and hydrophobic tail structures of the surfactant, and therefore expresses the free energy of micellization as follows:

$$g_{mic} = g_{tr} + g_{def} + g_{int} + g_{st} + g_{dipole} + g_{ionic}, \qquad (2)$$

where g_{tr} represents the free energy for transfer of the tail from water to the hydrophobic core of the aggregate, g_{def} is free energy contribution resulting from the deformation owing to the conformational constraint on the surfactant tail, g_{int} is the free energy of formation of the aggregate core-water interface, g_{st} is the free energy contribution from the head group steric repulsion at the micelle surface, g_{dipole} is the contribution from head group dipole interactions (applicable for zwitterionic surfactants), and g_{ionic} accounts for the ionic interactions arising at the micellar surface (applicable for ionic surfactants).

Once the intramicellar interactions reflected in the micellization energy (g_{mic}) are known for these free energy models, various micellar properties (including critical micelle concentration (CMC), preferred micelle shape (sphere, cylinder, or laminar), and micelle size distribution) can be predicted by utilizing the equations of chemical equilibria.

Free energy models have been widely used for the prediction of micellization, phase behavior and phase separation of various micellar solutions and some surfactant mixtures. Formulations of commercial interests usually are mixtures of amphiphiles with presence of other additives. It is difficult to apply these free energy models to multi-component systems if the molecular parameters of one surfactant component or the additive are unavailable. Furthermore, the details of molecular arrangement within the micelles (particularly in the head group shell) are simply lost when the free energy models are applied.

Statistical molecular-based simulations provide an option to overcome these problems. However, full-scale simulation of micelles may be computationally expensive and almost intractable for industrially relevant systems (*26*). In an attempt to address these issues, a "complementary model" was developed that combines a molecular thermodynamic approach (e.g., the Blankschtein model) to predict the free energy of the surfactant micellar core, with a Monte Carlo simulation to predict amphiphile head group ordering and interaction energies (*27*). Reference 27 describes the "complementary model" in detail; here we will review the most salient features.

The Monte Carlo simulations were carried out on either a spherical or cylindrical shell filled, initially, with C_2-CTA$^+$ groups. The short $-CH_2-CH_2-$ segment modeling the start of the hydrocarbon tail was represented by two united-atom spheres directed towards the center of the core. The CTA^+ head group was represented as a central nitrogen atom with tetrahedrally arranged spheres representing united-atoms of CH_3. The Sal^- ion is internally rigid. The size and energy parameters for the atoms and united-atom spheres were obtained from Rappe et al. (*28*). During the simulation the total number of molecules, pressure, temperature, and the chemical potential of Sal^- ions are kept constant (semi-grand canonical Monte Carlo simulation). The cell volume is allowed to vary as the system relaxes. During the simulation both the C_2-CTA^+ and the Sal^- groups can rotate or translate. There is a penalty associated with any movement that causes the groups to penetrate the core, or that causes the C_2-CTA^+ groups to protrude from the system (which would, in the real material, expose the hydrocarbon tail to water and increase the surface area of the micelle). There is also a swapping move in which the identity of C_2-CTA^+ group is changed to a Sal^- ion (or vice versa). Interactions with any counterions in the solution outside the micelle are calculated by the Poisson-Boltzmann equation (*29-31*).

Simulations are carried out at different chemical potentials of Sal^-. These are then used to calculate the free energy of a shell with respect to the pure C_2-CTA^+ system. Of course the distribution of the headgroups and the preferred locations for the Sal^- are also predicted by the simulations. Figure 1 illustrates the combination of Monte Carlo simulations and free energy model to create a complementary model. The free energy of micellization is defined as:

$$g_{mic} = g_{head,SIM} + g_{tail,comp} - g_{tail,Blank} + g_{CTAB,Blank} , \qquad (3)$$

where $g_{head,SIM}$ represents the free energy obtained from the headgroup Monte Carlo simulations for a given composition of CTA^+ and Sal^-, $g_{tail,comp}$ is the surfactant tail contribution in the complementary model, $g_{tail,Blank}$ is the tail contribution calculated by the Blankschtein model, and $g_{CTAB,Blank}$ accounts for the free energy of an amphiphile (including the head group and tail) in a pure CTAB micelle captured by the Blankschtein model. This complementary model can not only calculate the free energy, but the Monte Carlo simulations involved also provide molecular details (e.g., intramicellar order and molecular orientation) in the micelle shells and their effect on the free energy of the head groups.

In order to understand how the addition of inorganic or aromatic salts affects the behavior of aqueous solutions of cationic quaternary ammonium surfactants, and to test the effectiveness of the different molecular-thermodynamic models in the micellization prediction, we consider here the specific example of cetyltrimethylammonium bromide (CTAB) aqueous

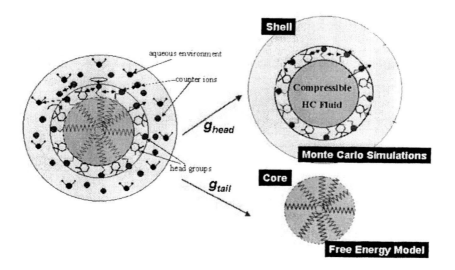

solutions with added NaSal. We would like to understand the ordering of the bound ions, the mixing nonideality of the surfactant head groups with the additive ions, and how these factors affect the interactions of head groups of CTAB surfactants with NaSal additives. At the same time, we wish to check the potential and limitations of the free energy models and the complementary model on the study of micellization of cationic surfactant-additive systems.

CTAB Surfactant Solution

The Krafft point of CTAB is about 25 °C. In aqueous solution, the critical micelle concentration (CMC) of CTAB measured by static light scattering at 35°C is 1.10 mM (32). Above the CMC, up to about 0.3 M, the surfactant exists as spherical micelles with an aggregation size of around 90 molecules (32, 33).

The CMC values of CTAB aqueous solution predicted by several models are shown in Table 1. We find that both the Nagarajan model and the Blankschtein model predict larger CMC values than measured experimentally. The complementary model gives a somewhat better estimate of the pure CTAB micellization. On the basis of the free energies calculated, the CMC for pure CTAB aqueous solution, as predicted by the complementary model (27), is 0.73 mM. This compares reasonably well with the experimental value of 1.10 mM.

Table 1. CMC values of CTAB aqueous solution predicted by different models at 35°C.

Method as defined in text	CMC, mM
Experiment	1.10
Nagarajan Model	7.48
Blankschtein Model	3.76
Complementary Model	0.73

CTAB-NaBr System

Imae *et al.* (32) investigated the formation of spherical and rod-like (cylindrical) micelles of CTAB in aqueous sodium bromide (NaBr) solutions. The data summarized in Table 2 demonstrate that adding NaBr lowers the CMC values and helps induce a spherical to rod-like micellar structure transition.

Table 2. Properties of CTAB micelles in NaBr solutions at 35°C measured by static light scattering by Iame *et al.* (32).

C(NaBr), M	CMC, mM	Shape	M, $\times 10^4$	Aggregate number	R_g, A
0	1.10	sphere	3.3	91	
0.1	0.55	sphere/cylinder	5.0/14.6	137/401	56
0.2	0.41	cylinder	66.2	1820	362
0.3	0.27	cylinder	197	5410	510
0.5	0.14	cylinder	347	9530	802

NOTE: M – molecular weight; R_g – radius of gyration.

Both the Nagarajan model and the Blankschtein model show that the CMC value of the CTAB aqueous solution decreases when adding salt, consistent with the experimental observations. This decrease is due mainly to the decrease in the thickness of the ionic atmosphere surrounding the ionic head groups (Debye length) in the presence of the additional electrolyte and to the consequent decreased coulombic repulsion between head groups (represented by the free energy contribution: g_{elec} for the Blankschtein model and g_{ionic} for the Nagarajan model). The quantitative effects of NaBr on micellization of CTAB surfactant in aqueous solutions are estimated using these models and the results are demonstrated in Figure 2. Although at the same electrolyte concentration the CMC value calculated by the Nagarajan model is larger than that predicted by the Blankschtein model, both models predict similar salt effects of additional NaBr electrolyte on the reduction of the CMC value. All the data predicted by the Nagarajan model are off-set downwards by a larger CMC value for the pure CTAB solution. Though both models specify the free energy contributions associated with the head groups, hydrocarbon tails, and counterions in a similar way, they calculate the CMC value based on the free energy of micellization slightly differently; this is the main contribution to the discrepancy of the CMC values predicted by these two models. The salt effects predicted by both models are slightly more pronounced than the experimental predictions. The preferred micelle shape remains a sphere when the Blankschtein model is used or a cylinder when the Nagarajan model is used throughout the NaBr concentration range investigated here. The micelle shape transformation from sphere to cylinder that occurs in the experiments has not yet been captured; this may be the result of incomplete parameterization. Since in CTAB/NaBr solution, the anions (Br⁻) form a double layer around the micelle and do not penetrate into the head group shell, the complementary model cannot show any improvement on the predictions of the free energy models. However, in the next section, the penetrating salicylate anion is addressed.

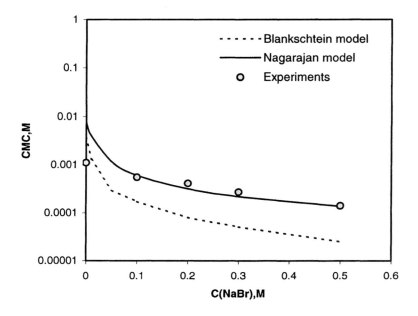

Figure 2. Effect of concentration of NaBr on CMC values of CTAB in NaBr solutions.

CTAB-NaSal System

Experimental studies (*33-36*) with NaSal suggest that NaSal is much more effective than NaBr in inducing the spherical to rod-like (cylindrical) micelle transition. It has been suggested that NaSal transforms the spherical micelles to rod-like micelles by insertion of the salicylate ion (Sal⁻) into the surfactant shell, which reduces the preferred mean curvature of the surfactant shell (*27 and references therein*). Thus it is possible to treat the Sal⁻ ion as another component of the micelle shell. The free energy models of surfactant mixtures (*16, 17, 20-25, 37-40*) might then be used or adapted, but some of the molecular constants and parameters required to calculate the free energy contributions of NaSal are unavailable. The complementary Monte Carlo and molecular thermodynamics approach can be used with little difficulty. In reference (*27*) using this complementary approach we showed that the arrangement of molecules, especially the Sal⁻ ions, within the micelle shell is highly structured. This suggests that even if interaction parameters were available, free energy models which make use of a simple mixing rule (assuming random arrangements of the molecules in the micelle shell) might not correctly model this type of system. To test this idea, we next compare the free energy predictions based on a simple mixing rule with the Monte Carlo simulation results.

A random mixing rule would allow, once the interactions between components of a mixture are known, the total interaction energy at any composition to be approximated as ideal:

$$g^{ideal} = x_1 \cdot g_{11} + x_2 \cdot g_{22} + kT(x_1 \cdot \ln x_1 + x_2 \cdot \ln x_2), \qquad (4)$$

where g_{ii} represents the interaction energy owing to contributions from pure i components and x_i is the mole fraction of component i in the mixture. The first two terms in equation 4 represent a linear interpolation between the energy values of the pure components, and the last term accounts for an ideal entropy of mixing. The energy owing to CTA^+-CTA^+ interactions (g_{11}) is calculated from the Monte Carlo simulation of a pure CTA^+ micelle. Sal^--Sal^- interactions (g_{22}) are calculated from the Monte Carlo simulation in a hypothetical state of a head group shell completely occupied by Sal^- ions. These calculated parameters based on the complementary model are presented in Table 3 for both spherical and cylindrical micelles.

Table 3. Interaction energies of the pure components predicted by the complementary model (in kT).

	Cylindrical micelle	Spherical micelle
g_{11} (CTA^+-CTA^+)	6.07	4.27
g_{22} (Sal^--Sal^-)	2.01	4.10

Based on the parameters in Table 3, the free energy contributions from head group interactions are calculated using the ideal mixing rule. The results are then compared with the Monte Carlo simulation data. Figures 3 and 4 demonstrate the comparison for cylindrical and spherical micelles, respectively.

In these two figures, the data points mark the Monte Carlo (MC) simulation results. The solid line represents the estimates from the ideal mixing rule. It is evident from Figure 3 that the estimates of the free energy based on a random mixing model for cylindrical micelles are much more positive than the simulation results. This verifies that the ordering of the molecules in the micelle shell, especially the orientation of the additive ions (Sal^-), deviates from the ideal mixing of the components. Indeed, the Monte Carlo simulations of reference (27) show that the additive ions have orientation preference along the all three Euler angles.

Figure 4 shows that for spherical micelles the estimates assuming random mixing are smaller than the simulation results, but the difference is not as significant as that for the cylindrical micelles, perhaps due to the weaker molecular ordering in the spherical micelles than in the cylindrical micelles as shown in reference (27).

The above results thus reiterate that the head groups in both the cylindrical and spherical micelles are ordered. The ordering causes the most important difference in the free energies, as evident in Figures 3 and 4. The results also suggest that the increased curvature in the spherical micelles frustrates molecular ordering and so reduces the difference.

In order to quantitatively characterize the mixing nonideality for the CTAB/NaSal binary mixtures, we also estimated the nonideal micelle formation using regular solution theory (41), in which the interaction energy at any composition depends on the interactions of the pure components, g_{11} and g_{22}, according to the following mixing model:

$$g = x_1 \cdot g_{11} + x_2 \cdot g_{22} + kT[x_1 \ln x_1 + x_2 \ln x_2 + \beta \cdot x_1 \cdot x_2]. \qquad (5)$$

The nonideality of mixing is represented by a term of the form $\beta \cdot x_1 \cdot x_2$, where β is a constant parameter quantitatively measured the interaction between the two component species in the mixed micelle. If $\beta = 0$, the mixing micelle formation is ideal. The larger the absolute value of β, the stronger the mixing nonideality and the corresponding synergism. Normally the value of the β parameter is calculated using the experimentally measured CMC values for the pure surfactants and the surfactant mixtures. (In this study, since the NaSal is an organic salt that does not form micelles on its own, estimating β based on the CMC values is not suitable. However, we can use the Monte Carlo simulation data based on the complementary model to estimate the mixing nonideality of the CTA$^+$/Sal$^-$ head groups in micelles.)

The free energy contributions from head group interactions are first obtained based on the complementary model for various compositions of CTA$^+$/Sal$^-$. Then the values of β parameter are calculated according to equation 5 for both the cylindrical and spherical micelles, as shown in Table 4.

Table 4. Values of the β parameter for various compositions of CTA$^+$/Sal$^-$.

	Cylindrical micelle			Spherical micelle	
x_1	g (kT)	β	x_1	g (kT)	β
0.6	1.04	-11.39	0.6	3.87	1.42
0.7	2.10	-10.20	0.7	3.98	1.77
0.8	2.62	-13.36	0.8	4.06	2.02
0.9	4.23	-12.32	0.9	4.13	2.24
Average:		-11.82			1.86

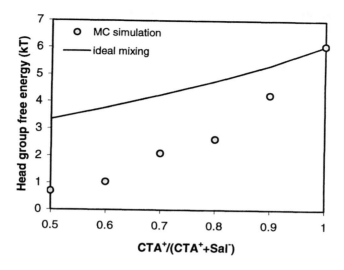

Figure 3. Comparing the head group free energy predictions from random mixing rule with simulation data for cylindrical micelles.

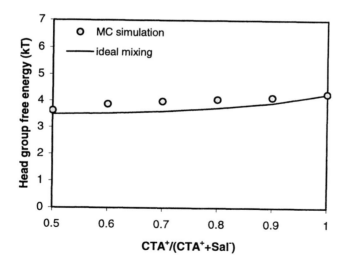

Figure 4. Comparing the head group free energy predictions from random mixing rule with simulation data for spherical micelles.

The negative values of β for the cylindrical micelle indicate synergism in mixed micelle formation. On the other hand, the positive values of β indicate antagonism for the spherical micelle. The non-zero values of β further verify that the molecular ordering in the micelle shell deviates the ideal mixing of the components. The much smaller absolute β values for the spherical micelles comparing to those for the cylindrical micelles suggest that the molecular ordering effects in spherical micelles are weaker and therefore the deviation from the random mixing becomes smaller.

Conclusions

The molecular-thermodynamic free energy models and the complementary model are used to investigate the effects of added salts on micellization of cetyltrimethylammonium bromide (CTAB) surfactant solutions. Both the Blankschtein model and the Nagarajan model show ability to predict the effects of NaBr on micellization of CTAB solutions. The complementary model predicts the micellization behavior of pure CTAB aqueous solutions with more reasonable accuracy comparing to the free energy models. Although it is not appropriate to use the complementary model to predict effects of nonadsorbing inorganic salts, it is necessary to account for specific ordering effects using the complementary model for systems with larger penetrating organic anions such as Sal$^-$. The complementary model demonstrates the great potential on the micellization investigation for the CTAB-NaSal mixtures. The Monte Carlo simulation data from the complementary model show that intramicellar molecular ordering is crucial to the estimation of free energies and that ordering is affected by the curvature of the micelles. The nonideality parameter of mixing for the CTA$^+$/Sal$^-$ in the micelle shell is calculated based on the regular solution theory and the complementary model. The results show that the ordering of molecules, particularly the salicylate ions, causes the deviation from the ideal mixing of the CTA$^+$ and Sal$^-$ ions, especially for cylindrical micelles.

Acknowledgement

We acknowledge support from the Industrial Partnership for Research in Interfacial and Materials Engineering (IPRIME), Center for Interfacial Engineering, and Supercomputing Institute at University of Minnesota. We are also grateful for the thoughtful advice of the monograph editor.

References

1. Tanford, C. *The Hydrophobic Effect;* Wiley: New York, NY, 1973.
2. Israelachvili, J. N., Mitchell, D. J., and Ninham, B. W. *J. Chem. Soc. Faraday Trans. II* **1976,** 72, 1525-1568.
3. Israelachvili, J. N. In *Physics of Amphiphiles: Micelles, Vesicles, and Microemulsions*; Degiorgio, V.; Corti, M. Eds.; Proceedings of the International School of Physics; Elsevier: New York, NY, 1985.
4. Israelachvili, J. N. *Intermolecular and Surface Forces;* 2nd Ed. Academic Press: London, 1991.
5. Evans, D. F.; Wennerstrom, H. *The Colloidal Domain;* VCH Publishers: New York, NY, 1994.
6. Lin, Z.; Zheng, Y.; Davis, H. T.; Scriven, L. E.; Talmon, Y.; Zakin, J. L. *J. Non-Newtonian Fluid Mech.* **2000,** 93, 363-373.
7. Lu, B.; Li, X.; Zakin, J. L.; Talmon, Y. *J. Non-Newtonian Fluid Mech.* **1997,** 71, 59-72.
8. Shikata, T.; Hirata, H.; Kotaka, T. *Langmuir* **1987,** 3, 1081-1086.
9. Shikata, T.; Hirata, H.; Kotaka, T. *Langmuir* **1988,** 4, 354-359.
10. Nash, T. *Nature* **1956,** 177, 948.
11. Gadd, G. E. *Nature* **1966,** 212, 1348-1350.
12. White, A. *Nature* **1967,** 214, 585-586.
13. Zakin, J. L.; Poreh, M.; Brosh, A.; Warshavsky, M. *Chem. Eng. Prog. Sysmposium* **1971,** 67, 85-89.
14. Rosen, M. *Surfactants and Interfacial Phenomena;* John Wiley & Sons: New York, NY, 1989.
15. Puvvada, S.; Blankschtein, D. *J. Chem. Phys.* **1990,** 92, 3710-3724.
16. Puvvada, S.; Blankschtein, D. *J. Phys. Chem.* **1992,** 96, 5567-5579.
17. Puvvada, S.; Blankschtein, D. *J. Phys. Chem.* **1992,** 96, 5579-5592.
18. Naor, A.; Puvvada, S.; Blankschtein, D. *J. Phys. Chem.* **1992,** 96, 7830-7832.
19. Carale, T. R.; Pham Q. T.; Blankschtein, D. *Langmuir* **1994,** 10, 109-121.
20. Shiloach, A.; Blankschtein, D. *Langmuir* **1998,** 14, 1618-1636.
21. Nagarajan, R.; Ruckenstein, E. *Langmuir* **1991,** 7, 2934-2969.
22. Nagarajan, R. *Langmuir* **1985,** 1, 331-341.
23. Nagarajan, R. *Adv. Colloid Interface Sci.* **1986,** 26, 205-264.
24. Nagarajan, R. In *Mixed Surfactant Systems;* Holland, P. M.; Rubingh, D. N., Eds.; ACS Symposium Series 501; American Chemical Society: Washington, DC, 1992.
25. Nagarajan, R. In *Structure-Performance Relationships in Surfactants;* Esumi, K.; Ueno, M., Eds.; Marcel Dekker: New York, NY, 1997.
26. Gelbart, W. M.; Ben-Shaul, A. *J. Phys. Chem.* **1996,** 100, 13169-13189.
27. Mohanty, S.; Davis, H. T.; McCormick, A. V. *Langmuir* **2001,** 17, 7160-7171.

28. Rappe, A. K.; Casewit, C. J.; Colwell, K. S.; Goddard, W. A. III; Skiff, W. M. *J. Am. Chem. Soc.* **1992**, 114, 10024-10035.
29. Evans, D. F.; Ninham, B. W. *J. Phys. Chem.* **1983**, 87, 5025-5032.
30. Mitchell, D. J.; Ninham, B. W. *J. Phys. Chem.* **1983**, 87, 2996-2998.
31. Evans, D. F.; Mitchell, D. J.; Ninham, B. W. *J. Phys. Chem.* **1984**, 88, 6344-6348.
32. Imae, T.; Kamiya, R.; Ikeda, S. *J. Colloid Interface Sci.* **1985**, 108, 215-225.
33. Olsson, U.; Soderman, O.; Guering, P. *J. Phys. Chem.* **1986**, 90, 5223-5232.
34. Rao, U. R. K.; Manohar, C.; Valaulikar, B. S.; Iyer, R. M. *J. Phys. Chem.* **1987**, 91, 3286-3291.
35. Clausen, T. M.; Vinson, P. K.; Minter, J. R.; Davis, H. T.; Talmon, Y.; Miller, W. G. *J. Phys. Chem.* **1992**, 96, 474-484.
36. Lin, Z.; Cai, J. J.; Scriven, L. E.; Davis, H. T. *J. Phys. Chem.* **1994**, 98, 5984-5993.
37. Sarmoria, C.; Puvvada, S.; Blankschtein, D. *Langmuir* **1992**, 8, 2690-2697.
38. Shiloach, A.; Blankschtein, D. *Langmuir* **1997**, 13, 3968-3981.
39. Reif, I.; Mulqueen, M.; Blankschtein, D. *Langmuir* **2001**, 17, 5801-5812.
40. Hines, J. D. *Langmuir* **2000**, 16, 7575-7588.
41. Rubingh, D. N. In *Solution Chemistry of Surfactants;* Vol. 1; Mittal, K., Ed.; Plenum:New York, NY, 1979.

Applications of Mesoscale Phenomena

Chapter 21

Formation of Multilamellar Vesicles from Ethylene Oxide–1,2-Butylene Oxide Diblock Copolymers

J. Keith Harris[1,2], Gene D. Rose[1], and Merlin L. Bruening[2]

[1]Corporate Research and Development, The Dow Chemical Company, Midland, MI 48674
[2]Department of Chemistry, Michigan State University, East Lansing, MI 48824

We recently developed ethylene oxide/1,2-butylene oxide (EO/BO) diblock copolymers that form stable, micron-diameter multi-lamellar vesicles (MLVs). EO/BO copolymers with EO blocks between 5 and 15 units long spontaneously form large MLVs upon simple mixing with water. EO lengths greater than 15 result in smaller vesicles or no vesicles at all. This behavior is consistent with a geometric argument that asserts vesicle formation is preferred only when the proper ratio of hydrophobe volume to the product of head group area and hydrophobe length is achieved. EO length is critical to vesicle formation, and polydispersity of EO block lengths also strongly affects structure formation. The EO/BO MLVs survive thermal cycling, mild sonication and moderate shear. In contrast, extrusion reduces vesicle sizes, but final structure diameter is independent of shear rate and the pore size used for extrusion. Structures were characterized using plane-polarized light microscopy, dynamic light scattering and cryo-SEM.

Vesicles

Vesicles form when selected surfactants self-assemble into bilayers that curve in on themselves to form hollow, spherical nano-structures. Naturally occurring vesicles typically form from phospholipids and generally consist of a single bilayer. Such structures are called uni-lamellar vesicles (ULVs) or liposomes. The surfactant bilayer is stable and resilient, making ULVs well-suited for isolating their contents from the dispersing medium. Hydrophilic materials are contained in the structure's aqueous core, while hydrophobic materials are sequestered in the bilayer interior. Additionally, varying surfactant composition or introducing additives can modify ULVs to achieve selective permeability, controlled release and compatibility with a given environment (1).

Phospholipids may also assemble in concentric bilayered spheres to form multi-lamellar vesicles (MLVs) (Figure 1). Compared to ULVs, these structures have a much larger hydrophobic volume and are capable of encapsulating greater amounts of hydrophobic agents. Multi-lamellar vesicles also offer advantages in controlled-release of water-insoluble agents because multiple barriers must be broken before release of the agent is complete (2,3).

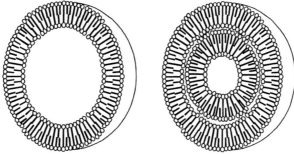

Figure 1. Diagram of a uni-lamellar vesicle (left) and a multi-lamellar vesicle (right). (Reproduced from reference 20. Copyright 2002 American Chemical Society.)

Although formation of both ULVs and MLVs from phospholipids is well known, lipids possess inherent limitations such as difficulty of formation, low capture efficiency and poor stability (4). To increase capture efficiency, Roux et al. have used controlled shear to convert surfactants from the lamellar phase (L_α) into MLVs, or "onions". These structures have average diameters of 1-5 μm (5). Florence and Cable also describe increased capacity through the use of non-ionic surfactant/ cholesterol mixtures. These mixtures can be processed in the same manner as phospholipids to produce non-ionic liposomes, or niosomes, for applications in drug delivery (6). These structures have many of the attributes of conventional liposomes in terms of size and stability.

Some limitations of phospolipids can also be overcome by using amphiphilic diblock copolymers as surfactants. Block copolymer syntheses can be simple processes that do not require extensive purification of the product. Thus, in many cases, copolymers are purer and much less expensive than phospholipids. Diblock copolymers can also be engineered to form vesicles that are compatible with specific environments. Ethoxylated copolymers, for example, have inherent biocompatibility, making them appealing as drug delivery agents (7). Additionally, nonionic copolymers demonstrate lower pH and saline sensitivity than ionic surfactants, (8) and vesicles produced from nonionic copolymers would be expected to show this same reduced sensitivity. The feature of copolymers which is of most interest to this work, however, is the ability to produce surfactants with varying geometric designs by appropriate selection of monomer type and block length.

Several studies confirm that diblock copolymers can be quite useful in nano-structure formation. Laibin and Eisenberg showed that polystyrene/poly(acrylic acid) diblock compositions form a variety of structures, including vesicles (9). Ding and Liu produced polyisoprene-*block*-poly(2-cinnamoylethyl methacrylate) vesicles in solvent mixtures (10), while Discher et al. formed ethylene oxide/ethylethylene vesicles and showed that these structures display mechanical stabilities 10 times greater than those of phospholipid-based structures (11). Other groups examined the self-assembly of ethylene oxide/1,2-butylene oxide (EO/BO) diblock copolymers in water, and although numerous structures were observed, no MLVs were reported (12-16).

This work demonstrates the formation of MLVs from EO/BO diblock copolymers upon simple mixing with water. Examination of EO/BO compositions with similar BO lengths, but increasing EO block lengths shows a marked effect of head group size on spontaneous MLV formation. The polydispersity of EO block lengths has an equally strong influence on structure formation. This chapter also explores how copolymer concentration, shear, sonication, thermal cycling and the addition of electrolyte affect structure formation and size. To better understand vesicle formation, we first review a simple geometric model of aggregate structures and discuss the properties of EO/BO diblock copolymers.

Geometric Packing Factor

Israelachvili described aggregate structure formation in terms of a simple critical packing factor, Φ, where, in its reduced form for a nonionic surfactant, $\Phi = V/(l_c a_o)$ (17). Here, V is the hydrophobe volume in the structure core, l_c is the hydrophobe chain length in the micelle, and a_o is the optimum cross-sectional area of the hydrophilic head group at the structure surface. While a_o represents the optimal head group area, other values of a may be used that represent metastable states for the head group since complete hydration of the hydrophile may be a time-dependent phenomenon. The model hypothesizes that changes in surfactant geometry, which are reflected in changes in V, l_c, or a_o, affect the type of aggregate that the amphiphile may form because of geometric constraints. For values of Φ from 0 to ⅓, the model predicts formation of spherical micelles. With ⅓ < Φ < ½, rod-like structures are predicted, and when Φ reaches ½, flexible bilayers form. At surfactant concentrations less than 25 wt %, these flexible bilayers generally curve in on themselves to form vesicles. As the packing factor increases further, the radius of curvature for the bilayers decreases, and the mean diameter of the vesicle increases. Eventually the curvature of the bilayers becomes negligible, and planar lamellae form at $\Phi = 1$. While this model is highly simplified, it provides important insights into the relationship between surfactant geometry and aggregate structures.

EO/BO Copolymers as Possible Vesicle-Forming Surfactants

EO/BO diblock copolymers offer special opportunities in the study of vesicle formation. The EO block constitutes the head group and forms hydrogen bonds with water. Larger EO blocks are more hydrophilic and require higher surfactant concentrations to initiate structure formation. However, increasing either EO lengths or surfactant concentrations can also lead to pronounced aggregate interactions and loss of discrete structures (18). Solvation of EO is driven by enthalpy, so the hydrophile has decreasing solubility with increasing temperature (19). EO/BO copolymers, like many ethoxylated surfactants, can form liquid crystalline structures that are temperature sensitive and have an obvious phase transition point (20, 21). Above this temperature, loss of hydrophilicity eventually leads to structure degradation. Increasing concentrations of selected salts or organic solvents also alter aggregate formation. Thus, variation of temperature, solvent, or ionic strength should allow manipulation of ethoxylated structures.

In spite of the versatility of the EO head group, few ethoxylated polymers form vesicles, and very few without additives. The relatively unique feature of EO/BO surfactants is the large volume of the BO block. Because of the ethyl substituent on the oligomer backbone, a BO block has a volume ~50 % larger than an *n*-aliphatic block of similar length. We selected EO/BO copolymers for vesicle formation based on the premise that the BO monomer would produce a hydrophobic block with the large volume to length ratio needed to achieve packing factors greater than 0.5. This strategy is similar to forming vesicles from phospholipids that contain two aliphatic tails per molecule. In the phospholipid case, the second aliphatic chain almost doubles the value of V for a given l_c as compared to a single-tailed surfactant.

Synthesis of EO/BO copolymers is a two-step, batch process that begins with the BO block (22). The BO monomer, unlike propylene oxide, another common hydrophobic monomer, is resistant to re-arrangement during polymerization, assuring a more uniform composition (23). Gel permeation chromatography (GPC) showed that the BO blocks typically had a polydispersity index between 1.06 and 1.10. The ethylene oxide group is highly reactive so producing EO blocks of uniform size is more challenging. However, variations in both EO and BO block lengths are reduced by using low monomer concentrations and only ethoxides (no added ethanol) as chain initiators. Figure 2 shows the structure of a typical EO/BO diblock copolymer.

The surfactants we examined had molecular weights ranging from 1100 to 1825 daltons. Compositions were determined using H^1 NMR spectroscopy and, in several cases, mass spectrometry. The copolymer purity and polydispersity were examined using GPC. Surfactants with short EO blocks (< 18 units) form clear liquids at room temperature. Increasing the EO block length leads to increasing order in the copolymer and higher viscosity. If the average EO block length exceeds 25, the copolymer is a wax-like solid due to the crystallinity of the EO chain.

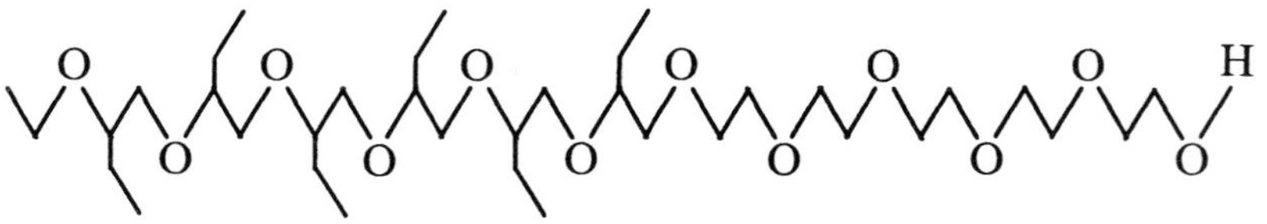

Figure 2. Structure of an ethoxide-initiated, hydroxy-terminated diblock copolymer (BO₆EO₅).

Structure Formation and Detection

Numerous methods have been developed for generating vesicles, many of which are labor intensive, have long equilibration times, or require extensive use of solvents (4). Mechanical processes include film casting and rehydration, dialysis, freeze/thaw cycles with ultra-sonication, extrusion of dispersions, shear and electrode-surface methods. In contrast, the process we used for MLV formation from EO/BO copolymers was simple hydration via hand shaking. The copolymer was added to a vial, followed by the desired amount of deionized water. (Generally the copolymer concentration was between 0.05 and 15 wt %, as higher levels tended to produce gels.) The vial was agitated by hand for 30 to 60 seconds to form an opaque dispersion of MLVs with average diameters of 1 to 5 μm.

Optical Microscopy

Because of the liquid crystallinity and large size of EO/BO MLVs, plane-polarized light microscopy (PLM) provides a simple tool to determine the extent of vesicular structure formation in surfactant dispersions. Under plane-polarized light, spherical lamellae produce a unique optical signature called a Maltese Cross extinction pattern, and many EO/BO dispersions display an abundance of these patterns (Figure 3). [Numerous micrographs of this structure can be found in the literature, and the optical pattern is sufficiently distinct as to allow facile identification (24-31).] The main drawback to this technique is that the Maltese cross pattern is difficult to resolve on structures with diameters less than about 1 μm.

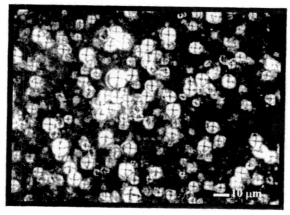

Figure 3. Plane-polarized light micrograph of MLVs in a dispersion of $EO_{11}BO_{11}$ (1 wt %).

We also used PLM in a screening test to identify copolymer compositions most likely to form MLVs (32). A drop of the copolymer was placed on a glass slide adjacent to a drop of water, and a cover slip was applied, forcing the two drops into contact and initiating hydration of the copolymer. Over the next 15 minutes, the hydration of the copolymer began to reach steady-state. A concentration gradient developed at the surfactant-water interface, and structure formation occurred as a function of distance from the interface. Generally,

EO/BO compositions favoring MLV formation produced lamellae (planar, spherical or tubular) at the interface within minutes of contacting the water.

Figure 4 shows an interesting phenomenon that frequently occurs during screening tests of EO/BO compositions. Lamellar tubes, called myelins, extend from the interface into the aqueous phase. Myelins lengthen as water diffuses up the tube to the tip (33). At the top of Figure 4 (slightly left of center), a vesicle can be seen forming at the end of the myelin as the copolymer becomes water saturated. The vesicle expands until it detaches from the myelin and floats away to be replaced by another growing vesicle. This cycle can occur several times, and these "budding" myelin are quite common. Based on these observations, and others, we refer to EO/BO MLV formation as "spontaneous".

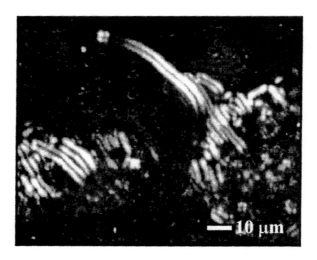

Figure 4. Plane-polarized light micrograph of a vesicle "budding" from a myelin tip at an $EO_{11}BO_{11}$-water interface.

Cryo-SEM

Cryo-SEM was used to examine the nature of EO/BO structures. Dispersions were prepared as described above and then flash frozen at -90 °C (22). Figure 5 shows a typical 6 µm-diameter structure. Several interior layers are visible through surface perforations, indicating that the structure is an MLV.

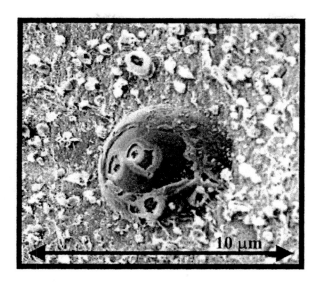

Figure 5. Cryo-SEM of a typical MLV structure in a dispersion of 1 wt %
$EO_{11}BO_{11}$. *(Reproduced from reference 20. Copyright 2002 American*
Chemical Society).

Fraction of Copolymer in Structures

Several samples of $EO_{11}BO_{11}$ and $EO_{19}BO_{11}$ formed structures that settled
out of the bulk phase after about a week. We estimate that the structures have a
density about 0.5 % greater than water, and this, in addition to the large particle
size, explains why sedimentation occurs (34). For the same reasons,
centrifugation provides a plausible means of isolating large structures to
estimate the fraction of copolymer included in the MLVs.

We centrifuged several 1.0 wt % dispersions of $EO_{11}BO_{11}$ at 3000 rpm for
24 h. Examination of the resulting sediment by PLM revealed only vesicular
structures; no other structures, such as planar lamellae, were visible. We
removed the supernatant and dried the sediment to determine the mass of the
MLVs and found that > 65 wt % of the copolymer was in these structures. In
contrast, the yield of MLVs from 2.0 wt % $EO_{19}BO_{11}$ dispersions was less than
10 %, showing the greater propensity for large vesicle formation by $EO_{11}BO_{11}$
(see Table 1 below).

Effect of EO/BO Composition on Structure Formation

Effect of EO Length

The critical packing factor can be re-written as $\Phi = (V/l_c)(1/a_o)$. For a given hydrophobe, e.g. BO_{11}, V/l_c is constant and independent of the head group. Vesicle formation then requires finding a hydrophile with the appropriate value of a_o to yield $0.5 < \Phi < 1$. For an ionic surfactant, where the hydrophile is generally short and composed of a few atoms, the head group is envisioned as a sphere whose diameter determines the head group area. For nonionic hydrophiles, which may be 20 times longer and consist of dozens of atoms, the head group "shape" is not as obvious, and determination of head group area is more complicated.

The relationship between EO block length and head group area for micellar structures is not first order as it is at the air-water interface (35). Instead, as the EO block is hydrated, it forms an expanding coil whose diameter is dependent on EO length (36). The most pronounced changes in coil size, and hence head group area, occur with increases to shorter EO blocks. Alexandridis et al. showed that the head group area for EO_2 to EO_{10} varies with the square root of the number of EO units (37). Thus, using an EO block as the hydrophile, one can generate various values of a_o simply by changing the block length during synthesis. Note that a_o is inversely proportional to Φ, so increasing the head group area is expected to produce smaller structures. Our research began with copolymers having short EO blocks (< 25 units) because we thought that such copolymers would be most likely to produce large MLVs.

Table 1 shows that increasing EO length produces slightly smaller structures, consistent with an increase in a_o. Continued increases in EO block length should eventually lead to a copolymer with a_o too large to produce vesicles. Consistent with this pattern, $EO_{24}BO_{10}$ did not form vesicles at any concentration, and dynamic light scattering (LS230, Beckman Coulter Inc, USA) showed the mean particle diameter in a 5 wt % $EO_{24}BO_{10}$ dispersion to be 14 nm, a value appropriate for a micelle formed from a copolymer of this size (22).

The length of the EO block also affects the concentration at which vesicles form. Shorter EO blocks should reduce the copolymer CMC and produce MLVs at lower concentrations. In agreement with this trend, Table 1 shows that the copolymers with larger EO blocks required higher concentrations to produce PLM-detectable MLVs. However, at lower concentrations, copolymers such as $EO_{19}BO_{11}$ may form vesicles that are simply too small to be detected with PLM. Dynamic light scattering from a 0.1 wt % dispersion of $EO_{19}BO_{11}$ does show

Table 1. Composition, molecular weight, and concentration range for PLM-detectable vesicle formation for EO/BO diblock copolymers

Copolymer [a]	M_n [b]	MLV Formation Concentrations	Ave. Diameter
EO_6BO_{11}	1104	0.05 – 10 wt %	5 μm [c]
$EO_{11}BO_{11}$	1322	0.05 – 10 wt %	5 μm [c]
$EO_{14}BO_{10}$	1384	0.5 – 15 wt %	3 μm [c]
$EO_{19}BO_{11}$	1661	2.0 – 25 wt %	1 μm [d]
$EO_{24}BO_{10}$	1824	none	----

a) H^1 NMR, b) mass spectrometry, c) 1 wt % dispersions, d) 5 wt % dispersions

particles with diameters of ~60 nm. This is significantly larger than the diameter expected for micelles, which is ~15 nm.

To confirm that the head group size is the primary reason for the structure-forming differences between copolymers, we attempted to induce micron-sized vesicle formation from a dilute $EO_{19}BO_{11}$ dispersion through chemical changes that affect head group size. One method for doing this is the addition of selected salts that compete with the hydrophile for free water (38,39). Thus, a saline solution should "dehydrate" the EO group, decrease the head group size and make large MLV formation more favorable (17). A 1.3 wt % sample of $EO_{19}BO_{11}$ was prepared using 12.5 wt % NaCl with gentle agitation. After 24 h the sample contained an abundance of multi-lamellar structures with diameters of ~ 1.0 μm. No MLVs were observed at the same concentration without salt.

Alternately, the head group size may be reduced using thermal dehydration. A 1.0 wt % dispersion of $EO_{19}BO_{11}$ was heated to 85 °C for 3 h, agitated while cooling and then examined using PLM. Figure 6 shows the changes in the dispersion due to heating. The large MLVs (> 3 μm dia.) were visible for up to 24 h after heating. (This provides some information as to the time required for complete rehydration of the EO block after thermal dehydration.) Based on these results, we conclude that the differences in structure formation between $EO_{11}BO_{11}$ and $EO_{19}BO_{11}$ are based primarily on head group size.

Ultra-Sonicated Structures

We also explored the effect of EO length on structure formation using a series of copolymers prepared by a sequential addition process. An $EO_{10}BO_{10}$ copolymer was prepared using a standard procedure. After copolymer formation, however, instead of quenching the entire reaction, only an aliquot was removed and quenched. A second reaction was performed to add another ~2 EO units to the remaining copolymer. An aliquot of this material was also

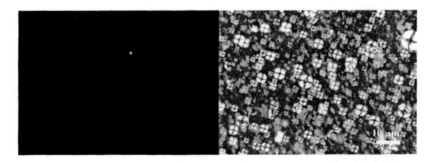

*Figure 6. PLM image of a 1.0 wt% $EO_{19}BO_{11}$ dispersion at room temperature
(left) and after heating at 85 °C for 3 hours and cooling for 45 min at room
temperature (right).*

removed and quenched, followed by another addition of EO_2 to the remaining copolymer. This procedure was repeated until a series of copolymers having the same BO block length and EO block lengths ranging from 10 to 24 was prepared.

Dispersions of 1 wt % surfactant were prepared from these copolymers. For copolymers with EO < 18, phase separation occurred after 5 days, forming a sediment of large structures and leaving a dilute, opaque, upper phase of structures less that 1 μm in diameter. After allowing the two phases to equilibrate for 3 weeks at room temperature, an aliquot of the upper phase was removed and ultra-sonicated. Ultra-sonication should yield smaller structures which should have a head group area, a, closer to the optimal area, a_o and reduce structure polydispersity (17). The structures were then evaluated using dynamic light scattering (90Plus, Brookhaven Instruments, USA).

Figure 7 shows structure diameter as a function of the EO block length. These data clearly show that sonicated structures produced from copolymers with shorter EO blocks are much larger than those produced from copolymers with EO greater than 16, which is in general agreement with the geometric model (17). Structures produced by $EO_{10}BO_{10}$ to $EO_{14}BO_{10}$ are too large to be spherical micelles and are probably ULVs. Conversely, the structures produced by EO_{20} to EO_{24} have diameters appropriate for spherical micelles formed from surfactants of this size. Particles produced from EO_{16} may represent transitional aggregate structures, and EO_{18} structures are likely swollen micelles.

Effects of Polydispersity on Vesicle Formation

The EO/BO copolymers synthesized for this work were not monodisperse, and copolymer polydispersity may impact structure formation. In a preliminary

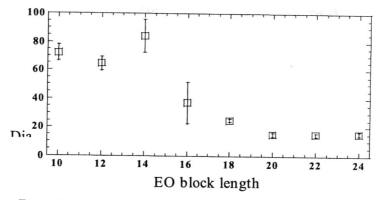

Figure 7. Mean ultra-sonicated structure diameters as a function of EO block length for EO_xBO_{11} copolymers. (Error bars represent standard deviations of 5 consective measurements.)

investigation of the effects of polydispersity, we examined BL50-1500, a commercial surfactant (The Dow Chemical Company, USA) described as having an average composition of $EO_{17}BO_{10}$. Based on the results described above, $EO_{17}BO_{10}$ should spontaneously produce MLVs at elevated concentrations. However at 5 wt %, dynamic light scattering (Brookhaven) showed that the copolymer produced micelles with diameters of 18 nm. The bulk copolymer was a waxy solid, suggesting that a significant portion of the surfactant had an EO length greater than 25, so we decided to fractionate the material. The copolymer was suspended in hexane and then centrifuged at room temperature to produce a two-phase sample (40). The clear, upper phase was decanted while the solid, lower phase, which contained the longer, insoluble EO chains, was removed after melting. The upper phase was re-centrifuged at -5 °C. The newly formed clear, upper phase was also decanted while the white solid was melted and removed. Finally, the hexane was removed from the upper phase to produce a clear liquid.

This clear liquid, which had an average composition of EO_8BO_{11} as shown by H^1 NMR, spontaneously produced 1 – 5 μm-diameter MLVs just as our $EO_{11}BO_{11}$ had, even at concentrations well below 1 wt %. This fraction accounted for ~ 30 wt % of the BL50-1500 copolymer, while the remaining material had an average EO length greater than 25. This result indicates that determination of the mean composition alone is not sufficient to indicate whether a given copolymer will form vesicles.

Our screening test with the EO_8BO_{11} copolymer also produced an unexpected structure. Upon hydration on a microscope slide, as described

340

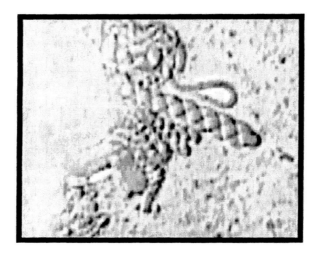

Figure 8. Plane-polarized light micrograph of twisted (helical) myelin formed from EO₈BO₁₁ upon hydration at a water-surfactant interface. (200X magnification).

above, the copolymer produced myelins, as frequently occurs with the other EO/BO surfactants we studied. However, many of these structures were twisted into helical shapes that persisted for hours (Figure 8). This may be an indication that a substantial fraction of these molecules have an average head group size at this level of hydration that is close to that required for planar lamellae structures, however further investigation is necessary. It should be noted that no vesicles were observed "budding" from the helical structures.

Stability of EO/BO Structures

Shear Stability

PLM showed that $EO_{11}BO_{11}$ formed the highest density of large, stable MLVs, so we examined structure stability primarily with this material. Figure 9 shows a typical particle size distribution for a spontaneously formed, 5.0 wt %, $EO_{11}BO_{11}$ dispersion. The light scattering data (Beckman Coulter) show that structure diameters range from 3.0 to 30 μm with a mean of 5.1 μm. Similar distributions occurred with 0.5, 1.0 and 2.0 wt % $EO_{11}BO_{11}$. None of the dispersions showed signs of sedimentation prior to testing (less than 48 h), and we observed no particles with diameters between 0.04 and 2.0 µm. (Based on copolymer length, spherical micelles would be too small to be detected using the Beckman Coulter.) Once formed, the particle size distribution was relatively unaffected by moderate speed magnetic stirring, vortex mixing (Thermolyne Vibrational Mixer, model 37615), or mild sonication (Branson Sonic Bath, model 2200). This stability is in contrast to phospholipid vesicles, for which many of these techniques result in a significant decrease in particle size (4).

We think that the enhanced stability of $EO_{11}BO_{11}$ MLVs relative to phopholipid vesicles results from contributions by both the EO and BO blocks. Hydration of the EO block produces a constrained, close-packed orientation, which leads to a liquid crystal structure (41). This crystalline "shell" can be robust and presents an energy barrier to vesicle disintegration. Collisions between vesicles are unlikely to penetrate this "shell" or have any impact on structural stability. The BO block contributes to the vesicle stability because of its length and branching. The free energy of micellization for a CH_2 group is about the same as for a BO unit that is 3 times longer than a CH_2 (21). Therefore, EO/BO copolymers that possess the same hydrophobic character as a phospholipid, have a BO length more than 3 times that of corresponding di-lipid tails. The additional BO length and ethyl branches allow for greater entanglement of hydrophobes, again slowing the process of disintegration.

Extrusion is also frequently used to reduce MLV diameters or convert MLVs to uni-lamellar vesicles (4). To evaluate the effect of extrusion on our EO/BO vesicles, a 5.0 wt % dispersion of $EO_{11}BO_{11}$ was forced through a 5µm Nylon® filter 7 times using a syringe pump. Light scattering data (Beckman Coulter) showed that the mean particle diameter decreased from 5 µm to 70 nm, and the monomodal particle size distribution became narrower. The same initial dispersion (5.0 wt % $EO_{11}BO_{11}$) was also extruded through a 0.45µm Nylon® filter with essentially the same results as with the 5.0 µm filter.

The particle size distribution was stable for at least 3 months with no reversion to larger structures. We repeated the experiment using the same initial dispersion and a 0.4 µm track-etched polycarbonate membrane. The results were the same; the new structures had a mean diameter of 74 nm. Generally, more than 95 wt % of the copolymer was recoverable after extrusion. The structure size after extrusion is consistent with that expected for a uni-lamellar structure (17).

Extrusion, in contrast to stirring, vortex mixing, and mild sonication, effectively reduced the mean particle diameter by almost two orders of magnitude. The success of extrusion is probably due to the way in which shear forces are applied. During extrusion, the MLVs experience elongational forces upon entering or exiting the pores as well as a steady, laminar flow field while passing through the pores. This may allow for sectional "delamination" of the lamellae from the MLV without disintegrating the liquid crystal structure. Once

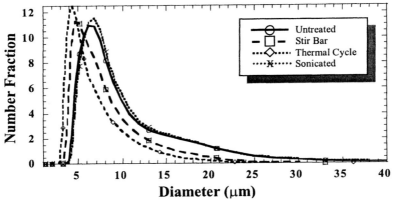

Figure 9. Particle size distribution in a 5 wt % $EO_{11}BO_{11}$ dispersion after various treatments.
(Reproduced from reference 20. Copyright 2002 American Chemical Society.)

outside the membrane pore, these sheets quickly close to form spherical lamellae. Others have shown that steady laminar flow can transform planar, lamellar liquid crystal structures into MLVs (42-44). Thus, the shear field presented by extrusion through 0.45 μm pores is more effective at separating the lamellae than sonication or high shear mixing. Furthermore, particle sizes after extrusion through both 5 and 0.45 μm-diameter pores suggest that 70 nm is the minimum stable size for vesicles prepared from this copolymer. This agrees with structure diameters produced by ultra-sonication (Figure 7) and may represent the diameter of uni-lamellar vesicles.

Thermal Stability of Liquid Crystalline Structures in EO/BO MLVs

The EO/BO vesicles are liquid crystalline structures as evidenced by the Maltese Cross pattern in PLM images. However, with elevated temperatures, the structures undergo a phase transition and lose this optical property. PLM examination of $EO_{11}BO_{11}$ and $EO_{19}BO_{11}$ structures using a hot stage showed that at about 38 °C, the Maltese Cross pattern disappeared. No liquid crystalline structures were evident between 38 and 65 °C, the maximum temperature examined. Non-polarized microscopy showed, however, that the structures maintained their spherical shape and did not coalesce. Upon cooling from 45 to 35 °C, the Maltese cross patterns reappeared within 5 minutes. Thus, there is no obvious redistribution of the block copolymers that make up the vesicles during this thermal cycling, and the formation of liquid crystalline structures in the vesicles is thermally reversible.

Conclusions

The use of BO in amphiphilic diblock copolymers results in hydrophobes with large volume-to-length ratios. This leads to EO/BO surfactants with compositions that spontaneously produce vesicular structures in water at room temperature. For a $BO_{11}EO_n$ copolymer, multi-lamellar vesicles spontaneously form when n is between 5 and 15 units, however when n is greater than 15, much smaller vesicles or micelles were observed. The differences in structure formation between various EO/BO compositions follow the general trend put forth in Israelachvili's geometric model of vesicle formation (17). Structure yields are more than 65 % of the total amount of copolymer. Temperature, surfactant concentration, solvent ionic strength and polydispersity of the EO block can alter the tendency for vesicle formation.

Cryo-SEM micrographs show that the structures formed are multi-lamellar spheres. These vesicular structures are mechanically stable and resistant to moderate thermal cycling or shear, but they can be extruded or ultra-sonicated to form smaller structures. Additionally, vesicles formed from both $EO_{11}BO_{11}$ and $EO_{19}BO_{11}$ underwent a reversible thermal transistion from a liquid crystalline to a non-crystalline phase at ~40 °C.

References

1. Lasic, D. D. in *Vesicles*; Rosoff M., Ed.; Surfactant Science Series #62; Marcel Dekker: New York, NY, **1996**; pp 448-458.
2. Allen, C.; Maysinger, D.; Eisenberg, A. *Colloids. & Surfaces B: Biointerfaces* **1999**, *16*, 3.
3. Ding, J.; Liu, G. *J. Phys. Chem. B* **1998**, *102*, 6107.
4. Gerasimov, O. V.; Rui, Y.; Thompson, D. H. in *Vesicles*; Rosoff M., Ed.; Surfactant Science Series #62; Marcel Dekker: New York, NY, **1996**; p 682.
5. Diat, O., Roux, D., Nallet, F. *J. Phys. II France* **1993**, *3*, 1427.
6. Florence, A. T.; Cable, C. in *Liposomes in Drug Delivery*; Gregoriadis, G.; Florence, A. T.; Patel, H. M., Eds.; Harwood Academic Publishers: Laghorne, PA, **1993**; pp 239-253.
7. Watson, K. J.; Anderson, D. R.; Nguyen, S. T. *Macromolecules* **2001**, *34*, 3507.

344

8. Rosen, M. J. *Surfactants and Interfacial Phenomena*, Wiley-Interscience: New York, NY, **1989**; pp 21-23.
9. Laibin, L.; Eisenberg, A. *J. Am. Chem. Soc.* **2001**, *123*, 1012.
10. Ding, J., Liu, G. *Macromolecules* **1997**, *30*, 655.
11. Discher, B. M.; Won, Y. Y.; Ege, D. S.; Lee, J. C-M.; Bates, F. S.; Discher, D. E.; Hammer, D. A. *Science* **1999**, *284*, 1143.
12. Pople, J. A.; Hamley, I. W.; Fairclough, J. P. A.; Ryan, A.; Komanschek, B. U.; Gleeson, A. J.; Yu, G-E.; Booth, C. *Macromolecules* **1997**, *30*, 5721.
13. Kelarakis, A.; Havredaki, V.; Yu, G-E.; Derici, L.; Booth, C. *Macromolecules* **1998**, *31*, 944.
14. Holmqvist, P.; Alexandridis, P.; Lindman, B. *J. Phys. Chem. B* **1998**, *102*, 1149.
15. Li, H.; Yu, G-E.; Price, C.; Booth, C.; Hecht, E.; Hoffmann, H. *Macromolecules* **1997**, *30*, 1347.
16. Alexandridis, P.; Olsson, U.; Lindman, B. *Langmuir* **1997**, *13*, 23.
17. Israelachvili, J. N. *Intermolecular and Surface Forces*, 2nd *Edition*, Academic Press: New York, NY, **1992**; pp 367–388.
18. Derici, L.; Ledger, S.; Mai, S-M.; Booth, C.; Hamley, I. W.; Pedersen, J. S. *Phys. Chem. Chem. Phys.* **1999**, *1*, 2773.
19. Bailey, F. E.; Koleske, J. V. In *Nonionic Surfactants: Physical Chemistry*; Shick, M. J. Ed.; Surfactant Science Series #23; Marcel Dekker: New York, NY, **1987**; pp 929-935.
20. Pople, J. A.; Hamley, I. W., King, S.M., Yang, Y.-W.; Booth, C. *Langmuir* **1998**, *14*, 3182.
21. Bedells, A.D.; Arafeh, R.M.; Yang, Z.; Attwood, D.; Heatly, F.; Padget, J.C.; Price, C.; Booth, C. *J. Chem. Soc. Faraday Trans.* **1993**, *89*, 1235.
22. Harris, J.K.; Rose, G.D.; Bruening, M.L. *Langmuir* **2002**, *18*, 5337.
23. Whitmarsh, R.H. in *Nonionic Surfactants: Polyoxyalkylene Block Copolymers*; Nace, V.M., Ed.; Surfactant Science Series #60; Marcel Dekker: New York, NY, **1996**; pp 13-19.
24. Wanka, G.; Hoffmann, H.; Ulbricht, W. *Macromolecules* **1994**, *27*, 4145.
25. Yu, G-E.; Li, H.; Fairclough, J. P. A.; Ryan, A. J.; McKeown, N.; Ali-Abid, Z.; Price, C.; Booth, C. *Langmuir* **1998**, *14*, 5782.
26. Kunieda, H.; Nakamura, K.; Davis, H. T.; Evans, D. F. *Langmuir* **1991**, *7*, 1915.
27. Gradzielski, M.; Muller, M.; Bergmeier, M.; Hoffman, H., Hoinkis, H. *J. Phys. Chem. B.* **1999**, *103*, 1416.
28. Dorfler, H. D.; Knape, M. *Tenside Surf. Det.* **1993**, *30*, 3.
29. Gray, G. W.; Windsor, P. A. *Liquid Crystals and Plastic Crystals;* Ellis Horwood: Sussex, England, **1974**; p 223.
30. Franses, E. I.; Hart, T. J. *J. Coll. & Interface Sci.* **1983**, *94*, 1, 1-13.

31. Collings, P. J. *Liquid Crystals*; Princeton University Press: Princeton, NJ, **1990**; p. 90.
32. Clint, J. H. *Surfactant Aggregation*; Chapman and Hall: New York, NY, **1992**; p 159.
33. Buchanan, M.; Egelhaaf, S.U.; Cates, M.E. *Langmuir* **2000**, *16*, 3718
34. Giddings, J. C. *Unified Separation Science*, Wiley-Interscience: New York, NY, **1991**; p 173.
35. Schick, M.J. *J. Coll. Sci.* **1962**, *17*, 801
36. Schick, M.J. *J. Am. Oil Chemists Soc.* **1963**, *40*, 680
37. Alexandridis, P.; Anthanassiou, V.; Fukuda, S.; Hatton, T.A. *Langmuir* **1994**, *10*, 2604
38. Lange, H.; Jeschke, P. In *Nonionic Surfactants: Physical Chemistry*; Shick, M. J. Ed.; Surfactant Science Series #23; Marcel Dekker: New York, NY, **1987**; p 14.
39. Rosen, M. J. *Surfactants and Interfacial Phenomena*, Wiley-Interscience: New York, NY, **1989**; p 194.
40. Yang, Y.-W.; Brine, G., Yu; G.-E.; Heatley, F.; Attwood, D.; Booth, C., *Polymer* **1997**, *38*, 1659
41. Gray, G. W.; Windsor, P. A. *Liquid Crystals and Plastic Crystals;* Ellis Horwood: Sussex, England, **1974**; p 4.
42. Zipfel, J.; Lindner, P.; Tsianou, M.; Alexandridis, P.; Richtering, W. *Langmuir* **1999**, *15*, 2599.
43. Muller, S.; Borschig, C.; Gronski, W.; Schmidt, C.; Roux, D. *Langmuir* **1999**, *15*, 7558.
44. Bergmeier, M.; Gradzielski, M.; Hoffmann, H.; Mortensen, K. *J. Phys. Chem. B.* **1998**, *102*, 2837.

Chapter 22

Cubosome Formation via Dilution: Kinetic Effects and Consumer Product Implications

Patrick T. Spicer

Complex Fluids Research, Corporate Engineering, The Procter & Gamble Company, 8256 Union Centre Boulevard, West Chester, OH 45069 (email: spicer.pt@pg.com)

Optical microscopy is used to study the mechanism of cubosome formation via dilution of ethanol solutions of the monoglyceride monoolein. When water is used for dilution, large cubosomes form that require further dispersion. When aqueous Poloxamer 407 solution is used for dilution, spontaneous emulsification occurs, forming numerous sub-micron particles as well as larger particles that slowly crystallize into cubosomes. Unique intermediate myelin structures are observed as emulsion droplets hydrate and form cubosomes. The formation of cubosomes by dilution of isotropic liquid phase is controlled by the kinetics of the relevant liquid crystalline phase transitions.

Cubosomes are dispersed particles of bicontinuous cubic liquid crystalline phase in equilibrium with excess water (*1*). Bulk cubic phase is formed by hydration of monoolein at levels between 20-40% w/w (*1-3*). Cubic phase is unique and desirable as a result of its mesoscale structure: a contorted lipid bilayer separating two continuous but non-intersecting water regions (*4, 5*). The tortuous structure of bulk cubic phase provides controlled release of solubilized active ingredients (*6*), while cubosomes exhibit burst release because of their sub-micron length scales (*7*). Cubosomes have been patented for use as active delivery vehicles (*8*), emulsion stabilizers (*9*), and pollutant scavengers (*10, 11*) in various pharmaceutical and personal care products (*12-15*).

Two main approaches are used to produce cubosome particles. The top-down approach applies high energy to fragment bulk cubic phase (*16-18*). The bottom-up approach forms cubosomes from molecular solution by, for example, dilution of an ethanol-monoolein solution (*19*). Top-down or high-energy techniques require formation of cubosomes prior to their use in a product. Bottom-up techniques avoid high-energy drawbacks and allow formation of cubosomes in use by a consumer or during product formulation. Both techniques require a colloidal stabilizer, like the tri-block copolymer Poloxamer 407 (*20*), to prevent cubosome aggregation. Cubosome formation by any method, even dispersion of bulk cubic phase, requires some time for the viscous cubic phase to crystallize from less-ordered precursors (*21*). Figure 1 shows a cryo-TEM image taken several days after ultrasonic treatment of bulk cubic phase (40% w/w water and 60% monoolein) in aqueous Poloxamer 407 solution (*19*). Well-formed cubosomes with regular cubic lattices are visible in Figure 1, as are less ordered cubosomes and simple vesicles, indicating the kinetic dependency of cubosome formation. The mechanism of cubosome formation by high energy dispersion is clearly the fragmentation of bulk cubic phase into smaller pieces. The dilution process produces sub-micron cubosomes in the absence of fluid shear by dilution of an isotropic liquid precursor, but the exact cubosome formation mechanism is not known (*19*).

Previous work has been limited in its ability to quantify cubosome formation mechanisms and rates because time-resolved observations with cryo-TEM are difficult. In this chapter we study the formation of cubosomes by dilution of monoolein-ethanol solutions with water and with aqueous Poloxamer 407 solution. When Poloxamer 407 solutions are used, the dilution process results in immediate interfacial turbulence of the type associated with spontaneous emulsification (*22*), producing numerous sub-micron particles. Some large isotropic liquid (L_1) phase droplets remain after the initial spontaneous emulsification, and their transformation into cubosomes is followed using optical microscopy. Unique intermediate myelinic particle morphologies form as water and Poloxamer 407 diffuse into the droplets and ethanol diffuses out. By characterizing the intermediate behavior of cubosomes formed by the dilution technique, we are better able to predict and understand the mechanisms and rate of cubosome formation. For cubosomes in consumer products and pharmaceuticals, the most commonly envisioned uses are diffusive uptake/release of materials and/or deposition onto skin or tissue. In some cases, the bottom-up dilution process (*19*) is favored so that cubosomes form only during consumer use, such as by dilution via ingestion or sweating. Cubosome performance in such applications is then a function of the mechanism and rate of cubosome formation by dilution.

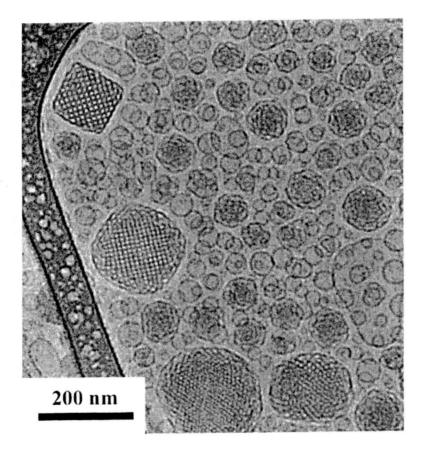

Figure 1. Cryo-TEM image of cubosomes formed by ultrasonic treatment of bulk cubic phase. (Reproduced with permission from reference 19. Copyright 2001 Am. Chem. Soc.)

Experimental

Materials

The system under investigation is a mixture of monoolein (Nu-Chek Prep), ethanol (Sigma Aldrich), Poloxamer 407 (PEO_{98}-PPO_{67}-PEO_{98}, with an average formula weight of 12,500, Spectrum), and deionized water (Millipore). The monoolein and ethanol are >99% pure.

Methods

Microscopy

All optical microscopy is carried out using a Zeiss Axioscop, equipped with differential interference contrast (DIC) optics, following alignment for Koehler illumination. Images are digitized using a Coolsnap camera (Photometrics) and a Flashbus MV frame grabber card (Integral Technologies) controlled by the software package Metamorph (v. 5.0, Universal Imaging Corp.).

Quantitative surfactant phase identification is difficult using only optical microscopy, although it is reasonable to assign some phases based on the previously established behavior of a given system and the optical defect textures exhibited by liquid crystalline phases (23, 24). For example, very viscous isotropic phases are assumed to be bicontinuous cubic phase. High viscosity is assessed by observations of low to no Brownian motion of tracer particles. The absence of birefringence between crossed polarizers indicates optical isotropy.

Cubosome Formation Experiments

Observations of cubosome formation are made directly on a microscope slide immediately following addition of either aqueous Poloxamer 407 solution to ethanol-monoolein solutions, or the opposite. Order of addition is not found to affect the general conclusions drawn for all cases examined. Cover slips are added to the sample immediately following cessation of the large amplitude interfacial turbulence and are not found to affect the conclusions drawn.

Results and Discussion

Cubosomes are formed by sufficient dilution of liquid precursors prepared in the binary monoolein-ethanol system. Such processes can be visualized using the ternary phase diagram determined by Spicer et al. (19), shown in Figure 2. In Figure 2, four single phase regions exist, including two cubic phases, one lamellar phase, and one isotropic liquid phase at high ethanol levels. Cubosomes form in the lower left portion of the phase diagram where cubic phase is in equilibrium with isotropic liquid phase. Dilution is represented by a straight line drawn from the starting composition toward the water apex of the triangle (Figure 2).

Cubosomes from the Ethanol - Monoolein -Water System

The formation of stable, discrete cubosomes is only possible when a colloidal stabilizer like Poloxamer 407 is used. The addition of 5 μL of 33% w/w monoolein solution in ethanol to 50 μL of deionized water on a microscope slide produces large pieces of cubic phase (Figure 3). Cubic phase forms when the monoolein-ethanol droplet contacts the water, ethanol diffuses out, and water diffuses in to hydrate the monoolein. In a matter of seconds, the system is essentially at equilibrium in the cubic phase-water region of the ternary phase diagram, the lower left portion of Figure 2. In this ternary system, dispersion of the cubosomes by fluid shear or another energy source is required to produce sub-micron particles. Less dispersion energy is needed than for the monoolein-water system because ethanol significantly reduces the cubic phase viscosity (19, 25). In most practical cubosome dispersion processes, a colloidal stabilizer like Poloxamer 407 is needed to provide steric stabilization against close approach and agglomeration. Beyond simple colloidal stabilization, however, the Poloxamer 407-monoolein-water system phase behavior significantly differs from that of the monoolein-water system (20). It is reasonable then to expect the amphiphilic polymer to affect the cubosome formation mechanism during its use in the dilution process.

Cubosomes from the Poloxamer 407- Ethanol-Monoolein -Water System

The inclusion of Poloxamer 407 in the ethanol-monoolein-water system has a significant effect on the mechanism of cubosome formation versus the ternary system without polymer. When 5 µL of 33% w/w monoolein solution in ethanol is added to 50 µL of 1.5% w/w aqueous Poloxamer 407 solution on a slide, violent interfacial turbulence immediately occurs that is visible unaided. The disturbance subsides in several seconds, leaving a dispersion of mostly sub-micron particles that can not be resolved well optically. Observations of the larger micron-scale particles provide insight into the driving force for the interfacial turbulence observed and thus the operative cubosome formation mechanism.

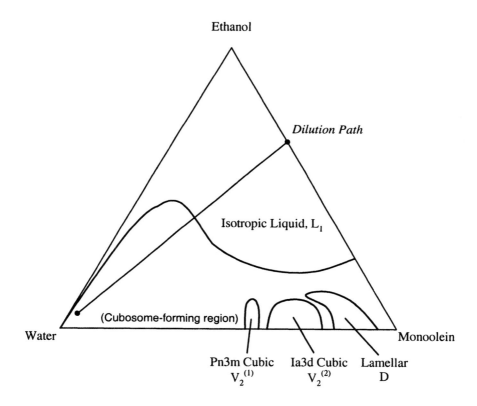

Figure 2. Equilibrium ternary phase diagram for the ethanol-monoolein-water system. The dilution path illustrates the process used to form cubosomes in this work. (Reproduced with permission from reference 19. Copyright 2001 Am. Chem. Soc.)

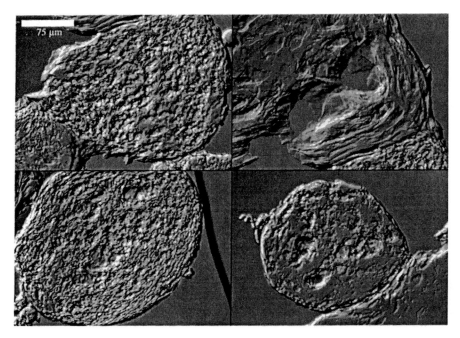

Figure 3. Optical micrographs of cubosomes produced without Poloxamer 407 polymer present.

The optical micrograph in Figure 4 shows a typical interface between the bulk solution and a droplet of concentrated ethanol-monoolein solution. Optical microscopy allows careful study of only the larger particles produced, but the background texture and Brownian motion observed indicate that the majority of the particles formed are sub-micron. The dilution process occurs with essentially zero fluid shear, so the production of such small particles must be the result of interfacial forces.

The interface in Figure 4 has surface protrusions resembling the myelin structures formed at surfactant-water interfaces far from equilibrium (26-28). Myelins are kinetically stable cylindrical structures, often appearing in surfactant-water systems with large miscibility gaps between the lamellar liquid crystalline phase and water (29, 30). Myelins are also observed when a surfactant-oil-water system crosses liquid crystalline phase boundaries during dilution, resulting in a type of spontaneous emulsification (31, 32). Very small emulsion droplets can be produced via myelinic spontaneous emulsification in the complete absence of shear (33). Droplets "bud" off of the tips of myelins,

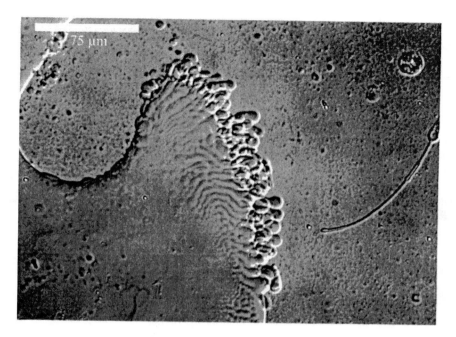

Figure 4. Interface between a droplet of monoolein-ethanol solution and aqueous Poloxamer 407 solution. Myelins are visible at the interface.

producing fine dispersions of droplets or liquid crystalline particles (*31, 34*). Myelin growth is only observed in the presence of Poloxamer 407 for the systems studied here. The associated spontaneous emulsification, caused by dilution through liquid crystalline phase regions, is apparently the mechanism of sub-micron cubosome formation.

Figure 5 shows four additional particles from the above quaternary system in various stages of myelin formation. The initial emulsification occurs almost instantaneously, forming numerous sub-micron particles. The larger droplets remaining continue to evolve, although more slowly as a result of their greater diffusion length scales and the formation of the more viscous cubic phase. The driving force for spontaneous emulsification and myelin formation is concentration gradients. As diffusion continues following the initial emulsification, the system heterogeneities decrease in magnitude and the driving force decreases for continued dispersion of particles. Nevertheless, all four particles in Figure 5 have myelinic tubules protruding from their surfaces, all in different stages of growth from the parent droplet. In some cases, such as the upper right and lower left images in Figure 5, smaller particles are visible that have just budded off of the parent myelin but retain their tubule morphology.

The myelins observed in Figures 4 and 5 are unique in that they form in a system with bicontinuous cubic phase compositionally adjacent to the lamellar liquid crystalline phase (Figure 2). Most studies examine myelins in the context of lamellar phase swelling toward equilibrium with water, while here a phase transition from lamellar to cubic occurs as the system continues to hydrate. As a result, the bilayer flexibility of the myelin structures decreases as the viscous cubic phase forms from the less viscous lamellar phase. This may explain why no coiled myelins are observed. Haran et al. (35) found that hydrotropic additives like toluene sulfonic acid enhance myelin coiling by increasing bilayer flexibility. Although ethanol possesses hydrotropic properties as well, any flexibility increase is likely offset by the formation of the viscous cubic phase, preventing coiling.

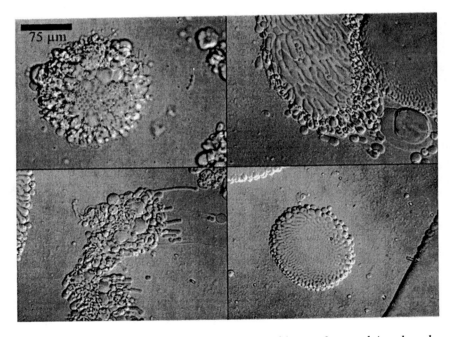

Figure 5. Examples of particles formed upon addition of monoolein-ethanol solution to aqueous Poloxamer 407 solution. All particles display myelin formation at various stages.

The order of addition is often a variable during non-equilibrium processes. For the example applications of the dilution process cited above, a preparation like ethanol-monoolein solution could be added to the mouth or skin where saliva or sweat induce dilution and cubosome formation. In such cases the order of addition is fixed: organic phase added to aqueous phase. In other applications, however, the order may be reversed. Application of organic phase to plant surfaces followed by rainfall is an example of the reverse case of aqueous phase added to organic phase. Figure 6 shows examples of particles formed by adding 50 µL of 1.5% w/w aqueous Poloxamer 407 solution to 5 µL of 33% w/w monoolein solution in ethanol. Spontaneous emulsification is again observed on the slide upon combination of the two phases. Comparison of the particles in Figure 6 with those in Figure 5 indicates no obvious differences. Most of the particles in Figure 6 exhibit some degree of myelin formation as a result of swelling by diffusion into the droplets.

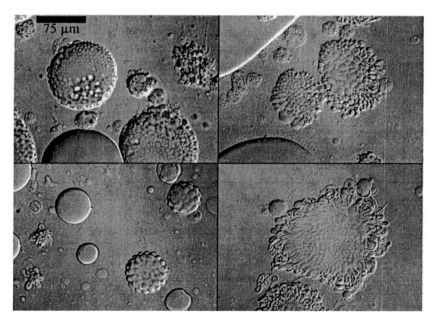

Figure 6. Particles formed by addition of aqueous Poloxamer 407 solution to monoolein-ethanol solution. Order of addition does not appear to affect particle morphology.

Some of the particles in the left hand side of Figure 6 exhibit a "raspberry" texture similar to particles observed by Buchanan et al. (29) when sheared lamellar phase particles, or onions, merged to form sponge, or L_3, phase. There is no evidence of such a transition here given the qualitative nature of these observations. However, a transition from isotropic, to lamellar, to sponge, to cubic phase is plausible on a structural basis. Sponge phase is simply disordered bicontinuous phase and is known to form in the Poloxamer 407-monoolein-water system (20) and may also form in the ethanol-monoolein-water system (19, 36). What is more likely is that the raspberry textures result from the early emergence of cylindrical myelins from the surface of a spherical droplet as hydration occurs. Buchanan et al. (29, 30) find that myelin formation proceeds via diffusive flux of water into myelin roots. They also calculate a packing density assuming the myelin cylinders order hexagonally. Figure 6 agrees with their finding as the upper left image shows the emerging myelins are ordered hexagonally. Also visible in Figure 6 are droplets that have not begun the transition through myelinic intermediates to cubic phase. It is useful to follow such a transition in order to better characterize the cubosome formation mechanisms and their rates.

Figure 7 shows a series of images taken at one minute intervals as the center droplet "crystallizes" from the isotropic liquid phase to the (presumed) cubic liquid crystalline phase. The first six images in Figure 7 illustrate development of the raspberry texture as myelins emerge from the droplet. Hydration occurs as water diffuses in between the myelins. After six minutes the droplet suddenly elongates, indicating a significant density or structural change, likely as a result of a phase transition. Image analysis indicates a linear change in droplet area with time and an area increase of 40% from the first to last image in Figure 7. Separate experiments with tracer particles support the formation of a very viscous, likely cubic, phase at the end of such a transformation. A similar change occurs in the particle partially visible in the lower right hand corner of each image, in this case the appearance changes less dramatically than the center particle but is also consistent with formation of a cubic phase. Clearly there is a distribution of times in which these transformations occur, consistent with the kinetic nature of transitions to the cubic phase (37). It is possible that the sub-micron particles also require some induction time between their spontaneous formation and their equilibration as cubosomes. The shorter length scales and thus diffusion times for smaller particles may speed things considerably, but cryo-TEM work indicates kinetic limitations to cubosome formation at the nanometer scale as well (21).

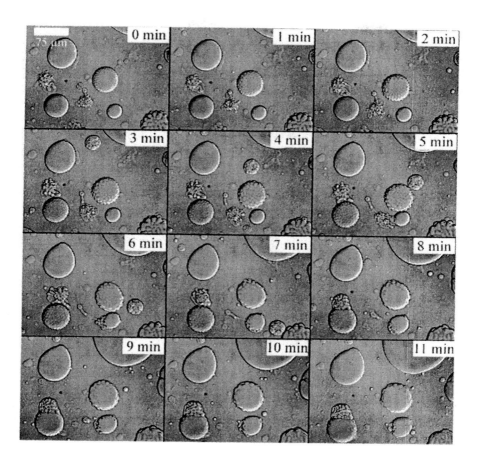

Figure 7. Montage of images of a droplet of isotropic liquid crystallizing into cubic phase.

Conclusions

Cubosome formation via dilution of the monoolein-ethanol system is described using the concepts of spontaneous emulsification and the associated myelinic interfacial instabilities. Upon dilution of the monoolein-ethanol system with aqueous Poloxamer 407 solution, spontaneous emulsification produces numerous sub-micron particles that form cubosomes and vesicles with time. Larger droplets with slower diffusion length scales form unique particulate structures as a result of myelinic interfacial instabilities that slowly transform into the viscous bicontinuous cubic liquid crystalline phase.

The formation of cubosomes via dilution is limited by the kinetics of the relevant phase transitions. Some of the most desirable properties of the bicontinuous cubic liquid crystalline phase, including complex microstructure and bioadhesion, will thus also be present only after an induction time characteristic of the cubic phase. The interfacial properties of the quaternary Poloxamer 407-ethanol-monoolein-water system need to be studied in more detail to better understand and control the kinetics of the relevant phase transitions.

References

1. Hyde, S. T.; Andersson, S.; Ericsson, B.; Larsson, K. *Z. Kristallographie* **1984**, *168*, 213-219.
2. Qiu, H.; Caffrey, M. *Biomaterials* **2000**, *21*, 223-234.
3. Laughlin, R. G. *The Aqueous Phase Behavior of Surfactants*; 1st ed.; Academic Press: San Diego, CA, 1994; p 138.
4. Mackay, A. L. *Nature* **1985**, *314*, 604-606.
5. Scriven, L. E. *Nature* **1976**, *263*, 123-125.
6. Anderson, D. M.; Wennerström, H. *J. Phys. Chem.* **1990**, *94*, 8683-8694.
7. Boyd, B. *International Journal of Pharmaceutics* **2003**, *unpublished*.
8. Landh, T.; Larsson, K. US Patent 5,531,925, 1996.
9. Ribier, A.; Biatry, B. US Patent 5,756,108, 1998.
10. Biatry, B. EP Patent App. 968704, 2000.
11. Biatry, B. EP Patent App. 1161938, 2001.
12. Schreiber, J.; Albrecht, H. DE Patent App. 10057769, 2002.
13. Schreiber, J.; Albrecht, H. DE Patent App. 10057768, 2002.
14. Schreiber, J.; Eitrich, A. DE Patent App. 10057767, 2002.
15. Schreiber, J.; Schwarzwaelder, C.; Cassier, T. DE Patent 10057770, 2002.
16. Gustafsson, J.; Ljusberg-Wahren, H.; Almgren, M.; Larsson, K. *Langmuir* **1996**, *12*, 4611-4613.
17. Gustafsson, J.; Ljusberg-Wahren, H.; Almgren, M.; Larsson, K. *Langmuir* **1997**, *13*, 6964-6971.
18. Ljusberg-Wahren, H.; Nyberg, L.; Larsson, K. *Chimica Oggi* **1996**, *14*, 40-43.
19. Spicer, P. T.; Hayden, K. L.; Lynch, M. L.; Ofori-Boateng, A.; Burns, J. L. *Langmuir* **2001**, *17*, 5748-5756.
20. Landh, T. *J. Physical Chemistry* **1994**, *98*, 8453-8467.

21. Almgren, M.; Edwards, K.; Karlsson, G. *Colloids and Surfaces A* **2000**, *174*, 3-21.
22. Sternling, C. V.; Scriven, L. E. *AIChE J.* **1959** 514-523.
23. Rosevear, F. B. *J. American Oil Chemists' Society* **1954**, *31*, 628-639.
24. Rosevear, F. B. *J. Society of Cosmetic Chemists* **1968**, *19*, 581-594.
25. D'Antona, P.; Parker, W. O., Jr.; Zanirato, M. C.; Esposito, E.; Nastruzzi, C. *Journal of Biomedical Materials Research* **2000**, *52*, 40-52.
26. Sakurai, I.; Kawamura, Y. *Biochim. Biophys. Acta* **1984**, *777*, 347-351.
27. Sakurai, I.; Suzuki, T.; Sakurai, S. *Biochim. Biophys. Acta* **1989**, *985*, 101-105.
28. Sakurai, I.; Suzuki, T.; Sakurai, S. *Mol. Cryst. Liq. Cryst.* **1990**, *180*, 305-311.
29. Buchanan, M.; Arrault, J.; Cates, M. E. *Langmuir* **1998**, *14*, 7371-7377.
30. Buchanan, M.; Egelhaaf, S. U.; Cates, M. E. *Langmuir* **2000**, *16*, 3718-3726.
31. Miller, C. A. *Colloids and Surfaces* **1988**, *29*, 89-102.
32. Miller, C. A.; Raney, K. H. *AIChE J.* **1987**, *33*, 1791-1799.
33. Walstra, P.; Smulders, P. E. A. In *Modern Aspects of Emulsion Science*; Binks, B. P., Ed.; Royal Society of Chemistry: Cambridge, 1998; pp 57-99.
34. Ekwall, P.; Salonen, M.; Krokfors, I.; Danielsson, I. *Acta Chem. Scand.* **1956**, *10*, 1146.
35. Haran, M.; Chowdhury, A.; Manohar, C.; Bellare, J. *Colloids and Surfaces A* **2002**, *205*, 21-30.
36. Engström, S.; Alfons, K.; Rasmusson, M.; Ljusberg-Wahren, H. *Progress in Colloid and Polymer Science* **1998**, *108*, 93-98.
37. Squires, A. M.; Templer, R. H.; Seddon, J. M.; Woenckhaus, J.; Winter, R.; Finet, S.; Theyencheri, N. *Langmuir* **2002**, *18*, 7384-7392.

Chapter 23

The Importance of Mesoscopic Structures in the Development of Advanced Materials

Cristina U. Thomas[1], Gregg Caldwell[1], Richard B. Ross[1], Sanat Mohanty[1], and Miriam Freedman[2,3]

[1]Materials Modeling, 3M Company, Maplewood, MN 55144–1000
[2]School of Mathematics, University of Minnesota, Minneapolis, MN 55455
[3]Department of Chemistry, University of Chicago, Chicago, IL 60637

As part of an industrial laboratory, we face the challenge of developing advanced materials by manipulating the relation between the chemical structure and the desired performance. Many of the processes governing these relations are clearly beyond the limit of atomistic scales. They involve length and time scales from nanometers to microns and from nanoseconds to microseconds. It is then imperative to understand the basic principles that govern these mesoscopic systems. Our laboratory contributes to the innovation process via the utilization of novel technologies, including computational modeling, to control the desired material behavior. We face many challenges in the area of blends/compatibilizers, miscibility, nanocomposites, colloidal suspensions, complex fluids, and surface modification. This chapter illustrates some industrially relevant systems used in our industrial, health care, and specialty materials markets.

Industrial scientists face the challenge of developing advanced materials by manipulating the relation between the chemical structure and the desired performance. Structure-property predictions have become a significant focus of computer modeling activities in the industrial world. In our laboratory, 3M Advanced Materials Technology Center (AMTC), we contribute to the innovation process via the utilization of mathematical and computational models that simulate material behavior. Predicting the final performance property requires an integrated approach among the various length and time scales of material behavior.

Industrial research and development organizations have the core responsibility of developing technologies that bring about innovative products and solutions. Our laboratory develops materials for 7 major businesses: 1.Consumer and Office, 2. Display and Graphics, 3. Electro and Communication, 4. Health Care, 5. Industrial, 6. Safety, Security and Protection Services; and 7. Transportation. Each business has its own technological requirements within various markets and business models. The AMTC drives strong 3M growth, short and long term, through building and applying expertise in chosen materials-related technology areas, anticipating new technology platforms aligned with 3M major growth opportunities. One of those key materials technologies is materials modeling. Molecular, mesoscopic, microscopic and macroscopic issues are studied and modeled. Indeed, this

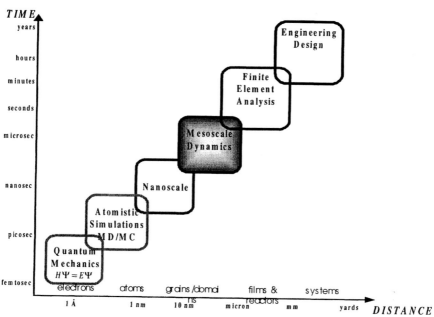

Figure 1. Multiscale approach for material behavior

group's charter is to accelerate 3M innovation by using mathematical and computational models of materials. We do this by using an integrated approach among the various length scales and time scales of material behavior (Figure 1). The latter is a requirement for any computational solution or research development in the complex world of real materials. At very small scales, on the order of angstroms, and time scales less than picoseconds, electronic interactions are driving the material behavior. Therefore electronic characterization and computational methods (Quantum Mechanics) are employed. Moving up to picoseconds/nanoseconds and nanometers, we are located in the atomistic scale or nanoscale. Here atomic and molecular analytical techniques and atomistic computational methods (Molecular Dynamics, Monte Carlo) are used. At the opposite side of the scales, millimeters/meters, and milliseconds/seconds, we find ourselves in the macroscopic world where continuum methods in structural mechanics, fluid dynamics and mass transport provide good solutions for describing the material behavior.

There is something in between - the mesoscale – which is very relevant and challenging for us as industrial scientists. In this symposium, we define the mesoscale as the length scale ranging from 1 nanometer to 1 micron. Here, "coarser" descriptions (relative to atomistic) are needed. It is the intent of this chapter to highlight the importance of mesoscopic scales in an industrial setting.

Due to the complexity of real materials, our group's workhorse is computer simulation. Traditionally, computer simulations have been used to test theories and to discriminate between these theories(1). Our group uses computer simulations together with real experiments to identify key variables. The computational model is created and validated against a set of known experimental results. Once the model is validated, we then use it to produce understanding or significant information that guides the development or design of new experiments and aids in interpreting the results. We strive to strengthen the connection between computational experiments and "real" experiments. This connection is vital to produce structure-property relations or predictions to relate mesoscopic details of the system to the desired performance properties of the material. This connection also aids us enormously when experiments are difficult and/or time-consuming to carry out. It is vital when the conditions of the experiment are severe, hazardous to the environment or to human health. We capitalize everyday on the information that is easily extracted from computational experiments, but difficult or impossible otherwise.

This chapter centers on illustrating major efforts for molecular and mesoscopic understanding of material behavior. The first corresponds to self-assembling materials (SAM), specifically fluorochemicals. The second gives a quick overview of how mesoscopic variables effect adhesive performance and adhesive development. It includes the use of fractals to capture how the complexities of meso- and micro- length scales effect adhesive-substrate interactions. The final effort is in complex fluids and in particular nanocomposites since they constitute an important class of industrial materials.

Self-Assembling Materials and the Mesoscale: Fluoromaterials

Perfluoroalkanes and polytetrafluoroethylene (PTFE) exhibit unique structural conformations and some remarkable properties. We use perfluoro-n-eicosane ($C_{20}F_{42}$) for purposes of illustrating the importance of mesoscopic structures in the resulting material behavior. Like other chain molecules, the crystal structure is composed of stacking layers(2-5). Within each layer, the molecules form a two-dimensional lattice. The phase transitions are first order and due to disorder resulting from rotational motion within each molecule or translational motion among molecules. Perfluoro-n-eicosane has three solid phases(6), **M** (monoclinic) for T < 146 K; **I** (intermediate) for 146 < T < 200 K, and **R** (rhombohedral) for T > 200 K. The crystal cell parameters are shown in Table 1. We used atomistic methods to predict the crystal structures of this SAM. The critical issue is to use a highly accurate classical force field (FF) to model the atomic interactions among different atoms. An accurate torsional FF has been derived to characterize perfluoroalkanes and PTFE based on *ab initio* quantum mechanical calculations(7). To capture the conformational characteristics classically, we used a standard Dreiding force field to describe covalent bonds, angles, and non-bond interactions(8). We tuned the torsional potential to best represent helical displacements for fluorinated systems. The unit cell in the simulation of the **M**-phase $C_{20}F_{42}$ consists of two molecules. One is left-handed, the other right-handed. After constant pressure-energy minimization using molecular mechanics, the predicted cell parameters for the low temperature crystal material are listed in Table 1 together with experimental data.

Notice the good agreement with the experimental values. Adding multiple molecules yields the SAM structure shown in color Plate 1. Averaging over time, the surface structure is maintained at 120K. This order is transformed to a high temperature SAM in agreement with experimental results above 200K. A 3x3 super-cell of the **M**-phase was constructed (Plate 1). The molecules are packed in a very ordered array. A molecular dynamics simulation was then run at 120 K (below the first transition) and at 240 K (above the second transition). After 70 ps of dynamics at the lower temperature, the well ordered crystal structure is maintained. When the temperature is increased, however, the crystal structure tends to disorder due to a change in the helicity of each molecule. The barrier to change from a left-handed helix to a right-handed helix is ~ 1 kcal mol^{-1}. During the higher temperature simulation a bond angle changes from +15° to -15°. This in turn causes the other bonds in the molecule to switch thus changing its

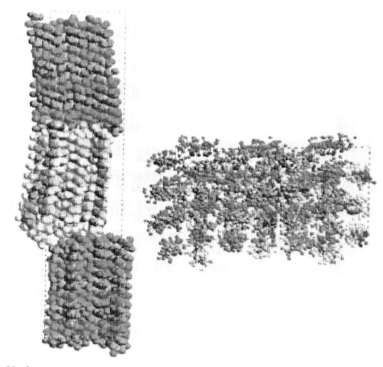

Plate 1. High-temperature structure of the SAM. (See page 10 of color insert.)

Table 1. Crystal Structures

	Experimental	Predicted
Low T	*Monoclinic*	*Monoclinic*
a	9.65 Å	9.73 Å
b	5.7 Å	5.63 Å
c	28.3 Å	28.71 Å
α	90°	88.9°
β	97°	100.3°
γ	90°	94.6°
High T	*Rhombohedral*	*Rhombohedral*
a	5.7 Å	5.61 Å
b	5.7 Å	5.69 Å
c	85 Å	169.8 (or 84.9) Å
α	90°	89.9°
β	90°	89.9°
γ	120°	119.2°

handedness. The adjacent molecule then changes handedness and the order of the entire crystal is changed.

Our calculations at 240K show a high temperature SAM (color Plate 2) whose unit cell is based on three molecules (comprising the three layers) with the molecules alternating between left and right-handed (Table 1). The results were a closer match to the experimental crystal structure after minimization. A recent paper(9) described the SAM formation of $C_{20}F_{42}$ by vapor deposition of the compound onto a glass slide. Meridional X-ray patterns of our simulated high temperature phase are in excellent agreement with the experimental X-ray patterns.

Adhesive Development and the Mesoscale

The development of adhesives or materials used to bond surfaces together constitutes a significant scientific and engineering challenge. This section is not intended to be a complete description of adhesive science. It only hopes to introduce the challenge of mesoscopic treatments for enhancing our ability to develop novel adhesives for more and more complex substrates. It summarizes learning done at 3M in collaboration with many adhesive experts, in particular with Al Pocius whose book(10) is a good reference that expands the concepts introduced here.

There are numerous advantages of adhesive bonding. There is no stress concentration (in contrast to mechanical joints) which improves fatigue resistance. Most adhesives are polymers, hence light-weight structures and corrosion resistant solutions are implicit in the technology. Due to their viscoelastic behavior, adhesives can join and seal simultaneously. Also, many adhesives can be formulated at low cost.

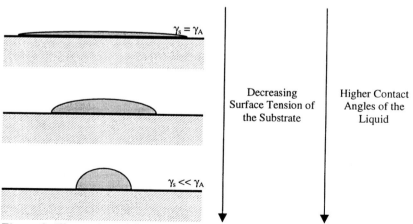

Figure 2. Illustration of drop formation of a material on surfaces with different critical wetting values

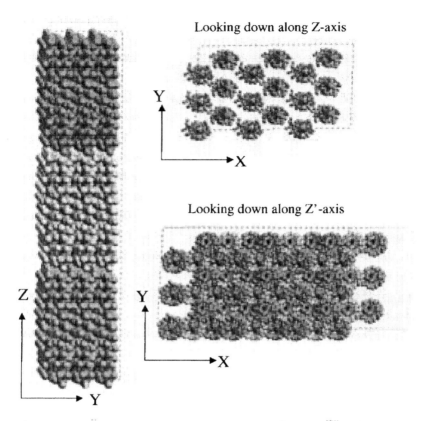

Plate 2. Low-temperature structure of the SAM. (See page 11 of color insert.)

Adhesion and Wetting

A necessary criterion for adhesion is that the adhesive and the substrate make intimate contact, that is, the adhesive can wet the substrate, i.e., it can spread on the surface such that it maximizes interfacial contact. Contact angle measurements are the most widely used tools for wetting studies. These measurements are very sensitive to surface characteristics, probing intermolecular forces including the short-range Van der Waals forces. A broadly utilized relationship based on work from Young(*11*) and Zisman(*12*), $\cos \Theta = 1 + b(\gamma_C + \gamma_{LV})$, determines critical wetting. Θ is the contact angle, γ_{LV} is the interfacial tension of the liquid in saturated air and γ_C is the critical surface tension for wetting. Extrapolated to zero contact angle, it predicts the liquid surface tension that would spontaneously wet the solid surface. If we add a drop of a material (Figure 2) to a surface with a critical wetting of γ_S, the material will spontaneously spread if its surface tension is equal to or less than γ_S. It will "bead up" or form a drop as the critical wetting tension of the surface decreases. A very relevant subject is wetting of the self-assembling materials described earlier. How these coatings behave on other polymer substrates depends on their chemical structure and on their ability to wet or spread on these polymers. These are macroscopic phenomena driven by intermolecular forces at the atomistic and mesoscale levels.

Adhesion and Solubility

When two materials A and B are in close contact (Figure 3), diffusive bonding may take place. If it does, it creates an interface with properties of A progressively changing into properties of B. When the adhesive solubilizes materials A and B (Figure 3), it produces adhesive welding. These molecules interdiffuse on mesoscale time and length scales. The diffusive bonding is enhanced if the two materials have similar solubility parameters (another good criterion for adhesion). The bond fracture will be cohesive, i. e, within the adhesive region when the solubilities match (Figure 4). Otherwise, the bond line will fracture in adhesive failure.

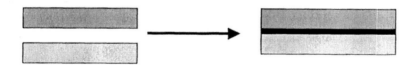

Figure 3. Adhesive Bonding via Diffusion

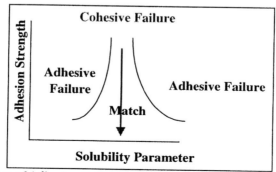

Figure 4. Types of Adhesive Failure and the Importance of Solubility Matching between adhesive and substrate

It is of great value to be able to estimate the cohesive energy density (CED) and the the related Hildebrand solubility parameter ($\delta = $ (CED)$^{0.5}$) for any given solvent or polymer from first principles. These predictions are also useful for selection of polymers in blending, for understanding polymer kinetics and monomer distribution compositions in copolymers, or for the proper selection of solvents in new formulations and synthesis processes.

To assess the likelihood of adhesive failure between two materials, we calculate and compare their solubility parameters. We use a simple hierarchical approach. First, accurate charges are obtained for each molecule using quantum mechanics. Atomic charges are calculated via Hartree-Fock studies (6-31G** basis set) employing Jaguar(*13*). Both Mulliken and electrostatic derived atomic (EDA) charges are employed. Molecular dynamics simulations are then carried out on an ensemble of the molecules using Cerius2(*14*) employing the Dreiding force field. Figure 5 illustrates the optimization strategy. The cohesive energy density (CED) of the bulk material is calculated from the difference in energy between molecules in the bulk state (Figure 5) and in isolation. A CED parameter analysis was carried out in collaboration with Blanco *et al* (*15*).

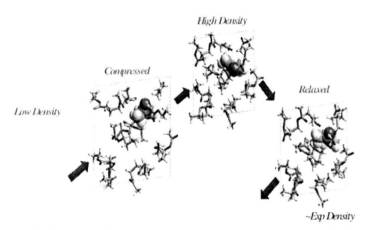

Figure 5. Molecular Dynamics sequence for optimizing a model of a bulk material used for calculation of CED and solubility parameters

Adhesives and Interfaces: Block-Copolymers

Block-copolymers are polymers with at least two chemically distinct blocks of polymers. They are well known for their ability to form interesting and useful mesoscopic morphologies. When different blocks repel each other they microphase separate. Industry uses these materials, in particular in the development of compatibilizers and adhesives. Because of the distinct blocks, they are used to join dissimilar polymers. The block-copolymer material migrates to the interface, as illustrated in Figure 6.

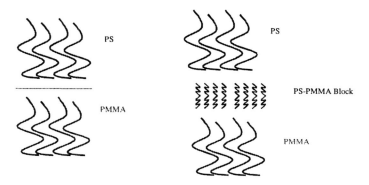

Figure 6. Adhesion and formation of interfaces via block-copolymers

The phase separation dynamics at the mesoscopic level may be simulated using the MesoDyn program incorporated in the Cerius2(*14*) This approach is a dynamic variant of the mean-field density functional theory(*16,17*) and is described in the chaper by Sevink, Zvelindovsky, and Fraaije earlier in this book. The polymer chains are modeled as ideal Gaussian chains consisting of beads, each bead representing a number of monomers of the real polymer (characteristic ratio). The current implementation of MesoDyn restricts our simulations to the same bead size and bead mobility. Figure 7 contains our simulation results for a 50/50 diblock copolymer system in which polymer A is not miscible with polymer B. Zero interaction energies were assumed for the pure bead interaction and a value of 4 kJ/mol for the unfavorable interaction between the A and B beads. Each block is modeled by 5 beads. The parameters used in our simulations for A and B were: 300 Å^3 for molecular volumes, 10^{-7} cm^2/s for the diffusion coefficients, 11.97Å for the Kuhn lengths of the beads. The simulation was conducted at room temperature and a total time of 2 ms was used. In Figure 7, we can clearly observe the microphase separation that has occurred between A and B since it shows the interface location. We reproduce the well-known behavior of this system(*18*). This work is useful in analyzing the effects of additives in this system as well as the use of more complex copolymer architectures than the diblock copolymer system shown in Figure 7.

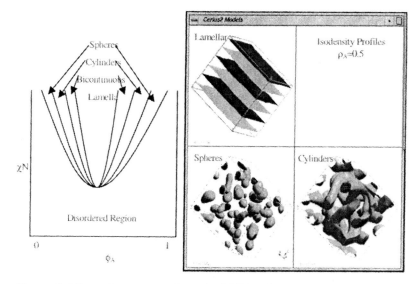

Figure 7. Microphase separation of a 50/50 Diblock-copolymer system

Adhesives and Fractals

Design of the pavement markings used to control traffic on our roads require a complete understanding and control of the adhesion of the various products to different pavements. A complete description of these materials is available elsewhere(*19*). It is important then to understand road adhesion for a large variety of pavements. Topographical data are required for characterization of the surface roughness. (These data have the form of height (z) as a function of area (x,y) and are gathered using a profilometer). We rely on the use of fractal methods developed by Brown et al(*20*) to assign physical interpretation to the topographical parameters. These patchwork methods have been patented(*21*) and software is available to perform the analysis. The main hypothesis here is that the adhesive strength is related to the relative area at the scale of interaction.

This adhesive strength depends on a finite number of bonds requiring a certain space and each having a certain strength. Therefore, the overall strength is a function of the single bond strength divided by the projected area. The number of interactions is also a function of the actual area at one interaction divided by the area of a single bond. The overall strength is proportional to the product of the relative area and the characteristics of a single bond. The patchwork method creates tile surfaces with triangles of constant area (x,y,z).

The tiling is produced with a different triangle sizes. At the end, the relative area is the product of the number of triangles of specific size, the triangle area, and the projected area. A log-log plot of relative area versus triangle area is produced (Figure 8). Inspection of this plot produces two very important parameters, the slope (related to the fractal dimension) and the crossover (related to the scale at which the surface becomes fractal).

Figure 9 contains data obtained with a laser profilometer from an asphalt road and from polymer replicas obtained from the asphalt road. One can see from the agreement of the data that the replica material does a good job at

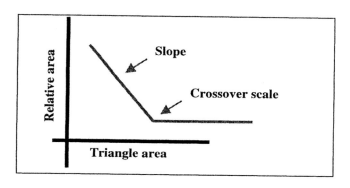

Figure 8. Fractal analysis of rough surfaces

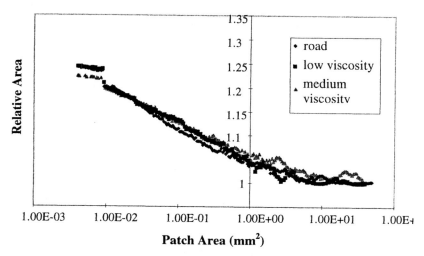

Figure 9. Fractal analysis of asphalt

372

reproducing that surface. Evaluating the two parameters for a variety of surfaces help us understand the characteristics of the adhesive which can wet this surface

Complex Fluids/Nanocomposite Materials and the Mesoscale

Surfactant assemblies are useful in lubricants, cleaning agents and other applications that take advantage of their amphiphilic properties as well as templates for design of other mesostructured materials (Figure 10). The practical formulation of surfactant products often requires additives that help regulate physical properties such as micelle shape and size. For example, mixed cationic/anionic and nonionic/ionic surfactants exhibit considerable micellar growth compared to single component surfactant systems. Additions of certain electrolytes and organic additives can be used to engineer micelles of specific shapes and sizes. Despite the industrial importance of such systems, little progress has been made in understanding the intramicellar order and structure, or in using this understanding to predict free energies and phase behavior. There is little understanding of why additives change the sizes and shapes of micelles. In a study by Mohanty et al (*22,23*), an algorithm is developed that can simulate the behavior of complex mixed systems of amphiphiles and micelles and predict the size, shape and configurations of the micelles that form under different conditions - temperature and concentrations of amphiphile, etc.

In this study, sodium salicylate (an aromatic additive) is added to aqueous solutions of cetyl trimethyl ammonium bromide (CTAB) to engineer the shapes

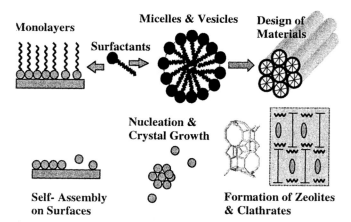

Figure 10. Self assembling materials forming various micelles and showing complex fluid behavior

and sizes of micelles. Monte Carlo simulations of realistic micellar systems are prohibitive, given the number of atoms and the size of the system that must be considered. Free energy models fail for anything more complex than single component amphiphiles (or mixtures of similar amphiphiles) with electrolytes. Using a model that combines aspects of free energy models with Monte Carlo simulations, this model spans length scales to predict the behavior of these complex systems correctly. It shows why intramicellar order affects the shape and the free energy of the micelle significantly and highlights the effect of the additives on the order. Further information is given in the chapter by Lin, Mohanty, McCormick, and Davis earlier in this book.

Engineered nanoparticles have caused a paradigm shift in materials engineering owing to their ability to control the properties of mixtures and composites at the nanoscale. However, the properties of such mixtures and composites strongly depend on the behavior of the particles in the matrix. Do they aggregate? Do they precipitate? Do they form stable emulsions? The energetic and entropic interactions of these particles are often not predictable and lead to surprises. Hence, modeling of the behavior of engineered nanoparticles promises to play a critical role in development of nanoparticle based materials.

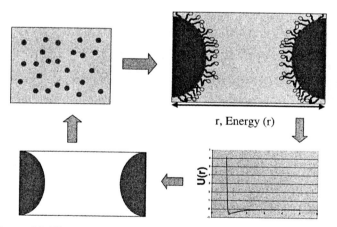

Figure 11. Illustration of the Mesoscopic Algorithm to Treat Nanoparticles

The problem of modeling most mesoscale systems is that the models must span multiple length scales. This is true for materials using nanoparticles as well. One has to include interactions between atoms on the surface and the solvent. In the case of surface modified nanoparticles, the surface groups include a wide range of functionality and interactions. In addition, the model must account for the nano-sized particles and the interactions between these particles. The large number of atoms needed to predict any useful trend makes the problem even more difficult. Most simulations overcome these problems by simplifying the system or using model 'beads'. In the industrial setting, however, simulations have to predict behavior of real systems. Thus, we have developed an algorithm that "coarse-grains" the interactions between nano-particles to predict the behavior of surface modified nanoparticles(24).

The algorithm (Figure 11) works by calculating interaction energies between two nanoparticles of interest at increasing distances between the particles. This is not a difficult problem and can be solved atomistically, accounting for the presence of solvent, ions in the solvent, as well as complex interactions between the functional groups on the particles. Once the interaction energy function between particles of a certain kind in a specific solvent has been obtained, Monte Carlo simulations of larger numbers of particles are run with the interaction between these particles in a certain solvent completely defined by the energy function.

Conclusions

Mesoscopic scales are directly related to the properties of advanced materials. Characterization techniques as well as computational techniques are and will keep playing a key role in the understanding and development of new materials. Advances in mesoscopic simulations and probing techniques are vital to industrial scientists developing advanced materials.

References

1. Allen and Tildesley, *Computer Simulation of Liquids,* Claredon Press, 1994.
2. Piesczek, W., et al *Acta Crystallegr Sect. B* **1974**, *30*, 1278.
3. Strobl, G., et al. *J. Chem. Phys.* **1974**, *61*, 5259
4. Ewen, B.; Strobl, G.; Richter, D. *Faraday Discuss* **1980**, *69*, 19.
5. Schwickert, H. *J. Chem. Phys.* **1991**, *95*, 2800.
6. Schwickert, H. J.; Strobl, G.; Kimmig, M., *Chem. Phys.* **1991**, *95*, 2800.
7. Caldwell et al. In preparation **2003**
8. Mayo S.L.; Olafson B.D.; Goddard W.A., *J. Phys. Chem.,* **1990** *94*, 8897.
9. Nishino T., et al *Langmuir,* **1999**, *15*, 4321
10. Pocius, A. V. Adhesion and Adhesives Technology. Hanser Gradner Publications, Inc. **1997**
11. Young, T. *Trans. Roy. Soc.,* **1805**, *95*, 65
12. Zisman, W. A., in: Polymer Science & Tech. Vol 9A, L.H. Lee (Ed.) **1975**

13. http://www.schrodinger.com/Products/jaguar.html (January 7, 2003)
14. http://www.accelrys.com/cerius2/index.html (August 28, 2001)
15. Blanco, M. et. al. submitted to J. Phys. Chem. B. 2002
16. Fraaije, J. G. E. M. *J. Chem Phys.* **1993**, 99, 9202.
17. Fraaije, J. G. E. M. et al *J. Chem. Phys.* **1997**, 106, 4260-4269
18. Bates et al., *Macromolecules* .**200**1, *34*, 6994
19. http://www.3m.com/us/safety/tcm(January 7, 2003)
20. Christopher Brown et al, Wear (1993)
21. Brown et. al. Patchwork Method (Patent- US 5 307 292)
22. Mohanty et. al. *Langmuir*, **2001**, vol 17, 7160-7171
23. Mohanty et. al. J. Chem Phys., **1992**, v 96, pp5579-5592
24. Mohanty et. al. AIChE, Annual Conference, Session 395g. **2002**

Chapter 24

Impact of Mesoscale Structure and Phase Behavior on Rheology and Performance in Superwetting Cleaners

Guy Broze[1] and Fiona Case[2]

[1]Colgate Palmolive R&D Inc., Milmort, Belgium
[2]Colgate Palmolive Technology Center, 909 River Road,
Piscataway, NJ 08855

Some macroscopic properties of multi-component systems such as personal care products and household cleaners are better described by the structure of their assemblies than by the chemical functions or molecular structures of the individual ingredients. This is particularly true for the rheology and solubilizing behavior of surfactant containing systems. This paper shows how we are beginning to correlate mesoscale structure predicted using Dissipative Particle Dynamics (DPD) to rheology and macroscopic phase behavior. By studying phase behavior (the phase diagram), and understanding how it reveals the mesoscale properties of the system, we can design products with optimum properties and performance. For example, an extraordinary increase in soil lifting kinetics can be obtained by careful control of the mesoscale structure and phase behavior of a system containing only three ingredients. A mixture of water, an ester and a hexanol polyether formulated in the vicinity of a thermodynamic tri-critical point lifts tar soil in only a few seconds, while compositions further from the tri-critical point act much more slowly, even if they are richer in "actives".

Introduction

Most personal care and cleaning products - soaps, dish liquids, laundry detergents, shampoo, household cleaners - are formulated from aqueous solutions of amphiphilic molecules (i.e. surfactants). They also contain oils (fragrances or actives), or interact with oils during their use. The physical properties of these products depend on their constitutive elements, the formulation. However, it is hard, if not impossible, to predict the behavior of a surfactant containing product directly from the molecular structures. Surfactants assemble in solution, interacting with the oil and/or water to form mesoscale structures ranging from a few nanometers to microns in size. These mesoscale structures determine the finished product performance profile (cleaning efficacy, rheology, active deposition, etc.).

There are a number of experimental techniques that can be used to characterize mesoscale structure in fluid systems[1] and these are being increasingly applied to characterize the types of oil/water/amphiphile systems used in consumer products. Mesoscale structure in surfactant systems may also be predicted using modeling methods[2,3], or monitored by following the macroscopic changes in phase behavior that result from changes in mesoscopic structure[4]. This insight can direct us towards optimal formulas for particular applications.

Example 1: Predicting Mesoscale Structure in Surfactant Solutions and Control of Rheological Behavior

A useful albeit very simple way to predict the mesoscale structure adopted by a particular surfactant solution is to compare the volume of the surfactant molecule (V) to the product of the surfactant extended chain length (l_c) with the effective area of the head group (a_0) [5,6]. If $\dfrac{V}{l_c a_0}$ is about 1/3, spherical micelles are obtained. If the ratio is about 1/2, elongated rod-like micelles are expected. Between 1/2 and 1 a flexible bilayer or vesicle is predicted. Finally, if the ratio is about the unity, rigid bilayers are most probable. This approach is applied to interpret experimental data in the earlier in this book.

Molecular thermodynamic methods[7-10] elaborate on this further by calculating the contribution of molecular scale structure to the energetic stability of particular mesoscale structures (spherical or rod micelles, lamellae, vesicles). They can be used to predict the thermodynamic equilibrium structures for systems containing several different surfactant species.

However, within the personal care industry we are often interested in non-equilibrium thermodynamic behavior. In use our products are diluted or sheared, they may form bubbles, and remove or deposit materials on surfaces. Simulation methods such as Dissipative Particle Dynamics (DPD) may be capable of predicting this behavior. The DPD method is described and referenced in detail in Chapter 15 *(11,12)*. The soft DPD bead represents many atoms (typically 20-100). This could be several small molecules, or a fragment of a larger molecule. The system is represented by a large number of these beads (in our simulations at least 240000, at a bead density of 3.0) in a 3D simulation box, which is replicated in space using periodic boundary conditions. There are three forces on each bead. A dissipative force (F^D_{ij}) which proportional to the relative velocity of two beads (i and j), a random force (F^R_{ij}) which can be thought of as the effect of random collisions between the atoms represented by the bead and other atoms in the system (within the cutoff radius), and a conservative interaction (F^C_{ij}). The conservative interaction (F^C_{ij}) has maximum magnitude a_{ij}. This parameter captures molecular scale interactions of the structures represented by the beads (van der Walls attraction/replusion, hydrogen bonding etc.). An a_{ii} value of 25.0 with a bead density of 3 has been shown to model water *(13)*. Larger values model more repulsive interactions.

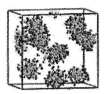

Figure 1 – mesoscale structure predicted for a 12% v/v solution of a surfactant with a strongly repulsive head group. Left images, snapshot of the surfactants and the tail density distribution for a HT surfactant model. Right images, results for a HTT (longer tail length). Water is not displayed.
(See page 11 of color insert.)

Figure 1 and color Plate 1 show mesoscale structure predicted using DPD for a 12% v/v solution of surfactants with a highly repulsive head-head interactions and more attractive head-water interactions. The structures on the left are for a surfactant comprised of one DPD head bead and one tail bead (HT), the structures on the right are predicted for a surfactant with a longer tail (HTT). The parameters used to describe the interactions between the surfactant head groups (h), tails (t), and water (w) are $a_{hh} = 35$, $a_{hw} = 15$, $a_{ht} = a_{tw} = 80$, $a_{ww} = a_{tt} = 25$. These are similar to a published DPD model sodium decyl sulfate (SDS) *(14)*, except that in our case more moderate tail-tail interactions are assumed.

Using the order of magnitude analysis in this paper (fitting to self-diffusion rates for the micelles) in which one DPD time step = 0.44 ± 0.08 ns, the systems in Figure 1 have been equilibrated for 3.3 ms (7500 DPD time steps). They have reached equilibrium, as judged by the leveling out of the surfactant self diffusion rate as a function of time.

In a simple surfactant shape analysis the head repulsion in ionic surfactants is modeled by increasing the effective head size (leading to a small value of $\frac{V}{l_c a_0}$) spherical micelles are predicted. Increasing the surfactant tail length does

not change the $\frac{V}{l_c a_0}$ ratio (both the tail volume and the tail length increase) –

and so would not be predicted to change the micelle shape. DPD predicts the increase in micelle aggregation number and reduction in CMC that results from the increase in surfactant tail length (there are far fewer HTT monomers than HT monomers)

External conditions (water temperature, presence of electrolytes or other low molecular weight components) can affect the surfactant parameters and accordingly the mesoscale structure. For example, the addition of salt to an anionic surfactant solution such as an alcohol ethoxy sulfate screens the electrostatic repulsion between head groups. Addition of a nonionic cosurfactant will also screen the ionic headgroups. Although the DPD method we use does not explicitly model electrostatic interactions, the effect of reducing the head group repulsion may be predicted by reducing a_{hh} and increasing a_{hw} (Figure 2, Plate 2, $a_{hh} = a_{hw} = 27$, $a_{ht} = a_{tw} = 80$, $a_{ww} = 25$, $a_{tt} = 25$)

The rod-like structure forms, breaks and reforms over time. Both the average rod length and the breaking time are needed to understand the rheological behavior of these systems *(15)*.

Figure 3 shows the experimentally measured effect of this change in mesoscale structure in a sodium dodecyl sulfate (SDS) solution with increasing salt, and varying amount of nonionic (lauryl dimethyl ethanolamide, LDEA) co-surfactant

The formation of elongated, rod-like micelles with increasing salt is responsible for the development of viscoelastic behavior due to their entanglement (picture the behavior of cooked spaghetti). Further increase in salt results in the formation of branches or small lamellae – and a reduction in micelle entanglement. The substitution of SDS with the non ionic LDEA makes it easier for the rod like structures to form (requiring less salt).

Figure 2 –DPD simulation of a system with less repulsive head group interactions. Predicted evolution of structure over 10,000 DPD time steps starting with a random distribution of surfactant molecules. Individual surfactant molecules (head = red, tail = green) can be seen in color Plate 2.

(See page 12 of color insert.)

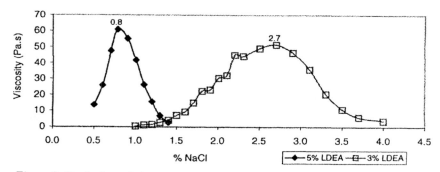

Figure3. Evolution of viscosity with increasing salt concentration for aqueous solutions of 9% SDS + 3% LDEA, and 7% SDS + 5% LDEA

Example 2: Effect of Micelle Shape on Oil Uptake Capacity

The micelle structure controls the capacity of a surfactant system to entrain oil. Spherical micelles exhibit a limited oil uptake capacity. This is because the sphere has a minimum surface/volume ratio. Increasing the volume by oil uptake necessarily implies an increase of surface, inducing a higher interfacial energy. Rod-like micelles however absorb oil quite readily. DPD simulations predict that in contact with oil, one single rod-like micelle breaks into several globular micelles, each containing a significant quantity of oil (Figure 4 and color Plate 3).

The effect of mesoscale structure on oil uptake is particularly evident in a microemulsion system containing sodium dodecyl sulfate (SDS), monoethoxy hexanol (C6E1) and brine (0.15 sodium chloride).

Figure 5 shows the experimental contour plot of dodecane uptake as a function of the SLS and C6E1 concentrations. As was discussed in the last section, substituting SDS with a nonionic co-surfactant facilitates the transition from spherical to rod-like micelle. This allows the micelles to accommodate more oil.

The oil uptake capacity depends not only on the surfactant/co-surfactant ratio, but also on the oil itself. Generally, the lower the molecular weight of the oil, the higher its "solubility", as illustrated in Figure 6. This suggests that in addition to structural factors, the entropy of mixing is also important. With lower molecular weight oil, more molecules are in a constant volume, leading to a greater number of configurational states of equal energy. This means that the micelle shape (as predicted by simple surfactant shape analysis) is far from sufficient to fully describe the behavior of a given system.

382

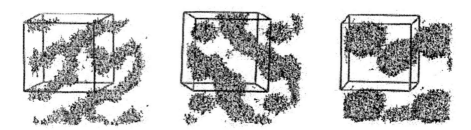

Figure 4 – Predicted structure for 12% HT surfactant (parameters as in Figure 3) with 0%, 5% and 10% oil.
(See page 12 of color insert.)

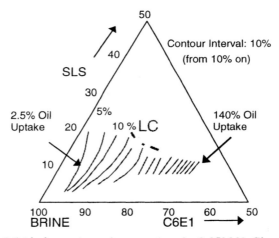

Figure 5. Oil (dodecane) uptake capacity of a 0.15M NaCl aqueous solution of SLS and C6E1. LC indictes the liquid crystal phase.

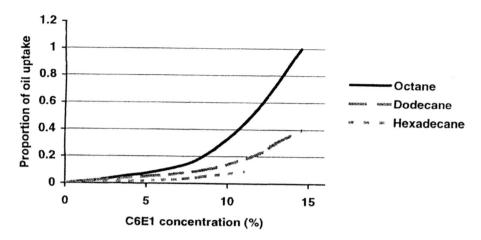

Figure 6. Effect of oil molecular weight and co-surfactant concentration on the oil uptake of a microemulsion. Proportion is given by weight of the initial microemulsion. Total surfactant + co-surfactant is 20%.

Example 3: Mesoscale Structure in 3-Component Oil/Water/Amphiphile Systems - Control of Kinetic Properties and Cleaning

Properties such as rheological behavior and oil uptake capacity are not the only ones to be affected by molecular topology. Surprisingly, adjusting a composition according to a phase diagram at thermodynamic equilibrium can enhance kinetic factors such as soil removal. Furthermore, it is possible to obtain very fast cleaning of tough stains such as tar with a system containing no conventional surfactant. An approach to designing these systems, based on prediction of mesoscale structures and analysis of phase behavior (the phase diagram), is explained here.

Three-Component Phase Diagram and Critical Point

A single point in an equilateral triangle can represent any three-component composition. The vertices represent the pure components and the binary compositions are represented on the edges. For any point in the triangle, the concentration of a given component is proportional to the perpendicular distance to the edge opposite the vertice corresponding to that component.

Figure 7 is an example of a phase diagram for a mixture of two immiscible liquids (such as water and oil) with a third liquid (such as a amphiphile or surfactant). There is a miscibility gap between the two immiscible liquids (A and B). The introduction of the third component (C) reduces the size of the miscibility gap, which may close totally if the amount of amphiphile is sufficient (any mixture of A and B will form a single phase solution).

A composition located in the miscibility gap, X, will separate into two phases, Y and Z. The separation proceeds according to the tie-lines, which are not necessarily parallel to the A-B side. In the case represented in Figure 7, the compatibilizing agent is in greater proportion in the B-rich phase than in the A-rich phase

The tie-lines converge on a point at which the compositions of the two co-existing phases become identical. This point is a critical point, sometimes called "plait point".

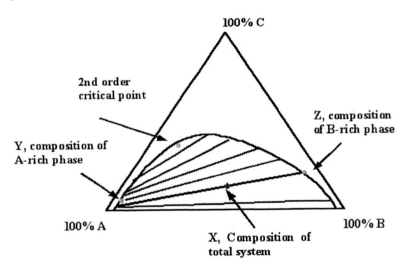

Figure 7. Three-component phase diagram for two immiscible liquids (A and B) with a compatibilizing agent (C).

The mesoscale structure corresponding to this phase behavior can be predicted using DPD. In Figure 8 and color Plate 4 the predicted A component isodensity surface, and a snapshot showing the amphiphile distribution, are shown for a mixture of 45% A and 45% B with 10% amphiphile (composition X on the phase diagram in Figure 7). The system is predicted to separate into two phases (Y and Z on Figure 7). In color Plate 4 the A beads (red) and B beads (green) can be distinguished. The phase separation is assymmetric because the amphiphile is constructed from one A and two B beads (ABB) – and thus is more soluble in the B phase.

Figure 9 shows the predicted behavior for only 6% amphiphile. Much less of A is "solubilized" in the B-rich phase. In both these simulation $a_{AA} = a_{BB} = 25$, $a_{AB} = 80$. The systems were sheared for 5000 DPD time steps, and then allowed to relax for a further 10,000 DPD time steps.

Analysis of these structures (16) shows that with 6% amphiphile, 0.7% of A is taken into the B-rich phase, 10% amphiphile takes 4% A into the B-rich phase. Virtually no B, or compatibilizer, is found in the A-rich phase at these compatibilizer concentrations.

Optimizing the Three-Component Phase Behavior

To optimize the mesoscale structure and phase behavior for cleaning we must vary one of the components in the oil/water/amphiphile system. As shown in Figure 7 we can control behavior of a mesoscale system by changing the nature of the oil (there are further examples later in this paper). The mesoscale structure can also be controlled with water-soluble additives (which change the properties of the solvent) (17). But, the most flexible control is obtained by altering the hydrophilic/hydrophobic balance of the amphiphile. This can be achieved by changing the relative amounts of more hydrophobic and hydrophilic surfactants in a mixed surfactant system (18,19), by varying the amount of salt in an ionic surfactant system (20), or for some surfactants (such as ethoxylated alcohols) by varying temperature.

The change in phase topology with temperature for a system containing water and oil and ethoxylated alcohol is shown in Figure 10. At low temperature, represented by the bottom of the prism, the ethoxylated alcohol is more soluble in water than in oil. The predicted mesoscale structure would be as shown in Figure 8 and 9 assuming that the red beads are hydrophobic and the green hydrophilic. At high temperature, the amphiphilic ethoxylated alcohol is more soluble in the oil, due to the thermally induced dehydration of the ethylene oxide chain. At this point the mesoscale structure would also look something like that shown in Figures 8 and 9 – assuming now that the red beads are hydrophilic (you could also rotate the graphic so that the new oil rich phase was on the top – but in fact there is no gravity in a DPD simulation!)

In the temperature domain between these two extremes, T_l and T_u, the ethoxylated alcohol partitions almost equally between oil and water, and a third phase (usually called "middle phase") appears.

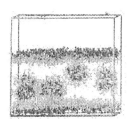

Figure 8 – DPD simulation of of 45% A, 45% B, 10% C (ABB amphiphile).
Predicted distribution of A (isodensity surface averaged over 500 time steps),
and a snapshot of the surfactant distribution.

(See page 12 of color insert.)

Figure 9 - DPD simulation of of 47% A, 47% B, 6% C (ABB amphiphile).
Predicted distribution of A (isodensity surface averaged over 500 time steps),
and a snapshot of the amphiphile distribution.

(See page 12 of color insert.)

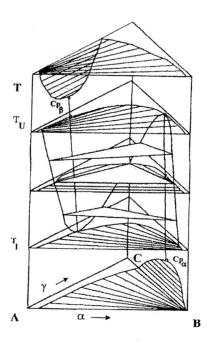

Figure 10. Phase behavior evolution with temperature of a water (A) – oil (B) –
ethoxylated alcohol (C).

What is the mesoscale structure in this new phase? In the DPD simulation of this system (using the same interaction parameters as for the simulations in Figures 8 and 9), for most concentrations between 4% and 35% of the balanced (AB) amphiphile in equal amounts of A and B, disordered bicontinuous structures are predicted (Figure 11d). At the lower concentrations these structures phase separate on shearing. At higher concentrations they appear stable – when sheared they form lamella-like structures (Figure 11e), but revert to disordered bicontinuous structures again when the shear field is removed (11f). These bicontinuous structures are consistent with NMR self-diffusion results that show rapid (approximately 60% the rate in the neat liquid) diffusion rates for both oil and water in systems found in this region of the phase diagram *(21,22)*. For particular concentrations, for example at 6% weight fraction AB, a more ordered structure is predicted (Figure 11b). This does agree with schematic drawing that have been proposed for this phase *(23)*, but it may be an artifact of the artificial periodicity in the simulation. At concentrations of AB greater than 40% stable lamella structures are predicted (Figure 11c).

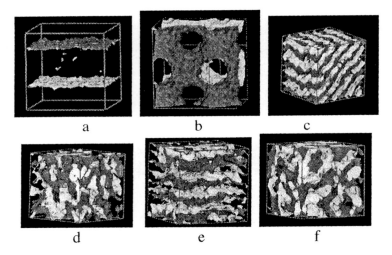

a   b   c

d   e   f

Figure 11 – DPD predictions of mesoscale structure for a 50/50 mix of A and B with varying amounts of a balanced AB amphiphile. a) 2%, b) 6%, c) 60%, d) 40% before shear, e) 40% during shear, f) 40% after shear.

The temperature domain between T_l and T_u (called "T lower" "T upper") is particularly useful in practice because the interfacial tension between oil and water reaches a minimal value. It is also between T_l and T_u that a minimum amount of amphiphile agent is required to compatibilize equal amounts of oil and water.

Tri-critical Points

T_l and T_u can be changed by modifying one of the three components. As shown in Figure 6 reducing the oil molecular weight increases its solubility in a given surfactant. The decrease in chain length for a linear hydrocarbon will reduce T_l and T_u and also reduce the difference between them. The three-phase domain will exist over a narrower temperature domain. Further reduction would lead to the merger of T_l and T_u in this system. However, the experiment cannot be conducted under atmospheric pressure due to the volatility of the hydrocarbon.

Modification of the chemical structure of the oil can produce equivalent results. In the alkyl benzene series T_l and Tu become identical between hexyl benzene and heptyl benzene. When this happens the three co-existing phases merge simultaneously into one single phase. This is a tri-critical point.

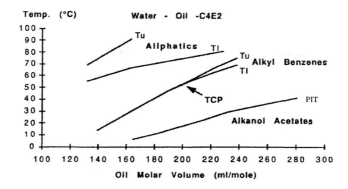

Figure 12. *Effect of oil molar volume on T_l and T_u for three different types of oils in a water-oil-diethylene glycol monobutyl ether system. A tri-critical point is visible in the alkyl benzene series.*

A tri-critical point may be reached in several different ways – by changing oil molecular weight or polarity, by modifying the nature of the amphiphile or amphiphile mixture, or even by adjusting the properties of water through addition of electrolytes or hydrotropes. Examples of these approaches have been studied by Kalweit and Stein *(24,25,26)*

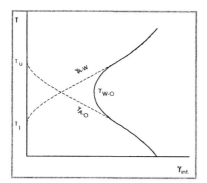 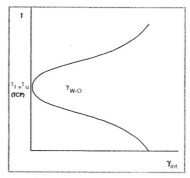

Figure 13. *Evolution of water-oil interfacial tension as function of temperature. Left: for a system showing a three-phase behavior. Right: for a system exhibiting a tri-critical point.*

In the temperature domain between T_l and T_u, the interfacial tension between the oil and water is strongly reduced, as illustrated on the left side of Figure 13. As the phase behavior of the system is changed so that T_l and T_u

approach each other, the minimum interfacial tension between the oil-rich and water-rich phases decreases, and eventually vanishes at the tri-critical point, as shown on the right side of Figure 13.

Effect of a Tri-critical Point on Soil Removal Kinetics

When two phases are in thermodynamic equilibrium, the number of molecules of one particular component (for instance, oil) moving from the water-rich phase to the oil-rich phase during a fixed time is exactly equal to the number moving in the opposite direction (from the oil-rich to the water-rich phase). Intuitively, the closer a system is to a critical point, the greater the rate of this exchange. One may even predict that this rate is infinite at the critical point. Of course, this does not make sense in practice as the two exchanging phases are identical. It remains that in the vicinity of a critical point, and *a fortiori* of a tri-critical point, phase exchange becomes very fast.

The results of andexperiment validating this point are illustrated in Figure 14 *(27,28)*. We measured the time required to lift tar dried on a glass surface with different proportions of water, decylacetate (oil) and diethylene glycol monobutyl ether (amphiphile). This system is close to a tri-critical point.

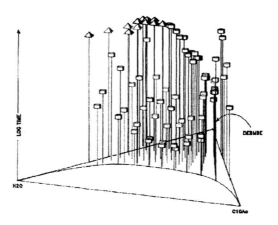

Figure 14. Time required to remove tar from a glass surface treated with different proportions of water, decylacetate and diethylene glycol monobutyl ether. Time is in log scale.

The cleaning time is the shortest for the compositions closest to the tri-critical point, and increases if the composition changes in any direction. The change is particularly significant as the proportion of amphiphile is increased.

For practical reasons, it was desirable to develop a composition with a tri-critical point richer in water. Using the insight we had gained into the effect of surfactant structure on mesoscale structure, and the effect of mesoscale structure on phase behavior, we could do this. We know that oil will be more effectively solubilized by a more hydrophobic amphiphile, moving the cloud point (critical point) into a higher water regime. Accordingly, we developed a system based on a longer amphiphile: triethylene glycol monohexyl ether. In this case, the oil is d-limonene. We achieved the tri-critical point at composition with 85% water.

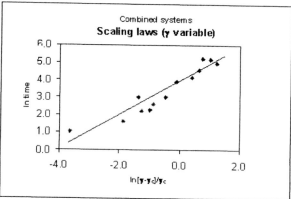

Figure 15. Relationship between the cleaning time and the "distance" from the tri-critical point for the water-decylacetate-diethylene glycol monobutyl ether and the water-limonene-triethylene glycol monohexyl ether systems.

The kinetic behavior appears to follow a scale phenomenon (27,28). As shown on Figure 15, the tar removal kinetics for the two systems follows the same equation:

$$t(s) = 56\left(\frac{\gamma - \gamma_c}{\gamma_c}\right)^{1.02}$$

γ is the proportion of amphiphile and γ_c is the proportion of amphiphile at the tri-critical point. The exponent is very close to the unity, a normal value for a critical exponent.

Acknowledgements

The data in Figures 14 and 15 was originally presented by Louis Oldenhove, Colgate Palmolive, Milmort, Belgium at the XXXI Jornadas anuales del Comite Español de la Detergencia, Barcelona in 2001.

Joan Gambogi, Colgate Palmolive, Piscataway provided the exprimental data in Figure 3.

References

1. *Mesoscale Phenomena in Fluid Systems;* Case F. H.; Alexandridis P., Eds.; ACS Symposium Series, Oxford University Press, **2003**. Chapters 1-12
2. *Mesoscale Phenomena in Fluid Systems*; Case F. H.; Alexandridis P., Eds.; ACS Symposium Series, Oxford University Press, **2003**. Chapters 13-20
3. Shelley, J. C.; Shelley, M. Y. *Current Opinion in Colloid and Interface Science* **2000**, 5, 101-110.
4. Sinoda, K.; Friberg, S. E.; *Emulsions and Solubilization*; Wiley-Interscience: New York; **1986**
5. Israelachvili, J; Mitchell, D. J.; Ninham, B. W. *J. Chem. Soc. Faraday Trans.* **1976**, 2, 1525.
6. Nagarajan, R. *Langmuir* **2002**, 18, 31-38
7. Nagarajan, R.; Ruckenstein, E. *Langmuir* **1991,** 7, 2934-2969.
8. Nagarajan, R.; Ruckenstein, E. *In Equations of State for Fluids and Fluid Mixtures*; Sengers, J. V.; Kayser, R. F.; Peters, C. J.; White, H. J.; Eds, Elsevier Science; Amsterdam, **2000**, Chapter 15, pp 589-749
9. Reif, I.; Mulqueen, M.; Blankschtein, D; *Langmuir,* **2001**; 17(19); 5801-5812
10. Hines, J. D.; *Langmuir* **2000**, 16, 7575-7588.
11. Noro M. G.; Frederico; Warren, P. B., In *Mesoscale Phenomena in Fluid Systems*; Case F. H.; Alexandridis P., Eds.; ACS Symposium Series, Oxford University Press, **2003**; Chapter 15.
12. DPD program supplied by Accelrys inc. San Diego, CA
13. Groot, R. D.; Warren, P. B. *J. Chem. Phys.* **1998**, 107, 4423
14. Groot R.D.; *Langumir* **2000**, 16 , 7493-7502
15. Cates M. E.; Candau, S. J.; *J. Phys. Cond. Matt.* **1990**, 2, 6869-6892 .
16. Esprit MesoViewer 1.0, available from Accelrys Inc. San Diego, CA
17. Alexandridis P., Chapter 5, *Mesoscale Phenomena in Fluid Systems* Case, F. H.; Alexandridis, P. Eds, Oxford University Press, **2003**
18. Johansson, I; O Boen Ho, In *Mesoscale Phenomena in Fluid Systems* Case, F. H.; Alexandridis, P. Eds, Oxford University Press, **2003**, Chapter 25.

19. Soderman O.; Johansson I., *Current Opinion in Colloid and Interface Science* **1999** , 4(6), 391-401.
20. Strey, R., *Current Opinion in Colloid and Interface Science* **1996**, 1, 402-410.
21. Olsson, U.; Wennerström, H. *Adv. Colloid Interface Science.* **1994**, 49, 113-146
22. Olsson, U.; Shinoda, K.; Lindman, B. *J. Phys. Chem.* **1986**, 90, 4083-4088
23. Wennerström, H.; Söderman. O.; Olsson, U.; Lindman, B. Macroemulsions from the Perspective of Microemulsions, in *Encyclopedic Handbook of Emulsion Technology*, Ed. Sjöblom, J., Chapter 5, Marcel Dekker, New York, **2002**
24. Kahlweit, M; Strey, R. *Angew. Chem. Int. Ed.* **1985**, *24*, 654-668.
25. Kahlweit, M.; Strey, R. *J. Colloid Interface Sci.* **1987**, *118(2)* 436-453.
26. Kahlweit, M; Strey, R. *J. Phys. Chem.* **1988**, *92*, 1557-1563.
27. Oldenhove L.; Broze, G. Optimization of Aqueous Cleaning Compositions around Near-tricritical Points. *XXXI Jornadas anuales del Comite Español de la Detergencia, Barcelona*, March 28-30, **2001**. Proceedings pp. 111-123.
28. Oldenhove L. New cleaning method. *Detergents for the New Millenium AOCS Conference Proceedings*, Sanibel, USA October 14-17, 2001.

Chapter 25

Emulsions or Microemulsions? Phase Diagrams and Their Importance for Optimal Formulations

Ingegärd Johansson and O Boen Ho

Akzo Nobel Surface Chemistry AB, Stenungsund, Sweden

A series of practical examples taken from the literature and from our own work demonstrate how knowledge of phase diagrams and mesoscale structure may be applied to design formulas and processes to obtain stable emulsion structures, or to optimize cleaning.

Introduction

The mesoscale is defined as 10^{-9} m to 10^{-7} m. When mixing oil, water and suitable surfactants you can obtain droplets of oil in water (O/W) or water in oil (W/O) stabilized by the surfactants in the range of 3×10^{-9} m to 10^{-6} m or larger. If the droplets are 3-100 nm and thermodynamically stable they will constitute a microemulsion. If they are larger they will form an emulsion, which by definition is a non-equilibrium system prone to collapse and phase separate given enough time. Depending on your application you might wish for either of these two systems. How do you identify and control which system you obtain?

One way is to study the quasi-ternary phase diagram for oil, water and the surfactant(s), see Figure 1 (*1*). By keeping the oil/water ratio constant and varying the concentration and nature of the surfactant(s) the microemulsion phases, Winsor I, Winsor II, Winsor III and Winsor IV (*2*) can be located. In Winsor I, a water-rich microemulsion exists in equilibrium with an oil-phase, **2**. In Winsor II, an oil rich microemulsion-phase is separated from an excess phase of water, $\bar{2}$. In the Winsor III, phases of water and oil exist in equilibrium with a middle microemulsion phase [3]. Winsor IV consists of a one-phase microemulsion [1].

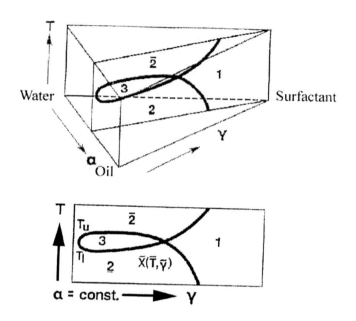

Figure 1: Phase diagram for oil, water and surfactant, tuned with temperature. α = oil/water ratio, γ = total surfactant concentration. (Reproduced with permission from reference 1. Copyright 1993.)

In the areas with more than one phase, an emulsion will be formed when the mixture is stirred. Winsor I will give a water continuous emulsion (oil in water, O/W) and Winsor II an oil continuous emulsion (water in oil, W/O). This depends upon the spontaneous curvature of the interfacial layer between the oil and water phases; convex towards water will give O/W, and convex towards oil will give W/O. Changing the curvature by varying temperature (for ethoxylated nonionics), salt (for ionics), addition of a co-surfactant, or dilution, will result in a switch over from one continuous phase to the other. The conductivity of the system can be used to follow this change since a water continuous emulsion will have a high conductivity and an oil continuous a very low(3). The bicontinuous phase will show an intermediate conductivity.

At the switchover or inversion of the emulsion, the spontaneous curvature of the oil-water interface is zero and the interfacial tension is at a minimum. Knowing this you can:

1. Make small droplet emulsions with little shearing force.
2. Optimize formulations for particular applications, for example cleaning.

This paper will present practical application examples within the two areas.

Making small droplet emulsions with little shearing force via phase inversion

Starting in the sixties Shinoda and his group used the concept of inverting emulsions to create small and stable droplets (3). Most published examples use ethoxylated nonionics as emulsifiers and temperature as the tuning factor, going from W/O to O/W emulsions by lowering the temperature. However, co-surfactant addition can give the same effect (4). It has also been shown that systems with a high tendency to form liquid crystalline phases give rise to stable emulsions (3).

a) Inversion with temperature, PIT (phase inversion temperature)

The path taken to achieve a particular composition can have a profound effect on the mesoscale structure that is obtained. An interesting example, with a practical potential to produce stable finely dispersed O/W emulsions without using high shear devices has been reported by Förster et al (5,6). In Figure 2, three different routes to create an emulsion are compared. The endpoint, determined by an earlier optimization step, is the same in all cases, but the path through the phase diagram is different.

In route 1 the whole formulation is heated until it reaches the microemulsion state, which in this case is a lamellar phase, L_α, and then cooled down to room temperature. In route 2 and 3 oil and surfactants are mixed with only 15% water, heated and diluted to the final formulation with cold water. Route 2 takes the mixture up to the bicontinuous microemulsion phase; route 3 only reaches the O/W area of the microemulsion. Route 1 and 2 give droplet sizes around 100 nm; route 3 results in much larger droplets, up to 1400 nm. It would be necessary to apply high shear mixing to obtain a finely dispersed stable emulsion from a system generated following route 3.

The conclusion is that the bicontinuous and/or lamellar structures can play a crucial role in the formation of finely dispersed and stable emulsions without high energy mixing.

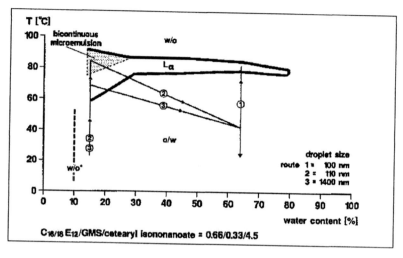

Figure 2. Phase diagram for $C_{16/18} E_{12}$ /Glyceryl monostearate/ cetearyl isononanoate in the proportions 0.66/0.33/4.5 in water. Route 1,2 and 3 refer to the preparation order for the emulsion. Reprinted with permission from ref 6.

When using the PIT techniques it has also been shown that the rate of cooling has an effect on the resulting emulsion stability. For example in the work by Ozawa, Solans, and Kunieda *(7)* the goal was to make highly concentrated O/W emulsions. This was achieved by passing from a W/O microemulsion through a very unstable oil/reverse bi-continuous phase and a multi-phase area containing lamellar structures. When the temperature decrease was fast, the end-result was a more finely dispersed and stable emulsion.

b) Inversion at constant temperature with dilution

Another example is taken from the paper industry, where an alkyl ketene dimer (AKD) is used to increase the hydrophobicity of the paper surface *(8)*. AKD is an oil with melting point 40-60 C, which is reactive to both the cellulosic surface and water. It is thus important to make a finely dispersed emulsion as close to the paper production as possible, without using sophisticated and energy demanding equipment. Two steps were required to identify a suitable process. First the surfactant system was optimized to obtain a "close to self-emulsifying" situation (8), and then the pseudo ternary phase diagram was determined for water, AKD and the optimized surfactant mixture at 60 C (9), see Figure 3.

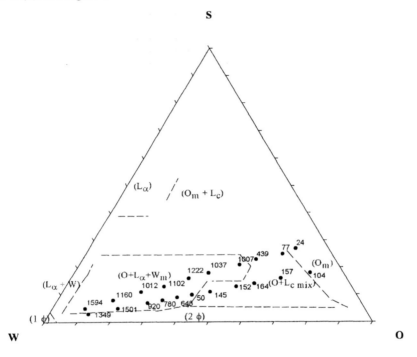

Fig. 3. Isothermal pseudo-ternary phase diagram for the system (AKD+dodecanol)/(C$_{11}$E$_6$+Ca-DBS)/water at 60 °C. Figures viewed in the diagram denote conductivity in • S/cm. Reprinted with permission from ref 9.

This knowledge was used to design different paths to arrive at the final composition at constant temperature, i.e. above the melting point of AKD. All the final compositions lie inside a three-phase region containing oil, micelles in water and a lamellar phase. Two strategies were identified, the first (path 1) starts with oil and surfactant, water is then added. In the second (path 2) the surfactants are added to the water phase and then diluted with the oil. In the first case seven different starting points along the surfactant-oil line were investigated, see Figure 4.

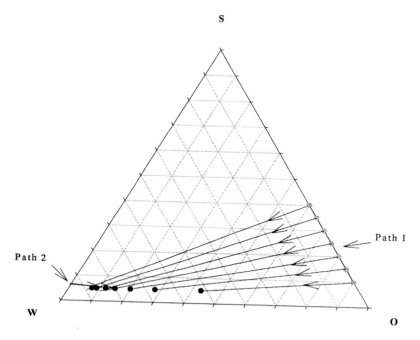

Fig. 4. Ternary plot showing emulsification path 1, i.e., addition of water to a mixture of oil and surfactant and path 2, i.e., addition of oil to a mixture of water and surfactant.

(Reproduced with permission from reference 9. Copyright 2003.)

In general the dilution with water (path 1) gave the smallest droplets and the most stable emulsions, see Figure 5. Since this route involves a transitional phase inversion the result was expected. However, one interesting exception was seen where, though droplet sizes were rather similar, the stability of the systems was different. The emulsion at R=0.08 was less stable than that at R=0.1. A tentative explanation can be given in terms of details of the phase diagram. The less stable emulsion had its inversion line closer to the end point. This means that the droplets are easily formed due to the low interfacial tension, but can coalesce equally rapidly. The more stable emulsion enjoys the stabilization from the lamellar phase at an earlier stage, and is thus easier to dilute without changing the droplet size. Salager and his group (10) have reported similar behaviour.

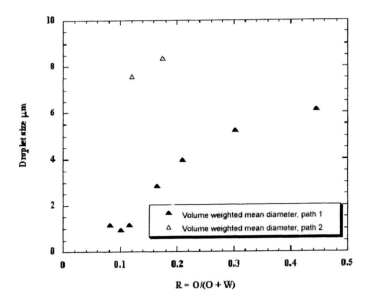

Fig. 5. Droplet size at 5 wt.% of surfactant as a function of R=O/(O+W) resulting from emulsification along paths 1 and 2 at 60 °C. , (Reproduced with permission from reference 9. Copyright 2003.)

Other investigations have shown the importance of the kinetics of the emulsification process, as well as the presence of liquid crystals and low interfacial tension, when forming nano-emulsions without high shear or high energy devices using dilution with water as the inversion technique (*11,12*). In the recommended sequence of steps the starting point is in an isotropic W/O phase. During dilution with water the natural curvature of the surfactant stabilized oil/water interface is changed in a continuous way from convex towards oil to convex towards water. The mixture passes through multiphase regions, and ends up in a three phase area containing excess oil and water phases, and a lamellar surfactant phase, which forms an O/W emulsion. The secret is said to be to combine low interfacial tension (easy to produce small droplets) with lamellar structures in the endpoint (stabilization of the droplets), and a continuous change of the interfacial curvature.

c) Inversion with addition of surfactant

The third example also concerns emulsifying at constant temperature, and is taken from our own laboratory. The shape of the interfacial layer of surfactant between the water and the oil is tuned by changing the surfactant composition during the emulsification. The components are coco benzyl ammonium chloride, a cationic surfactant with a C14 hydrocarbon tail used as a biocide in wood preservation, a more hydrophobic ethoxylate nonionic co-surfactant made from coco alcohol with an average of 4 mol of ethylene oxide, paraffinic oil and water. The surfactants were technical grade products from Akzo Nobel.

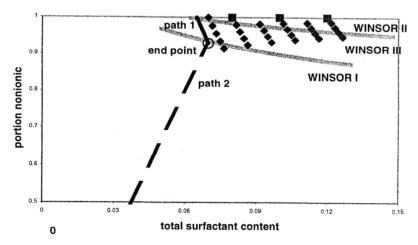

Figure 6. Equilibrium phase diagram, the squares and the diamonds, for paraffinic oil, coco benzyl dimethyl ammonium chloride and C1214E4. Inversion lines from conductivity scans with addition of the quaternary surfactant. Path 1 and path 2 show the ways of making the emulsion

First the optimal composition giving the most stable emulsion was identified (*13*). Next, the phase diagram was studied roughly, both with equilibrium samples and conductivity to find out where the inversion lines were and what phases that could be found in terms of Winsor I, II, III and IV. Finally, three different routes to make the emulsions were studied, to find out if the order of addition of the surfactants had any impact on the final emulsion. Figure 6 shows the phase diagram from the equilibrium points and the inversion line from the conductivity measurements.

The oil/water mixture with only the nonionic surfactant present as surfactant showed a Winsor II behaviour i.e. a W/O emulsion when stirred. Addition of small amounts of the cationic quaternary ammonium chloride surfactant (quat) resulted in a Winsor III situation. Further addition of this surfactant gave a very viscous mixture, which did not separate in definite phases. The upper inversion line from titration of the quat into the mixture where the nonionic is present appears below the upper border line of the three-phase region. Usually they coincide, but here there may be a delay due to the viscosity increase.

The endpoint for the emulsion is at 0.5% alkyl quat and 6.5% of the alkyl ethoxylate in a 1/1 oil/water medium. The three routes investigated were

1. Add the quat after the other components were thoroughly mixed
2. Add the nonionic after the other components were mixed
3. Add all components at the same time and mix.

The premix was stirred during a couple of minutes with a turbine stirrer, and then the whole amount of the second surfactant in route 1 or 2 was added as quickly as possible via a syrinx. The stirring continued for a few minutes, and then the droplet size was measured. The results are shown in Figure 7. The stability was judged by eye after a few hours and found to follow the drop size order (systems with larger initial drop size were less stable).

There is an obvious difference in droplet size due to addition order. Path 1 gives the smallest droplets and is also the only procedure that goes via a phase inversion. Path 2 is the worst and path 3 in between. The experiment is to be seen as a single test without optimization of the procedure. None of the final droplet sizes is in the nanometer size range. However, we see this as an example of the positive adjustments that can be made to emulsification procedures, starting with knowledge about the phase diagram and taking into account the evident effect of inverting the emulsion during the production.

A more detailed study of the emulsion changes along a transitional inversion made by changing surfactant composition can be found in ref (*14, 15*).

Making small droplet emulsions with little shearing force via phase inversion: Conclusions

In conclusion can be said that when aiming for a stable emulsion with low energy input and small droplet size, the information from a phase diagram covering the different components, and an understanding of the variations in interfacial curvature (mesoscale structure) controlled by temperature, salt or surfactant composition, is very useful. Both a low interfacial tension (to facilitate the formation of small droplets), and some stabilization mechanism (to prevent droplet coalescence) must be present. It is important to know both where the transition/inversion line is, and which phases the mixture might pass through and end up in, on the continuous emulsification route.

Optimizing formulations for a particular application - cleaning

Another use of the knowledge about interfacial curvature, minimal interfacial tension and the possibility of forming microemulsions can be found in applications involving cleaning. Much work has been done on the mechanism for detergency or cleaning generally speaking in the early nineties. Optimum cleaning has been found to occur:

- around the phase inversion temperature (PIT) for oily soils and cleaning formulations containing ethoxylated nonionics (16, 17,18)
- close to zero curvature for the interfacial layer between the oily soil and the cleaning solution (19),
- when solubilization into liquid crystalline phases or three phase microemulsions can take place (16,18,20).

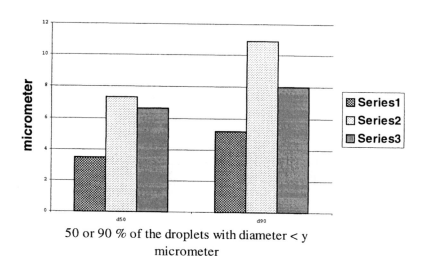

Figure 7. Droplet size, series 1 corresponds to the path with inversion, series 2 without inversion and series 3 all ingredients mixed at the same time.

404

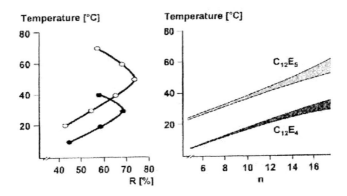

Figure 8. Detergency effects of C12E4 and C12E5 against hexadecane as a function of temperature and the corresponding three-phase ranges for these surfactants as function of the number n of carbon atoms of various n-alkanes

Several broad reviews have been published (*21,22,23*). In most cases systems containing ethoxylated nonionics and their changes with temperature have been studied, see Figure 8.

Since the polyhydroxyl nonionic surfactants, based on glycerol, glucose or sacharose, are gaining more and more interest due to their excellent environmental properties, we have been studying the mechanism for cleaning with these product types and related the results to their phase behavior.

It was pointed out by H.Kunieda at the ACS conference in Boston 2002 (*24*) that the ethoxylated nonionic surfactants become much more hydrophobic with increased temperature i.e. show a cloud point, than the poly hydroxyl based, for instance the alkyl glucosides, do. However, for very specific alkyl glucoside mixtures a mixing gap can be found that may look like a very steep well in the temperature dependent binary phase diagram (*25*) or a closed loop (*26*). For pure alkyl mono glucosides a mixing gap with vertical borders is seen for alkyl chains containing ten carbons or more (*27*). In most cases no phase change appears in the lower concentration range when the temperature is increased. Temperature can thus not be used to change the spontaneous curvature of an interfacial layer consisting of these surfactants.

Strey and his group (*28,29*) have made a thorough investigation into how the interfacial curvature of hydrocarbon /water interfaces stabilized by alkyl glucosides varys with the interfacial surfactant composition. The shape of the interface, and the mesoscale structure that is formed, can be controlled by including hydrophobic cosurfactant, in the same way as temperature controls structure in systems containing conventional nonionics.

The challenge was to see whether a formulation with optimal cleaning power could also be found for alkyl glucosides, when the interfacial curvature at the interface to the oily soil is close to zero, interfacial tension at its minimum, and a microemulsion phase can be formed at the soil surface. This would open the door for less temperature sensitive cleaning formulations.

We used the following chemicals,

- Hexadecane as a model dirt, radioactively labeled in the cleaning test,
- Decyl glucoside (Simulsol SL10 from SEPPIC)
- 2-propylheptyl glucoside (lab product from Akzo Nobel Surface Chemistry AB)
- 1-dodecanol (Merck)

Both glucosides are industrial products made according to the Fischer procedure and have a Degree of Polymerization (DP) of around 1.6.

Hard surface cleaning was measured as follows:

A defined amount of radioactive labelled hexadecane was deposited on a small aluminium plate, which was fixed to the rotor in a Terg-O-Tometer. The rotor was then submerged into the cleaning solution and rotated with a normalized speed for 5 minutes. The dirt remaining on the aluminium plates was extracted into a scintillation liquid and quantified using a tri-carb 1900CA scintillation analyzer.

The phase diagrams have been mapped out by weighing samples in closed vials, letting them equilibrate, and analyzing the phases present by visual inspection using crossed polarizing filters.

The results are given with the nomenclature introduced by Kahlweit and Strey (30). The mass and volume fraction of the oil ([oil] / [oil + water]) is denoted α. The mass fraction of the amphiphile mixture in the quaternary system is denoted γ([glucoside + alcohol] / [glucoside+ alcohol + oil+ water]). Finally the mass ratio of the alcohol in the amphiphile mixture is denoted δ([alcohol] / [alcohol+ glucoside]). The microemulsion phase behaviour is presented at constant mass fraction of oil/water and in a diagram of δ versus γ, so- called "fish-cut" diagram. Two different oil/water ratios have been investigated for both surfactants, 1:1 and 1:4.

The four phase diagrams are shown in Figure 9, a,b,c,d. In general the branched decyl glucoside shows a broader Winsor III area than the straight decyl glucoside. The same trend was seen in a more extensive investigation of the phase behavior of straight versus branched octyl, decyl and dodecyl glucosides, with dodecane and dodecanol, presented by Strandberg and Johansson at the Cesio conference 2000 (31). The actual amount and ratio of the surfactants present in the interfacial layer was calculated and the largest difference was

406

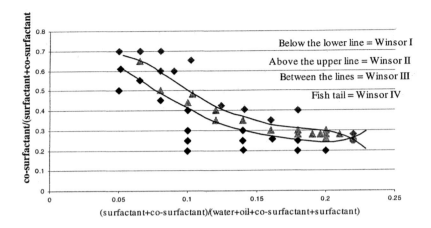

Figure 9a. Phase diagram for decyl glucoside with dodecanol as co-surfactant, water/oil ratio equals 1.

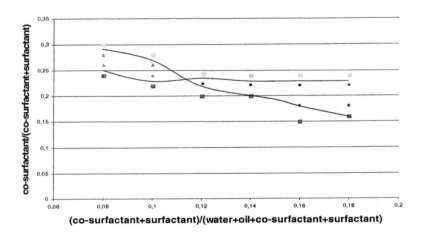

Figure 9b. Phase diagram for decyl glucoside with dodecanol as co-surfactant, water/hexadecane ratio equals 4:1

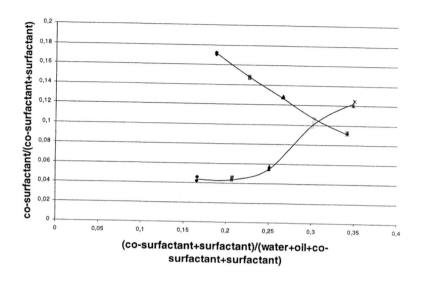

Figure 9c. Phase diagram for propyl heptyl glucoside with dodecanol as co-surfactant. Water/hexadecane ratio equals 1:1

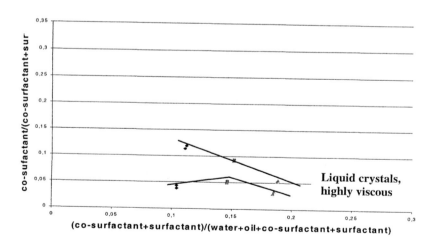

Figure 9d. Phase diagram for propyl heptyl glucoside with dodecanol as co-surfactant. Water/hexadecane ratio equals 4:1.

found between the straight chain decyl glucoside and its branched equivalent, the propyl heptyl glucoside. It is clear also in this investigation that the propyl heptyl glucoside needs much less hydrophobic co-surfactant than the decyl glucoside to create a zero curvature interface (see the fish tail point). This can be understood as a consequence of the structure, see Figure 10.

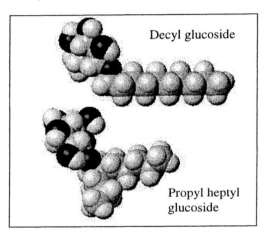

Figure 10: Space filling models of the structures of decyl glucoside and propyl heptyl glucoside

In systems with less oil, less co-surfactant as well as less total amount of surfactant is needed to form the one phase microemulsion, Winsor IV (compare Figures 9a with 9b, and 9c with 9d). This is because less emulsifier is needed to stabilize the smaller amount of oil, and less co-surfactant is lost into the oil phase. In a cleaning situation the amount of oily soil to be detached from the surface is very small compared to the volume of the cleaning solution. Consequently even less co-surfactant would be needed in a real cleaning situation. Even though the fish tail point cannot be detected in Figure 9d, due to highly viscous liquid crystalline phases, a rough extrapolation shows that hardly any co-surfactant would be needed to reach the zero curvature for low amounts of oil in the propyl heptyl glucoside case.

Formulations with different relative amounts of cosurfactant (δ) based on the two glucosides were then tested as cleaning agents, with hexadecane as model dirt according to the procedure described above. As can be seen in Figure 11, the cleaning results given as percent removed hexadecane can be improved by small additions of dodecanol for the straight decyl glucoside. The result became worse when too much dodecanol was added. For the propyl heptyl glucoside, Figure 12, the surfactant with no added dodecanol gives the most efficient dirt removal.

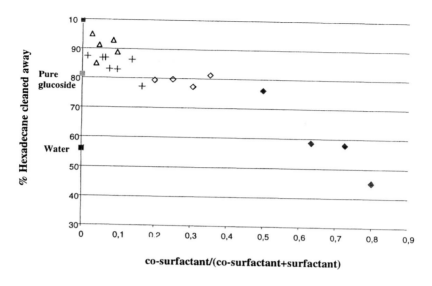

Figure 11. Cleaning results for decyl glucoside on hexadecane at an aluminium surface with different amounts of dodecanol added as co-surfactant

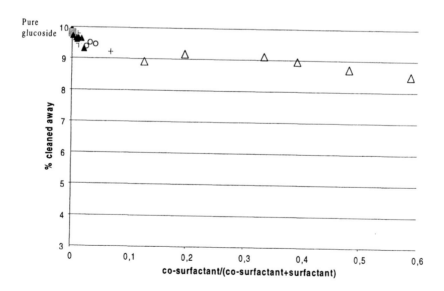

Figure 12. Cleaning of hexadecane from an aluminium surface by propyl heptyl glucoside with different amounts of dodecanol as co-surfactant.

Optimize formulations for particular applications, for example cleaning: Conclusion

We obtained an understanding of the effects of nonionic alkyl glucoside surfactant structure on the curvature of surfactant stabilized oil/water interfaces. Using this understanding, and phase diagrams for the oil/water/surfactant system with varying surfactant structures and varying amounts of cosurfactant, we designed systems with optimized cleaning ability. So far the results compare well with those seen for nonionic ethoxylate surfactants (where temperature effects are used to tune the interface shape and cleaning performance) (Figure 8). However, the optimum is not as pronounced as for the ethoxylates. This may be the result of an unrealistically easy cleaning situation in the current study. Continued work will be done with more difficult dirt, as well as other formulation parameters such as electrolytes and pH. Whether the solubilization into liquid crystalline structures at the interface(16, 18) plays a role in the cleaning procedure will also be investigated.

Overall conclusions

All in all the increased knowledge about the meso structures in water solutions of surfactants with or without additives like oil, co-surfactants, salt etc makes it possible to fine tune formulations and production procedures in ways that rely less on trial and error. This will in the future make product development and work with applications for surface active material more efficient, and open up for a better use of academic knowledge and more efficient co-operation between industry and the academic world.

References

1. Kahlweit, M.; *Tenside Surf. Det.* **1993,** *30,* 83.
2. Winsor, P.A.; *Solvent Properties of Amphiphilic Compounds.* Butterworth: London; **1954.**
3. Shinoda, K.; Friberg, S.; *Emulsions and solubilization,* John Wiley & Sons, Inc. **1986.**
4. Shinoda, K.; Kunieda, H.; Arai, T.; Saito, H.; *J.Phys.Chem.* **1984,** *88,* 5126.

5. Förster, T.; von Rybinski, W.; Wadle, A.; *Advances in Colloid and Interface Science,* **1995,** *58,* 119-149.
6. Wadle, A.; Förster, T.; von Rybinski, W.; *Colloids and Surfaces A,* **1993,** *76,* 51-57.
7. Ozawa, K.; Solans, C.; Kunieda, H.; *J. Coll. Interface Sci* **1997,** *88,* 275-281.
8. Mohlin, K.; Leijon, H.; Holmberg, K.; *J.Dispersion Sci Tech,* **2001,** *22(6),* 569-581.
9. Mohlin, K.; Holmberg, K.; Esquena, J.; Solans, C.; *Colloids and Surfaces A,* to be published **2003.**
10. Salager, J. L.; Marques, L.; Pena, A. A.; Rondon, M.; Silva, F.; Tyrode, E.; *Ind. Eng. Chem. Res.* **2000,** *39,* 2665-2676.
11. A.Forgiarini, J.Esquena, C.Gonzalez, C.Solans, *Progr Colloid Poly Sci* **2000,** *115,* 36-39.
12. Forgiarini, A.; Esquena, J.; Gonzalez, C.; Solans, C.; *Langmuir,* **2001,** *17,* 2076-2083.
13. Boen Ho, *Macro emulsions stabilized by an ethoxylated fatty alcohol and an alkyl quat.* In Proceedings of The third World Conference on Emulsions, Lyon, **2002**
14. Brooks, B. W.; Richmond, H. N.; *Colloids Surf* **1991,** *58,* 131-148.
15. Brooks, B. W.; Richmond, H. N.; *Chem engineering Sci* **1994,** *49 (7),* 1053-1064.
16. Raney, K. R.; Benton, W. J.; Miller, C. A.; *J. Coll Interface Sci* **1987,** *117(1),* 282-290.
17. Benson, H. L.; Cox, K. R.; Zweig, J. E.; *Household Personal Products Industry, HAPPI,* **1985,** *22:3,* 50.
18. Raney, K.; Miller, C. A.; *J.Coll Interface,* **1987,** *Sci 119(2),* 539-549.
19. Malmsten, M.; Lindman, B.; *Langmuir,* **1989,** *5,* 1105-1111.
20. Schambil, F.; Schwuger, M. J.; *Coll Polymer Sci ,* **1987,** *265 ,* 1009-1017.
21. Rubingh D. N.; in *Cationic surfactants, physical chemistry,* ed by Rubingh, D. N.; Holland, P. M.; *Surfactant Science Series;* Marcel Dekker **1991;** vol 37, pp 469-507.
22. Miller, C. A.; Raney, K. H.; *Colloids and Surfaces A,* **1993,** *74,* 169-215.
23. von Rybinski, W.; chap.3 in *Handbook of Applied Surface and Colloid Chemistry,* vol1, edited by Holmberg, K.; John Wiley & sons, **2001**
24. Kunieda, H.; Abstract 287, COLL, *ACS National Meeting,* Boston, 18-22 August **2002.**
25. Balzer, D.; *Langmuir* **1993,** *9,* 3375-3384.
26. Whiddon, C.; Soderman, O.; Hansson, P.; *Langmuir,* **2002,** *18(12),* 4610-4618.
27. Nilsson, F.; Soderman, O.; Hansson, P.; Johansson, I.; *Langmuir,* **1998,** *14 (15),* 4050-4058.

28. Sottman, T.; Kluge, K.; Strey, R.; Reimer, J.; Söderman, O.; *Langmuir*, **2002**, *18*, 3058-3067.
29. Kluge, K.; Stubenrauch, C.; Sottmann, T.; Strey, R.; *Tenside, Surfactants, Deterg.* **2001**, *38*, 30.
30. Kahlweit, M.; Strey, R., *J. Phys. Chem.* **1987**, *91*, 1553.
31. Strandberg, A.; Johansson, I.; *Microemulsions with medium chain alkyl polyglucosides* , CO6 in Proceedings , CESIO conference in Florence, **2000**

Indexes

Author Index

Alexander-Katz, Alfredo, 279
Alexandridis, Paschalis, 60
Bell, Richard C., 191
Berg, Mark A., 177
Bissig, Hugo, 143
Broze, Guy, 376
Bruening, Merlin L., 328
Caldwell, Gregg, 360
Case, Fiona, 376
Cipelletti, Luca, 161
Coveney, Peter V., 206
Cowin, James P., 191
Davis, H. T., 313
Dong, J., 2
Findenegg, Gerhard H., 117
Fraaije, J. G. E. M., 258
Fredrickson, Glenn H., 279
Freedman, Miriam, 360
Fütterer, Tobias, 117
Goldbeck-Wood, Gerhard, 227
Grigoras, Stelian, 290
Harris, J. Keith, 328
Hellweg, Thomas, 117
Hill, R. M., 2
Ho, O. Boen, 394
Iedema, Martin J., 191
Johansson, Ingegärd, 394
Kindt, James T., 298
Lin, B., 313
Manley, Suliana, 161
Mao, G., 2
McCormick, A. V., 313
McGrother, Simon, 227
Meneghini, Frederico, 242

Ming, Lam Yeng, 227
Mohanty, Sanat, 313, 360
Moreira, André G., 279
Nobbmann, Ulf, 44
Noro, Massimo G., 242
Pashkovski, Eugene, 161
Penfold, J., 96
Rojas, Luis, 143
Romer, Sara, 143
Rose, Gene D., 328
Ross, Richard B., 360
Scheffold, Frank, 143
Schlossman, Mark L., 81
Schmidt, J., 17
Schurtenberger, Peter, 143
Sevink, G. J. A., 258
Somoza, Mark M., 177
Spicer, Patrick T., 346
Staples, E., 96
Stilbs, Peter, 27
Stradner, Anna, 143
Talmon, Y., 17
Terech, P., 17
Thomas, Cristina U., 360
Thomas, R. K., 96
Tikhonov, Aleksey M., 81
Trappe, Veronique, 143
Tucker, I., 96
Vavrin, Ronny, 143
Wang, Hanfu, 191
Warren, Patrick B., 242
Weitz, David, 161
Zvelindovsky, A. V., 258

Subject Index

A

Adhesion
adhesive bonding via diffusion, 367*f*
block copolymers, 368–369
complex fluids/nanocomposite
materials and mesoscale, 372–374
critical wetting, 366–367
description, 366–367
formation of interfaces via block
copolymers, 369*f*
fractal analysis of asphalt, 371*f*
fractal analysis of rough surfaces,
371*f*
fractals, 370–372
illustration of mesoscopic algorithm
to treat nanoparticles, 373*f*
interfaces, 368–369
microphase separation of 50/50
diblock copolymer, 370*f*
molecular dynamics sequence for
optimizing model, 368*f*
self-assembling materials (SAM)
forming micelles and showing
complex fluid behavior, 372*f*
solubility and, 367–368
solubility matching, 367*f*
types of failure, 367*f*
wetting, 366–367
See also Advanced material
development
Adhesives, development and
mesoscale, 365
Advanced material development
adhesion and solubility, 367–368
adhesion and wetting, 366–374
adhesive development and
mesoscale, 365
adhesives and fractals, 370–372

adhesives and interfaces: block
copolymers, 368–369
complex fluids/nanocomposite
materials and mesoscale, 372–374
computer simulations, 361
crystal structures, 364*t*
efforts for molecular and mesoscopic
understanding, 362–363
fluoromaterials, 363, 365
fractal analysis, 371*f*
high temperature self-assembled
material (SAM), 366
illustration of drop formation on
surfaces, 365*f*
intermediate (I) phase, 363
materials modeling, 361–362
mesoscale, 362
monoclinic (M) phase, 363
multiscale approach for material
behavior, 361*f*
rhombohedral (R) phase, 363
self-assembling materials and
mesoscale, 363, 365
structure-property predictions, 361
super-cell of M-phase, 364
Aggregate
alkyl chain order and dynamics, 36
associative thickeners, 36
cross section of fiber, 135, 137–
139
micro-viscosity from spin relaxation
data, 37
Aggregate structure
geometric packing factor, 330–331
solid/liquid interface by soft-contact
AFM, 3
Aggregation
complex fluids, 145
destabilized suspensions, 149–158

417

rheology and diffusing wave spectroscopy (DWS), 153*f*
See also Colloidal suspensions
Aggregation and fragmentation, Becker–Döring model, 219–220
Aging
definition, 162
jammed systems, 165–166
See also Soft glassy materials (SGM)
Algorithms
dissipative particle dynamics (DPD), 243–245
external potential dynamics, 264
illustration of mesoscopic algorithm to treat nanoparticles, 373*f*
Metropolis Monte Carlo, 299
polydisperse insertion (PDI), 306
reversible growth of clusters, 300–301
n-Alkanol monolayers
disorder in chain, 89
electron density in head-group region, 90–91
electron density profile for triacontanol monolayer at water-vapor interface, 88*f*, 90
electron density profile normal to interface for triacontanol at water-hexane interface, 87, 88*f*
fitting parameters for triacontanol monolayers, 89*t*
interfacial tension vs. temperature for, at water-hexane interface, 86*f*
monolayers at water-hexane interface, 85–90
nomenclature for interfacial layers, 84*f*
oscillations in reflectivity, 86–87
sample preparation, 83
study at water-hexane interface, 82–83
temperature variation of monolayers at water-hexane interface, 91–92
triacontanol monolayer at water-vapor interface, 90–91

X-ray reflectivity normalized to Fresnel reflectivity vs. temperature, 92*f*
X-ray reflectivity vs. wave vector transfer normal to interface, 87*f*
See also Water-hexane interface
Alkyl chain order, aggregates, 36
Amphiphiles
dissipative particle dynamics, 216–217
lattice-Boltzmann methods, 211, 213, 216
lattice gas automaton models, 210–211
models of fluid dynamics, 208–217
self-assembled structures, 61
solvents and self-assembly, 61–62
statistical mechanics, 209*f*
surfactants, 207
See also Mesoscale fluid systems; Surfactants; Vesicles
Anionic surfactants. *See* Surfactant mixtures
Anthracene
calibration and validation of nanoviscometer, 180–181
probe object, 179
rotation times of, in various solvents, 181*f*
structure and van der Waals dimensions, 179*f*
See also Nanoscale vs. macroscale friction
Asphalt, fractal analysis, 371*f*
Associative thickeners, aggregation, 36
Atomic force microscopy (AFM). *See* Trisiloxane surfactants
Atomistic simulations
ladder structures, 291–293
See also Sol-gel chemistry
Autocorrelation function
analysis by inverse Laplace transformation, 122–123
dynamic light scattering, 48–49

field, for different contributions, 125, 127*f*
intensity measured and fits, 125, 126*f*
intensity time, 122
normalized electric field, 121–122

B

Becker–Döring model
aggregation and fragmentation process, 219
coarse-graining procedure, 220, 221*f*
fixed points of renormalization mapping in two-dimensional space, 222*f*
kinetic equations, 219–220
microscopic rate processes, 222–223
predictions of coarse-grained, of self-reproducing micelles, 223*f*
rate coefficients, 221–222
renormalizability, 224
See also Mesoscale fluid systems
Binding studies, self-diffusion, 33–34
Blankschtein model
critical micelle concentration (CMC), 319
free energy of micellization, 314–315
Block copolymers
adhesives and interfaces, 368–369
effect of polar organic solvents on phase behavior of $EO_{37}PO_{58}EO_{37}$ in water, 74*f*
effects of solvents on self-assembly, 75, 77
ethanol effects, 75
generalized indirect Fourier transformation (GIFT) method, 64
glycerol addition to poly(ethylene oxide)–poly(propylene oxide) [PEO–PPO], 71, 74
glycerol effects, 75
lyotropic liquid crystalline structures, 67*f*
micellization boundary for $EO_{37}PO_{58}EO_{37}$, 71*f*

PEO–PPO, 69–71
percent variation of lattice parameter at different polar organic solvent contents, 76*f*
predicting phase diagrams for, in water, 232
self-assembly, 62
small angle scattering techniques, 62
solvents and self-assembly, 61
structural information from SANS of $EO_{37}PO_{58}EO_{37}$ solutions vs. organic solvent content, 72*f*, 73*f*
temperature and PEO–PPO, 70–71
See also Ethylene oxide/1,2-butylene oxide (EO/BO) diblock copolymers; Small angle neutron scattering (SANS); Small angle x-ray scattering (SAXS)
Bragg diffraction peaks, positions for space groups, 68–69
Brownian motion
changing to subdiffusive motion, 152
light scattering, 47–48
rotation time, 178

C

Calibration, nanoviscometer, 180–181
Cetyltrimethylammonium bromide (CTAB) surfactant
addition of inorganic or aromatic salts, 316, 318
addition of sodium salicylate (NaSal), 314
Blankschtein model, 324
comparing cylindrical and spherical micelles, 323*f*
complementary model, 316
critical micelle concentration (cmc) values of CTAB solution by different models, 318*t*
CTAB–NaBr system, 318–319
CTAB–NaSal system, 320–324
CTAB solution, 318

effects of NaBr on micellization, 319, 320f
interaction energies of pure components by complementary model, 321t
models for free energy of micellization, 314–316
Monte Carlo (MC) simulations, 315–316
Nagarajan model, 315
nonideality of mixing, 322
properties of CTAB micelles in NaBr by static light scattering, 319t
schematic of complementary model, 317f
total interaction energy, 321
values of β parameter for various compositions of CTA^+/Sal^-, 322t
Chemical potentials, definition of intrinsic, 262–263
Cholesterol monohydrate crystal, cryo-transmission electron microscopy (TEM), 20f
Cleaning process
decyl glucoside cleaning hexadecane, 409f
detergency effects of C12E4 and C12E5 against hexadecane, 404f
formulations, 377
hexadecane as model dirt, 405
measuring hard surface cleaning, 405
model dirt, 405
optimizing formulations for applications, 403–405, 408
optimum cleaning conditions, 403
phase diagram for decyl glucoside with dodecanol as co-surfactant, 406f
phase diagram for propyl heptyl glucoside with dodecanol as co-surfactant, 407f
propyl heptyl glucoside cleaning hexadecane, 409f
space filling models of decyl glucoside and propyl heptyl glucoside, 408f

See also Superwetting cleaners
Cluster formation, suspensions, 144f
Cluster growth and fragmentation, Becker–Döring model, 219–220
Coarse-graining procedure
Becker–Döring model, 220, 221f
dissipative particle dynamics, 216
Colloidal particles
irreversible aggregation, 145
properties, 144
Colloidal suspension
3D cross correlation (3DDLS) experiments, 146
aggregation and gelation in complex fluids, 145
aggregation and gelation in destabilized suspensions, 149–158
averaged short-time diffusion coefficient, 150
correlation functions during aggregation and gelation, 150–151
critical cluster radius, 145
crystallization, 147
diffusing wave spectroscopy (DWS), 148–149
diffusion limited cluster-cluster aggregation (DLCCA), 145
DWS and rheological measurement combination, 154–156
DWS for characterization of dynamics of gels, 157–158
dynamic structure factor, 149
elastic modulus of gel, 154
electrostatic repulsion, 147–148
fractal character, 149
gel point and property changes, 152–153
hard sphere potential, 147
in situ variation of ionic strength, 149–150
interaction between colloidal particles, 146
limiting Brownian excursions of scatterers, 151–152
linking microstructural and macroscopic properties, 145–146

mean square displacement vs. time for gels, 151*f*

neutron scattering intensity and resulting structure factor for polystyrene (PS) suspension, 148*f*

nonergodicity limitation of light scattering in concentrated systems, 151–152

particle mean square displacement in sol and gel, 151*f*

reaction limited cluster-cluster aggregation (RLCCA), 145

rheology and DWS during aggregation and sol-gel transition, 153*f*

SANS–DWS experiment, 154–156

schematic of set-up for simultaneous light and small angle neutron scattering (SANS), 156*f*

shortcomings of DWS–SANS, 155–156

short time dynamics, 154

sol-gel transition, 151*f*

spring constant, 154

static structure factors S(qa) of deionized suspensions of polystyrene latex particles, 147*f*

structure and dynamics of stable suspensions, 146–149

study of aggregation and gelation, 145

supercooled liquids, 146

suppressing contributions from multiple scattering, 146

time dependence of mean square displacement, 152

time evolution of q-dependence of scattered intensity of particle suspension, 155*f*

time evolution of storage and loss moduli for destabilized latex particles, 150*f*

time evolution of suspension and aggregate structure during sol-gel process, 156–157

transition from liquid to solid state, 151*f*

volume fraction and dynamic properties of gel, 153–154

Colloidal systems, particles, 118

Complementary model

free energy of micellization, 316

interaction energies of pure components, 321*t*

schematic, 317*f*

Complex fluids

mesoscale, 372–374

properties depending on micelles, 61

Composition, ethylene oxide/1,2-butylene oxide (EO/BO) diblock copolymers, 337*t*

Compressed emulsions, dynamic structure factor decay, 166, 167*f*

Computer simulations, advanced materials, 362

Concentration range, ethylene oxide/1,2-butylene oxide (EO/BO) diblock copolymers, 337*t*

Confined polymer melts, diblock copolymer, 268–269

Continuum hydrodynamics, Stokes–Einstein–Debye (SED) model, 178

Contour length, particles, 123

Contrast matching experiment, particle characterization, 133–134

Copolymers. *See* Block copolymers; Ethylene oxide/1,2-butylene oxide (EO/BO) diblock copolymers

Core-corona form factor

block copolymers, 64

fits to scattering results, 66–70

Creep

creep compliance for xanthan paste, 172*f*

time-domain creep compliance, 163

xanthan paste, 170

Critical cluster radius, 145

Critical micelle concentration (cmc)

addition of salt, 319

calculation, 233

dynamic light scattering, 55

422

Critical micelle temperature (cmt), poly(ethylene oxide)–poly(propylene oxide) [PEO–PPO], 71
Critical packing factor
  aggregate structure formation, 330–331
  ethylene oxide (EO) length, 336
Critical point, three-component oil/water/amphiphile systems, 383–385
Cryo-scanning electron microscopy (SEM), ethylene oxide/1,2-butylene oxide (EO/BO) structures, 334, 335f
Cryo-transmission electron microscopy (TEM)
  cholesterol monohydrate crystal, 20f
  combination with indirect techniques, 18–20
  cubosomes by ultrasonic treatment of bulk cubic phase, 348f
  digital imaging, 19
  poly(paraphenylene) oligomers, 120
  small angle scattering of X-rays (SAXS) or neutrons (SANS), 19–20
  See also Sodium lithocholate (SLC)
Crystallizability, proteins, 53–54
Cubosomes
  cryo-TEM image of, by ultrasonic treatment, 348f
  dilution process with Poloxamer 407 solution, 347–348
  droplet of isotropic liquid crystallizing into cubic phase, 356, 357f
  equilibrium ternary phase diagram for ethanol-monoolein-water system, 351f
  ethanol-monoolein-water system, 350
  experimental, 349
  formation experiments, 349
  formation methods, 347

inclusion of Poloxamer 407 and formation mechanism, 350–351
interface between droplet of monoolein-ethanol solution and aqueous Poloxamer 407 solution, 353f
materials, 349
methods, 349
microscopy method, 349
myelins forming in system with bicontinuous cubic phase, 354
myelin structures, 352–353
observations at surface-water interfaces, 352–353
optical micrograph of, without Poloxamer 407, 352f
order of addition and morphology, 355
particle formation, 347, 353–354
Poloxamer 407-ethanol-monoolein-water system, 350–356
raspberry texture, 356

D

trans-Decalin, orientation of principal axes of rotation, 37f
Decyl glucoside
  cleaning formulations, 405
  cleaning results for, on hexadecane, 409f
  phase diagram for, with dodecanol as co-surfactant, 406f
  space filling model, 408f
  See also Cleaning process
Degree of binding, equation, 34
Density functional theory (DFT), MesoDyn simulation, 230–231
Detergency effects
  cleaning against hexadecane vs. temperature, 404f
  See also Cleaning process
Detergents
  filtration, 55
  hydrodynamic radius, 56t

light scattering, 55–57
Development. *See* Advanced material development
Diblock copolymer
confined polymer melt, 268–269
simulation for dissimilar substrates at top and bottom of simulation box, 271*f*
tapping-mode scanning force microscopy (SFM) images of polystyrene-*b*-polybutadiene-*b*-polystyrene (SBS) films, 270*f*
Diblock copolymers. *See* Ethylene oxide/1,2-butylene oxide (EO/BO) diblock copolymers
Dielectric constants, porous glasses, 294, 295*f*
Diffusing wave spectroscopy (DWS)
aggregation and sol-gel transition for latex particle suspension, 153*f*
characterizing dynamics of gels, 157–158
measurements for xanthan paste, 170–171
multiple scattering, 148–149
small angle neutron scattering (SANS)–DWS experiment, 154–156
xanthan paste, 168, 170
See also Colloidal suspensions
Diffusion
adhesive bonding via, 367*f*
coefficients for translational and rotational, 123–124
translational and rotational, 123
Diffusion limited cluster-cluster aggregation (DLCCA), 145
Digitial imaging, cryo-transmission electron microscopy (TEM), 19
Dilution
inversion at constant temperature with, 397–400
See also Cubosomes; Phase inversion
Dipole interaction, light scattering, 46
Dissipative particle dynamics (DPD)
basic algorithm, 243–245

bottom-up approach, 242
coarse-grained dynamics method, 229
continuous-time Langevin scheme, 216–217
dissolution of surfactants, 247–248
equilibrated mono-lamellar vesicle, 252
flat aggregate coexisting with larger vesicle, 252
free-standing bilayer, 255
homogeneous solution of micelles, 252
hydrodynamics, 242
many-body DPD, 245–246
membranes, 253–255
mesoscale structure predictions, 378–379
modeling method, 231
parameterization, 231–232
particle-based simulation, 243
plot of fluctuation spectrum as function of wave vector, 255*f*
polymers, 246–247
predicting non-equilibrium thermodynamic behavior, 378
snapshots of DPD starting from patch of crystalline bilayer, 249, 250
three-component oil/water/amphiphile systems, 385, 386*f*
top-down approach, 242–243
two small micelles coexisting with larger vesicle, 252
vesicle formation and growth, 248–253
See also Mesoscale modeling
Dodecane uptake, effect of micelle shape, 381, 382*f*
Drug delivery
MesoDyn simulation, 236*f*
mesoscale modeling, 233, 235, 237
transmission electron microscopy (TEM) image of pluronic in water, 235*f*
See also Mesoscale modeling

424

Dynamic density functional theory (DDFT)
predicting morphology, 259–260
See also Phase separation dynamics
Dynamic light scattering
amplitudes, 129
autocorrelation function, 49f, 121
autocorrelation technique, 48f
Brownian motion, 47–48
concentration dependence of mean fiber length, 127
critical micelle formation (cmc), 55
crystallizability, 53–54
dependence of mean length of fiber aggregates on polymer concentration, 128f
detection limit, 52–53
detergents, 55–57
exponential decay function, 49
field autocorrelation function of each contribution, 125, 127f
hydration and shape of proteins, 53
hydrodynamic radius, 49–50
hydrodynamic radius from globular proteins, 52f
hydrodynamic radius of detergents, 56t
hydrodynamic radius of surfactant micelles, 127–128
indirect Fourier transformation method, 118–119
intensity autocorrelation function, 49
intensity correlation function, 124
intensity time autocorrelation function, 122
intermediate scattering function, 121
inverse Laplace transformation, 118
inverse Laplace transformation for size distribution, 50–51
mean fiber aggregate length and polymer concentration, 125–127
measured intensity autocorrelation function and fits, 125, 126f
measuring diffusion of particles, 48
micelles, 54–55
molecular weight of proteins, 53

normalized electric field autocorrelation function, 121–122
normalized fiber contribution, 124–125
persistence length and contour length, 123
polymers, 57
protein shape and structure, 51–54
rotational diffusion, 123
scattering vector, q, dependence of relative amplitudes, 128–129
sensitivity curve, 52f
static model describing q dependence of relative amplitudes, 129–130
stretched exponential functions (Kohlrausch–Williams–Watts, kww), 122–123
structure factor, 129
surfactant Triton X-100, 55f
techniques, 48f
translational and rotational diffusion coefficients, 123–124
viruses, 57
See also Light scattering; Static light scattering
Dynamics
stable suspensions, 146–149
See also Phase separation dynamics
Dynamic structure factor
compressed exponential shape, 165
decay for compressed emulsions, 166, 167f
jammed systems, 165–166

E

Effective correlation length, polymer solutions, 287, 288f
Effective temperature, definition, 163
Electric field, ions, 200–201
Electric field alignment
dynamics, 266–268
scattering intensity vs. azimuthal angle and time, 267
Electron density profile

equation, 84
head-group region for alkanols, 90–91
triacontanol monolayer at water-hexane interface, 87, 88*f*
triacontanol monolayer at water-vapor interface, 88*f*, 90
Electron diffraction, combining imaging with, 19
Electronic industry, sol-gel chemistry, 291
Electrophoretic nuclear magnetic resonance (NMR), polymer and surfactant systems, 35–36
Emulsification
ternary plot showing paths, 399*f*
See also Cubosomes
Emulsions
dynamic structure factor decay of compressed, 166, 167*f*
inversion at constant temperature with dilution, 397–400
inversion with addition of surfactant, 401–402
inversion with temperature, 396–397
small droplets with little shearing force, 396–403
Winsor phases, 394–395
See also Phase inversion
Equilibrium polymers. *See* Grand canonical Monte Carlo (GCMC) simulation
Ethanol, changes of micelle structure with addition, 74
Ethylene oxide/1,2-butylene oxide (EO/BO) diblock copolymers
composition, molecular weight, and concentration range for vesicle formation, 337*t*
cryo-scanning electron microscopy (cryo-SEM), 334, 335*f*
delamination of lamellae, 341–342
effect of EO/BO composition on structure formation, 336–340
effect of EO length, 336–337

effect of extrusion on EO/BO vesicles, 341–342
effects of polydispersity on vesicle formation, 338–340
formation of multi-lamellar vesicles (MLVs), 329
formation of uni-lamellar vesicles (ULVs), 329
fraction of copolymer in structures, 335
mean ultra-sonicated structure diameters vs. EO block length, 339*f*
optical microscopy, 333–334
particle size distribution after various treatments, 341, 342*f*
plane-polarized light micrograph of MLVs in dispersion of $EO_{11}BO_{11}$, 333*f*
plane-polarized light micrograph of vesicle budding from myelin tip at $EO_{11}BO_{11}$-water interface, 334*f*
plane-polarized light microscopy (PLM) image of $EO_{19}BO_{11}$ dispersion at room temperature and after heating at 85°C, 338*f*
PLM image of twisted myelin, 340*f*
possible vesicle-forming surfactants, 331–332
shear stability, 340–342
stability of EO/BO structures, 340–343
structure formation and detection, 332–335
structure of ethoxide-initiated, hydroxy-terminated diblock copolymer, 332*f*
synthesis by two-step batch process, 332
thermal stability of liquid crystalline structures in EO/BO MLVs, 342–343
ultra-sonicated structures, 337–338
External fields, nanotechnology, 264–265

External potential dynamics, algorithm, 264

F

Fiber aggregates
  cross section, 135, 137–139
  mean length and concentration dependence, 127
Field theoretic polymer simulation (FTPS), field theory, 280
Field theory, polymer solutions, 280
Film voltage
  experimental 90%, fall-off temperature vs. water coverage for 3-methylpentene (3MP) films, 200f
  measurement, 193
  simulation of, vs. temperature, 195–196
  temperature evolution of, for Cs^+ migration through 3MP film, 199f
  See also Nano-scale liquid interfaces
Finite element simulation, polymer blends, 237–238
Flory–Huggins parameter
  dissipative particles dynamics (DPD) of polymers, 246–247
  mixing free energy, 232
Fluctuation-dissipation, plot for xanthan paste, 171, 173
Fluctuation-dissipation theorem (FDT)
  dissipative particle dynamics (DPD), 231
  simulating field theory, 283
  violation, 162–163
Fluidity
  free surface, 197
  interfacial, 193–197
  parameterizing interfacial effects, 196–197
  shielding ions from interfacial fluidity perturbations, 198
  surface-enhanced, 196
  See also Nano-scale liquid interfaces

Fluoromaterials, self-assembling materials, 363, 365
Form factor, block copolymers, 63–64
Formulations
  optimizing for cleaning, 403–405, 408
  See also Cleaning process; Personal care industry
Fourier transform pulsed-gradient spin-echo (FT–PGSE). *See* Pulsed-gradient spin-echo (PGSE)
Fractals
  adhesives and, 370–372
  analysis of asphalt, 371f
  analysis of rough surfaces, 371f
Free energy, finding minimum, 263–265
Free energy functional, definition, 260
Free energy of micellization
  Blankschtein model, 314
  complementary model, 316
  Nagarajan model, 315
Friction
  viscosity, 178
  See also Nanoscale vs. macroscale friction

G

Gaussian thread model, polymers in coarse-grained fashion, 280–281
Gelation
  complex fluids, 145
  destabilized suspensions, 149–158
  suspensions, 144f
  See also Colloidal suspensions
Gel point, property changes, 152–153
Gels, applications, 145
Generalized indirect Fourier transformation (GIFT)
  block copolymer micelles, 64
  block copolymers, 64
Generalized Stokes–Einstein relation (GSER)

study of colloidal systems by violating, 163
violation using light scattering study, 164
Geometric packing factor, aggregate structure formation, 330–331
Glucosides. *See* Cleaning process
Glycerol, changes of micelle structure with addition, 74
Grand canonical Monte Carlo (GCMC) simulation
acceptable probability, 302
advantages, 299
algorithms for reversible growth of clusters, 300–301
application of polydisperse insertion (PDI) method, 304–305
chemical potential vs. monomer concentration, 307*f*
configuration-bias Monte Carlo (CBMC) approach, 300
degree of control, 299
equilibration of system composition, 300
equilibrium polymer self-assembly definitions, 304
mean chain length vs. monomer concentration, 307*f*
methods for self-assembled network, 308–309
middle-up approach, 299
PDI, 300, 302–303
PDI approach advantage, 303–304
phase behavior of equilibrium network model, 310*f*
polydisperse, self-assembled systems, 300–306
probabilities, 301–302
schematic of PDI MC step, 303*f*
self-assembled network, 306, 308–309
semiflexible equilibrium polymers, 304–306
snapshots of isotropic and nematic phases of model equilibrium polymer, 308*f*

top-down method, 299
transition probability, 302
weighting function definition, 301

H

Head group area, silica substrate, 14
Hertz-dipole profile, light scattering, 46
Hexadecane
cleaning results for decyl glucoside, 409*f*
cleaning results for propyl heptyl glucoside, 409*f*
model dirt, 405
See also Cleaning process
Hydration, proteins, 53
Hydrodynamic radius
comparing surfactant micelles in surfactant solution with value in presence of polymer, 127–128
ion hydration, 201
Stokes–Einstein equation, 49–50
visualization, 50*f*
Hydrodynamics
anthracene rotation as viscometer, 181
dissipative particle dynamics (DPD), 242–243
generalizing to molecular levels, 178–179
rotation time, 178
Stokes–Einstein–Debye (SED) model, 178
See also Dissipative particle dynamics (DPD)

I

Indirect Fourier transformation
analysis of fiber cross section, 137–139
scattering length density distribution, 118–119

Intensity, light scattering, 47
Intensity correlation function,
  contributions by three species,
  124–125
Interaction, colloidal particles, 144
Interfaces
  adhesives and, 368–369
  See also Nano-scale liquid interfaces;
    Surfactant mixtures; Water-hexane
    interface
Interfacial area, block copolymer,
  67–68
Interfacial tension, function of
  temperature for alkanols at water-
  hexane interface, 86*f*
Interlayer dielectric (ILD), electronic
  industry, 291
Intrinsic chemical potentials,
  definition, 262–263
Inverse Laplace transformation, field
  autocorrelation function, 118
Ion mobility, Stokes formula, 195
Ions
  collective electric field of, 200–201
  hydrodynamic radius due to
    hydration, 201
  motion in presence of trap, 202
  solvation, 200
  See also Nano-scale liquid interfaces

J

Jammed systems
  aging, 165–166
  soft glasses, 163–164
  violation of generalized Stokes–
    Einstein relation (GSER) using
    light scattering study, 164
Jump-in event, description, 7

K

Kinetics
  self-reproduction of micelles, 218

tar removal equation, 391

L

Lattice-Boltzmann methods
  complex fluids, 211, 213, 216
  cubic Schwarz "P" and lamellar
    mesophases, 214, 215
  H-theorem, 213
Lattice gas automaton models
  arrest of domain growth, 214
  color flux and color field, 212
  immiscible mixtures, 210–211
  microemulsion phase formation, 214
  phase separation process, 212
Lattice parameter
  percent variation at different polar
    organic solvent contents, 76*f*
  structure information, 66
Light scattering
  Debye plot, 47
  dipole interaction, 46*f*
  dynamic, 47–51
  Hertz-dipole profile, 46
  ideal case of linearly polarized light,
    46
  intensity, 47
  isotropic scattering profile, 46*f*, 47
  schematic, 45*f*
  schematic setup for simultaneous
    light and small angle neutron
    scattering (SANS), 156*f*
  static, 46–47
  theory, 44–46
  violation of generalized Stokes–
    Einstein relation (GSER) using,
    164
  See also Dynamic light scattering;
    Static light scattering
Lipari–Szabo model, motional
  dynamics, 30
Liquid interfaces. *See* Nano-scale
  liquid interfaces
Long chain alkanol monolayers. *See*
  n-Alkanol monolayers

Loss modulus, suspension of
destabilized latex particles, 150*f*
Lyotropic liquid crystalline
microstructures, block copolymer
self-assembly, 67*f*

M

Macroviscosity
entanglement, 185
ordinary viscosity, 179
Magnetic glasses, aging, 162
Many-body dissipative particle
dynamics (DPD), requirements,
245–246
Mechanism
assembly of sodium lithocholate, 25
intramolecular spin relaxation, 30
Membranes
dissipative particle dynamics (DPD),
253–255
DPD simulation of free-standing
bilayer, 255
plot of fluctuation spectrum as
function of wave vector, 255*f*
MesoDyn simulation
coarse-grained dynamics method,
229
modeling method, 230–231
parameterization, 231–232
See also Mesoscale modeling
Mesoscale
adhesive development, 365
complex fluids/nanocomposite
materials and, 372–374
control of rheological behavior, 377–
379
definition, 394
predicting structure in surfactant
solutions, 377–379
self-assembling materials, 363, 365
See also Superwetting cleaners
Mesoscale calculations
polymeric matrix, 293–296
See also Sol-gel chemistry

Mesoscale fluid systems
arrest of domain growth, 211, 214
Becker–Döring model of aggregation
and fragmentation process, 219–
224
cell membrane showing lipid bilayer,
207*f*
color flux and color field definitions,
212
course-graining scheme for Becker–
Döring equations, 221*f*
cubic Schwarz "P" and lamellar
mesophases, 214, 215
depiction of surfactant mesophases,
208*f*
discrete models of amphiphilic fluid
dynamics, 208–217
dissipative particle dynamics, 216–
217
experimental measurement of self-
reproduction reaction kinetics,
218*f*
H-theorem, 213
interpretation of Boltzmann's
program for statistical mechanics,
209*f*
lattice-Boltzmann methods, 211, 213,
216
lattice gas automaton methods, 210–
211
micelle formation, 207–208
micelle-vesicle transition, 207*f*
microemulsion phase formation, 211,
214
modeling and simulation methods
across length- and time-scales,
209*f*
phase separation kinetics in binary
immiscible fluid mixture, 212
phase separation process of binary
mixture of two immiscible lattice
gas fluids, 212
predictions of coarse-grained
generalized Becker–Döring model
of self-reproducing micelles,
223*f*

renormalization methods in nucleation and growth processes, 217–224

self-reproduction of micelles, 217–218

stepwise formation of spherical micelles, 207*f*

surfactants, 207

two-dimensional triangular lattice, 210*f*

two-dimensional Voronoi tessellation of field of massive, point particles, 217*f*

See also Becker–Döring model

Mesoscale modeling

calculation of critical micelle concentration (cmc), 233

coarse-grained dynamics methods, 229

comparison of modeling techniques, 230*t*

core radii of self-assembled micelles in solution of pluronic in water vs. pluronic concentration, 237*f*

dissipative particle dynamics (DPD), 231

drug delivery, 233, 235, 237

effect of temperature, 233, 234

experimental transmission electron microscopy (TEM) image of 10% pluronic F127 in water, 235*f*

finite element methods, 231

Flory–Huggins parameter, 232

limitations, 228

MesoDyn, 230–231

MesoDyn simulation of pluronic F127 in water, 236*f*

MesoDyn simulation of polystyrene and polybutadiene mixture, 239

method description, 229–231

need for, 228

parameterization, 231–232

phase morphologies, 229

Pluronic 85 in water at 15°C and 70°C from MesoDyn simulation, 234

polymer blends, 237–238

predicted flux rates, 238, 240

predicting phase diagrams for block copolymers in water, 232

validation studies, 232–233

Mesoscale structure

aggregate micro-viscosity from spin relaxation data, 37

aggregation of hydrophobically modified polymers, 36

alkyl chain order and dynamics in aggregates from nuclear spin relaxation, 36

amplitude of spin-echo, 31–32

applications of electrophoretic NMR to polymer and surfactant systems, 35–36

associative thickeners, 36

degree of binding, 34

diffusion and NMR-based electrophoretic mobility in sodium dodecyl sulfate/poly(ethylene oxide) (SDS/PEO), 35*f*

dissipative particle dynamics (DPD) predictions for three-component system, 388*f*

dynamics in fluids through nuclear spin relaxation rates, 29–30

effective self-diffusion coefficient, 34

Fourier transform pulsed-gradient spin-echo (FT–PGSE), 28

FT–PGSE NMR self-diffusion studies for quantifying, 33

instrumental and methodological progress, 39–41

intramolecular spin relaxation mechanism, 30

issues in NMR characterization, 37–39

Lipari–Szabo model, 30

micellization of surfactants, 34–35

model-free model, 30

modeling random molecular reorientation processes, 30

nuclear magnetic resonance (NMR), 28
obstruction effects, 38–39
optimizing three-component, 385, 387–388
organization and dynamics in fluids through multi-component self-diffusion rates, 31–34
predicting in surfactant solutions, 377–379
problems in NMR investigations of complex fluids, 29
radio frequency pulse phase cycling, 31*f*
self-diffusion data, 33
sequence of proton FT–PGSE spin-echo spectra, 32*f*
simple binding studies, 33–34
Stejskal–Tanner relation, 31–32
three-component oil/water/amphiphile systems, 383-385
three-pulse stimulated echo PGSE variant, 32
trans-decalin and orientation of principal axes of rotation, 37*f*
two-pulse PGSE experiment, 31*f*
two-step model, 30
viscosity corrections, 37–38
See also Pulsed-gradient spin-echo (PGSE)
Mesoscopic regions, study, 82
3-Methylpentane (3MP). *See* Nanoscale liquid interfaces
Micelles
amphiphilic molecules, 61
core radii of self-assembled, 237*f*
effect of shape on oil uptake capacity, 381, 382*f*
formation, 54*f*, 70–71, 207–208
form factor of block copolymers, 63–64
generalized indirect Fourier transformation (GIFT) method, 64

light scattering, 54–55
onset in PEO–PPO block copolymers, 70–71
properties of self-assembled systems, 61
self-assembling materials forming, 372*f*
self-reproduction, 217–218
structure changes with addition of ethanol or glycerol, 74
See also Becker–Döring model; Cetyltrimethylammonium bromide (CTAB) surfactant; Surfactant mixtures
Micellization, surfactants, 34–35
Microemulsions. *See* Emulsions
Mixed dynamics, standard method, 263–264
Mixing, nonideality, 322
Mixtures. *See* Surfactant mixtures
Model-free model, motional dynamics, 30
Modeling
mesoscale, 228–229
random molecular reorientation processes, 30
solvation potential, 201
statistical mechanics, 209*f*
See also Mesoscale fluid systems; Mesoscale modeling
Models
scattering vector dependence in static light scattering experiment, 129
Stokes–Einstein–Debye model, 178, 182, 184
torsional relaxations, 187–189
See also Cetyltrimethylammonium bromide (CTAB) surfactant
Molecular packing parameter, definition, 14
Molecular weight
ethylene oxide/1,2-butylene oxide (EO/BO) diblock copolymers, 337*t*
proteins, 53

Monolayers of surfactants, water-oil interface, 82
Monte Carlo (MC) simulations
  algorithms for reversible growth of clusters, 300–301
  combination with free energy model, 315–316
  configuration-bias MC approach, 300
  free energy predictions by mixing rule, 321
  See also Cetyltrimethylammonium bromide (CTAB) surfactant; Grand canonical Monte Carlo (GCMC) simulation
Morphology
  formation in complex liquids, 259
  mesoscale modeling, 229
  order of addition for cubosomes, 355–356
  predictions using dynamic density functional theory (DDFT), 259–260
Multi-lamellar vesicle (MLV)
  formation, 329–330
  onions, 329
  See also Ethylene oxide/1,2-butylene oxide (EO/BO) diblock copolymers
Multiphase mixture, Pluronic surfactants, water, and oil, 168, 169*f*
Myelin
  formation in system with bicontinuous cubic phase, 354
  interface between droplet of monoolein-ethanol and aqueous Poloxamer 407 solution, 353*f*
  lamellar tubes, 334
  plane-polarized light micrograph (PLM) of twisted, 340*f*
  plane-polarized light micrograph (PLM) of vesicle budding from, 334*f*
  surfactant-water interfaces, 352–353

N

NaBr. *See* Cetyltrimethylammonium bromide (CTAB) surfactant
Nagarajan model
  critical micelle concentration (CMC), 319
  free energy of micellization, 315
Nanocomposite materials, mesoscale, 372–374
Nanoparticles
  smart, 272–276
  See also Smart nanoparticles
Nano-scale liquid interfaces
  collective electric field of ions, 200–201
  deposition of ions, 192–193
  effect of water on motion of ions across oil-water-oil interface, 199
  electric field effect on Cs^+ trapping/escape from $3MP/H_2O/3MP/Pt$, 201*f*
  experimental, 192–193
  experimental 90% film voltage fall-off temperature vs. water coverage for 3MP films, 200*f*
  film deposition method, 192
  film thickness, 196
  film voltage measurement, 193
  hydrodynamic radius and ion hydration, 201
  increased fluidity of free surface, 197
  interfacial effects on fluidity, 196–197
  interfacial fluidity, 193–197
  ion mobility with Stokes formula, 195
  ion motion in presence of trap, 202
  modeling solvation potential, 201
  potential across film ΔV, 193
  shielding ions from interfacial fluidity perturbations, 198
  simulation of film voltage vs. temperature, 195–196
  solvation near interfaces, 197, 199–202

spectrum for transport of D_3O^+ ions through 3MP film, 194*f*
surface-enhanced fluidity, 196
temperature evolution of film voltage for Cs^+ migration through 3MP/Pt, 199*f*
transport of D_3O^+ ions through 3MP film, 194–195
viscosity change for 3MP films, 198
viscosity via Vogel–Tamman–Fucher equation, 195
water for ion solvation, 200
Nanoscale vs. macroscale friction
anthracene as probe object, 179
calibration and validation of nanoviscometer, 180–181
capillary vs. anthracene viscometer, 181
dynamic correlation function, 189
end-flip and crankshaft mechanisms, 189
end-flips and relaxation mechanism, 188
entanglement and macroviscosity, 185
friction as viscosity, 178
macroviscosity, 179
macroviscosity and chain length, 184
macroviscosity and nanoviscosity vs. number of backbone bonds, 183*f*
methods to study motion, 180
model based on torsional relaxations, 187–189
nanoviscosity, 179
number of backbone bonds, 182, 184
PDMS behavior vs. chain length, 187–188
poly(dimethylsiloxane) (PDMS) and poly(ethylene) (PE), 185–187
rigid body motions, 187–188
rotation time, 178
rotation times of anthracene in various solvents, 181*f*
schematic of three types of polymer motion, 188*f*

SED (Stokes–Einstein–Debye) model breakdown and poly(isobutylene) (PIB) results, 186–187
small molecule to polymer conversion in PIB, 182, 184–185
solute rotation, 179
static correlation functions, 189
Stokes–Einstein–Debye (SED) model, 178
structures and van der Waals dimensions of anthracene, PIB, and PDMS, 179*f*
torsional kinetics of PIB and PDMS, 186
transition length ($_{SED}$), 182, 184
Nanotechnology, external fields, 264–265
Nanotubules
electron diffraction from nanotubes, 21–22
formation of single-molecular walled, 18
self-assembly of sodium lithocholate, 22
See also Sodium lithocholate (SLC)
Nanoviscosity
chain length, 184
effective viscosity, 179
See also Nanoscale vs. macroscale friction
Networks. *See* Grand canonical Monte Carlo (GCMC) simulation
Neutron reflectivity
hydrophilic liquid-solid interface, 105*f*
surfactant mixing, 97
theory, 98–100
Neutron scattering, surfactant organization, 82
Neutron scattering intensity, polystyrene (PS) suspension, 148*f*
Nonideality, mixing, 322
Nonionic surfactants. *See* Surfactant mixtures
Nuclear magnetic resonance (NMR)

issues in characterizing mesoscale structure, 37–39
mesoscopic structure and dynamics, 28
obstruction effects, 38–39
problems in investigating complex fluids, 29
tools, 28
viscosity corrections, 37–38
See also Pulsed-gradient spin-echo (PGSE)
Nuclear spin relaxation
alkyl chain order and dynamics in aggregates, 36
dynamics in fluids, 29–30
Nucleation process, study, 217

O

Obstruction effects, characterizing mesoscale structure, 38–39
n-Octadecyltrichlorosilane (OTS)
monolayer coverage, 3–4
substrate surface modification, 4
See also Trisiloxane surfactants
Oil uptake capacity, effect of micelle shape, 381, 382*f*
Optical microscopy
cubosomes, 352*f*
ethylene oxide/1,2-butylene oxide (EO/BO) multi-lamellar vesicles (MLVs), 333–334
Order-disorder transition (ODT)
electric field alignment, 266–268
polymer solutions, 280
Ordering, mesoscopic regions, 82
Oxygen diffusion rates, polymer blends, 238

P

Paper industry
alkyl ketene dimer (AKD) and hydrophobicity of paper, 397–398

See also Emulsions
Particles
colloidal systems, 118
See also Poly(paraphenylene)s (PPP)
Particle size distribution, ethylene oxide/1,2-butylene oxide (EO/BO) after various treatments, 341, 342*f*
Perrin factor, proteins, 53
Persistence length, particles, 123
Personal care industry
dissipative particle dynamics (DPD) method, 378
formulations, 377
mesoscale structure prediction using DPD, 378–379
non-equilibrium thermodynamic behavior, 378
See also Superwetting cleaners
Phase behavior
optimizing three-component, 385, 387–388
polar organic solvents and, of $EO_{37}PO_{58}EO_{37}$ in water, 74*f*
surfactants by dissipative particle dynamics (DPD), 247–248
Phase boundaries, solvent effects, 75
Phase diagrams
alkyl ketene dimer (AKD), water, and surfactant mixture, 398*f*
decyl glucoside with dodecanol as co-surfactant, 406*f*
equilibrium ternary, for ethanol-monoolein-water system, 351*f*
oil, water, and surfactant, tuned with temperature, 395*f*
paraffinic oil, coco benzyl dimethyl ammonium chloride, and C14E4, 401*f*
predicting for block copolymers in water, 232
propyl heptyl glucoside with dodecanol as co-surfactant, 407*f*
quasi-ternary, for oil, water, and surfactant, 394–395
ternary plot showing emulsification paths, 399*f*

three-component oil/water/amphiphile systems, 383–385

Phase inversion
addition of surfactant, 401–402
constant temperature with dilution, 397–400
droplet size and dilution with water, 400
isothermal pseudo-ternary phase diagram for water, alkyl ketene dimer (AKD) and optimized surfactant mixture, 398*f*
making small droplet emulsions with little shearing, 402–403
temperature, 396–397
ternary plot showing emulsification plot, 399*f*
See also Emulsions

Phase morphology, mesoscale modeling, 229

Phase separation, Flory–Huggins parameter mapping, 247

Phase separation dynamics
confined polymer melts, 268–269
dynamics of electric field alignment, 266–268
examples, 265
external fields, 264–265
external potential dynamics algorithm, 264
finding minimum of free energy, 263–265
free energy functional, 260–263
hard films, 269, 271*f*
intrinsic chemical potentials, 262–263
morphologies of $A_{N-M}B_M$ polymer surfactant nanodroplets, 273*f*
morphologies of $A_{N-M}B_M$ polymer surfactant nanogels, 273*f*
order parameter field, 262
scattering intensity as function of azimuthal angle and time, 267
shell position vs. initial radius for different layers, 275*f*

simulation of $A_3B_{12}A_3$ block copolymer film in simulation box, 270*f*
simulation of dissimilar substrates at top and bottom of simulation box, 271*f*
smart nanoparticles, 272–276
tapping-mode scanning force microscopy (SFM) phase images of thin polystyrene-*b*-polybutadiene-*b*-polystyrene (SBS) films on silicon, 270*f*
two-body mean field potential, 261

Phospholipids
formation of uni-lamellar and multi-lamellar vesicles (ULV and MLV), 329–330
limitations, 330

Photon correlation spectroscopy (PCS). *See* Dynamic light scattering

Plane-polarized light microscopy (PLM), ethylene oxide/1,2-butylene oxide (EO/BO) multi-lamellar vesicles (MLVs), 333–334

Pluronic surfactants, multiphase mixtures of, water, and oil, 168, 169*f*

Poloxamer 407 solution. *See* Cubosomes

Poly(butadiene)
MesoDyn simulation of polystyrene and, 239
simulation of polymer blends, 237–238

Poly(dimethylsiloxane) (PDMS)
behavior as function of chain length, 187–188
macroviscosity and nanoviscosity, 185–187
macroviscosity and nanoviscosity vs. number of backbone bonds, 183*f*
structure and van der Waals dimensions, 179*f*
torsional kinetics, 186
See also Nanoscale vs. macroscale friction

Polydisperse insertion (PDI)
  advantage, 303–304
  algorithm, 306
  application of PDI method, 304–305
  description, 300
  schematic, 303*f*
  See also Grand canonical Monte
    Carlo (GCMC) simulation
Polydispersity, effect on vesicle
  formation, 338–340
Poly(ethylene) (PE)
  macroviscosity and nanoviscosity,
    185–187
  macroviscosity and nanoviscosity vs.
    number of backbone bonds, 183*f*
  See also Nanoscale vs. macroscale
    friction
Poly(isobutylene) (PIB)
  breakdown of Stokes–Einstein–
    Debye (SED) model and PIB
    results, 186–187
  macroviscosity and nanoviscosity vs.
    number of backbone bonds, 183*f*
  small molecule to polymer transition,
    182, 184–185
  structure and van der Waals
    dimensions, 179*f*
  torsional barriers, 188–189
  torsional kinetics, 186
  See also Nanoscale vs. macroscale
    friction
Polymer blends
  MesoDyn simulation of polystyrene
    and polybutadiene, 239
  mesoscale and finite element
    simulation, 237–238
  oxygen diffusion rates, 238
  predicted flux rates, 238, 240
Polymer melts, diblock copolymer,
  268–269
Polymer motion
  crankshaft motions, 187–189
  end-flip motions, 187–189
  rigid body motions, 187–188
  types accommodating rotation, 188*f*

Polymeric systems. *See* Phase
  separation dynamics
Polymers
  dissipative particle dynamics (DPD),
    246–247
  electrophoretic nuclear magnetic
    resonance (NMR) of, and
    surfactants, 35–36
  light scattering, 57
  problems in nuclear magnetic
    resonance (NMR), 29
  See also Nanoscale vs. macroscale
    friction
Polymer solutions
  calculating reduced density profile
    across slit, 285
  complex diffusion equation, 282
  complex Langevin dynamics, 283
  cubic lattice, 284–285
  effective correlation length vs.
    parameter BC, 287, 288*f*
  field theoretic polymer simulations
    (FTPS), 280
  field theory, 280
  fluctuation dissipation theorem
    (DFT), 283
  partial function, 281
  reduced density profile across slit,
    285–287
  self-consistent field theory, 282
  simulation method, 283–285
  simulation schematic, 284*f*
  single-polymer partial function, 282
  theoretical background, 280–282
Poly(paraphenylene)s (PPP)
  absolute scattering curves of PPP(12)
    with C_8E_4 solution and C_8E_4 in
    D_2O, 136*f*
  amphiphilically substituted nonionic
    PPP oligomer, PPP(n_n), 119–120
  applications, 119
  arrangement of aromatic backbones
    of PPP(12) oligomers, 138–139
  comparing difference curve with
    scattering curve of mixture with

perdeuterated d-C_8E_4/PPP(12), 135, 136*f*

contrast matching experiment, 133–134

cross section of fiber aggregates, 135, 137–139

cryo-transmission electron microscopy (cryo-TEM), 120

difference curve for fiber aggregates of PPP(12), 137*f*

difference curve representing fiber contribution for PPP(12) plus C_8E_4 solution, 138*f*

dynamic light scattering, 121–130

gel permeation chromatography of PPP(12), 120

indirect Fourier transformation method, 137–139

radial excess scattering length density distribution function, 139*f*

small angle neutron scattering (SANS), 133–139

small angle neutron scattering (SANS) method, 120–121

static light scattering, 130–132

structure of oligomer PPP(12), 119*f*

total SANS intensity vs. mole fraction for PPP(12) solution and d-C_8E_4 in mixtures of H_2O and D_2O, 135*f*

Poly(styrene)

MesoDyn simulation of PS and polybutadiene, 239

simulation of polymer blends, 237–238

Porous glass

dielectric constants, 294, 295*f*

stress, 296

Young's modulus predictions, 294, 295*f*

See also Sol-gel chemistry

Propyl heptyl glucoside

cleaning formulations, 405

cleaning results for, on hexadecane, 409*f*

phase diagram of, with dodecanol as co-surfactant, 407*f*

space filling model, 408*f*

See also Cleaning process

Proteins

crystallizability, 53–54

dynamic light scattering, 51–54

hydration and shape, 53

hydrodynamic radius, 51, 52*f*

molecular weight, 53

sensitivity curve, 52*f*

See also Dynamic light scattering

Pulsed-gradient spin-echo (PGSE)

aggregate micro-viscosity from spin relaxation data, 37

aggregation of hydrophobically modified polymers, 36

alkyl chain order and dynamics in aggregates, 36

amplitude of spin-echo, 31–32

application of electrophoretic NMR to polymer and surfactant systems, 35–36

associative thickeners, 36

component-resolved (CORE) data analysis, 40

degree of binding, 34

diffusion-ordered spectroscopy (DOSY) approach, 40–41

extended application, 39–40

generalized rank annihilation method (GRAM), 40

instrumental and methodological progress, 39–41

micellization of surfactants, 34–35

obstruction effects, 38–39

organization and dynamics in fluids through multi-component self-diffusion rates, 31–34

quantifying mesoscopic structure, 33

radio frequency pulse phase cycling, 31*f*

self-diffusion coefficient, 34

sequence of proton FT–PGSE spectra of neopentanol in sodium dodecyl sulfate (SDS), 32*f*

simple binding studies, 33–34
Stejskal–Tanner relation, 31–32
three-pulse stimulated echo PGSE variant, 32
two-pulse experiment, 31*f*
viscosity corrections, 37–38

Q

Quasi-elastic light scattering (QELS). *See* Dynamic light scattering

R

Radius, critical cluster, 145
Rayleigh ratio, normalized scattering intensity, 130–131
Reaction limited cluster-cluster aggregation (RLCCA), 145
Regular solution theory (RST), surfactant mixing, 97–98
Renormalization, Becker–Döring model, 221–222
Rheology
   aggregation and sol-gel transition for latex particle suspension, 153*f*
   viscous and elastic properties of complex fluids, 20
   xanthan paste, 168, 170
Rotation, trans-decalin and orientation of principal axes of, 37*f*
Rotational diffusion, particles, 123
Rotation time, hydrodynamics, 178
Rough surfaces, fractal analysis, 371*f*
Rouse model, macroviscosity and chain length, 184

S

Scattering intensity, light scattering, 47
Scattering length sensitivities (SLD), block copolymers, 64–65

Scattering techniques
   characterization of colloidal dispersions, 118
   See also Poly(paraphenylene)s (PPP)
Self-aggregation. *See* Sodium lithocholate (SLC)
Self-assembled system
   equilibrium network, 306
   methods, 308–309
   properties depending on micelles, 61
   simulation results, 309, 310*f*
   See also Advanced material development; Grand canonical Monte Carlo (GCMC) simulation
Self-assembly
   block copolymer lyotropic liquid crystalline structures, 67*f*
   block copolymer poly(ethylene oxide)–poly(propylene oxide) [PEO–PPO], 69–71
   block copolymers, 62
   effects of solvents on, 75, 77
   effects of solvents on self-assembly, 75, 77
   mechanism for SLC, 25
   micellization boundary for $EO_{37}PO_{58}EO_{37}$, 71*f*
   sodium lithocholate (SLC) nanotubules, 22
   solvents and amphiphiles, 61–62
Self-consistent field theory, polymer solutions, 282
Self-diffusion
   binding studies, 33–34
   quantifying mesoscopic structure, 33
Self-reproduction, micelles, 217–218
Semiflexible equilibrium polymers
   application of polydisperse insertion (PDI) method, 304–305
   See also Grand canonical Monte Carlo (GCMC) simulation
Shape
   mesoscopic regions, 82
   proteins, 53
Shear, effect on mesoscale structure, 387, 388*f*

Shear stability, ethylene oxide/1,2-butylene oxide (EO/BO), 340–342

Silicone surfactants. *See* Trisiloxane surfactants

Simulations

computer workhorse for advanced materials, 362

static scattering intensity for clusters, fibers, and micelles, 132*f*

statistical mechanics, 209*f*

See also Dissipative particle dynamics (DPD); Mesoscale modeling; Polymer solutions; Sol-gel chemistry

Small angle neutron scattering (SANS)

absolute SANS intensity, 63

absolute scattering curves of PPP(12) plus C_8E_4 and C_8E_4 solution, 136*f*

arrangement of aromatic backbones for PPP(12) oligomers, 138–139

blank curve, 135

colloidal suspensions, 145

contrast matching experiment, 133–134

core-corona model, 64

cross section of fiber aggregates, 135, 137–139

cryo-transmission electron microscopy (TEM), 19–20

difference curve for fiber aggregates of PPP(12), 137

difference curve representing fiber contribution, 136*f*

difference curve representing fiber contribution for PPP(12) plus C_8E_4 solution, 138*f*

difference curves, 135, 136*f*

fits of core-corona model to SANS scattering intensities, 66*f*

form factor of block copolymer, 63–64

indirect Fourier transformation method, 137–139

micellar solutions, 61

poly(ethylene oxide)–poly(propylene oxide) [PEO–PPO] block copolymers, 62–65

PPO scattering length densities (SLD), 65

radial excess scattering length density distribution function, 139*f*

SANS–diffusing wave spectroscopy (DWS), 154–156

scattering intensity at given q value, 133–134

scattering length densities (SLD), 64–65

schematic of block copolymer micelles, 63*f*

schematic of lyotropic liquid crystalline structures, 67*f*

schematic setup for simultaneous light and SANS, 156*f*

structural information of $EO_{37}PO_{58}EO_{37}$ solutions vs. organic solvent content, 72*f*, 73*f*

structure factor, 65

surfactant mixing, 97

theory, 101–102

total SANS intensity vs. mole fraction for PPP(12) and d-C_8E_4 in H_2O and D_2O, 135*f*

volume fraction of polymer in core and corona, 65

volume fraction of polymer in core and polymer in corona, 65

See also Colloidal suspensions

Small angle scattering

study of self-assembly, 61

See also Small angle neutron scattering (SANS); Small angle X-ray scattering (SAXS)

Small angle X-ray scattering (SAXS)

bicontinuous cubic structure, 69

block copolymers polyethylene oxide–polypropylene oxide (PEO/PPO) , 67–69

Bragg diffraction peaks, 68–69

cryo-transmission electron microscopy (TEM), 18, 19–20

crystallographic space group of liquid crystalline phases, 68
diffraction pattern from micellar cubic samples, 70f
diffraction patterns from hexagonal samples, 68f
effects of solvents, 67
interfacial area per block copolymer, 67–68
lattice parameter, 67
micellar cubic samples from $EO_{37}PO_{58}EO_{37}$ glycerol system, 69
Smart nanoparticles
morphologies of $A_{N-M}B_M$ polymer surfactant nanodroplets, 273f
morphologies of $A_{N-M}B_M$ polymer surfactant nanogels, 273f
shell position vs. initial radius for different layers, 275f
simulation, 272–276
See also Phase separation dynamics
Sodium bromide (NaBr). See Cetyltrimethylammonium bromide (CTAB) surfactant
Sodium dodecyl sulfate (SDS). See Surfactant mixtures
Sodium lithocholate (SLC)
mature nanotubes, 21
mechanism of assembly, 25
nanotubules four hours after mixing, 22, 25
spontaneous self-assembly, 22
superposition of tubules, 22
time-resolved cryo-transmission electron microscopy (TEM), 22, 25
vitrified SLC dispersions four hours after mixing, 24f
vitrified SLC dispersions showing intermediate nanostructures, 22, 23f
Sodium salicylate (NaSal). See Cetyltrimethylammonium bromide (CTAB) surfactant
Soft condensed matter, properties depending on micelles, 61

Soft contact atomic force microscopy (AFM)
aggregate structures, 3
See also Trisiloxane surfactants
Soft glassy materials (SGM)
age as waiting time, t_w, 162
age dependence of characteristic time of final decay of dynamic structure factor, 167f
age-dependent relaxation process, 162
age of sample, t_w, 162
aging of jammed systems, 165–166
aging of magnetic glasses, 162
compressed emulsions, 166
creep compliance curves for xanthan paste, 172f
creep measurements of xanthan paste, 170
diffusing wave spectroscopy (DWS) measurements of xanthan paste, 170–171
dynamic structure factor, 165
final decay of dynamic structure factor by multispeckle dynamic light scattering (DLS), 167f
fluctuation-dissipation plot for xanthan paste, 171, 173
fluctuation-dissipation theorem (FDT) violation factor, 163
generalized Stokes–Einstein relation (GSER) violation, 163
jammed colloidal systems as "soft glasses", 163–164
mixture of Pluronic F108, water, and mineral oil, 169f
molecular mobility, 162
multiphase mixture of Pluronic surfactant, water, and oil, 168
scaled creep compliance curves and dependence of creep relaxation time vs. aging time for xanthan paste, 172f
time-domain creep compliance, 163
trap model, 163
violation of FDT, 162

violation of GSER using light
scattering study, 164
viscoelastic spectrum, 163
xanthan pastes, 168, 170–173
Sol-gel chemistry
atomistic simulations, 291–293
characterizing products, 291
effect of shape of inclusions in
polymeric matrix on elastic
properties, 294t
elastic constants along main
directions for ladder structures,
292t
electronic industry, 291
Hashin–Shtrikman model, 294
interlayer dielectric (ILD), 291
ladder structures, 292f
mesoscale calculations, 293–296
objective of work, 291
PALMYRA software, 293
predicted dielectric constants, 294,
295f
predicted values of Young's modulus
of porous glass, 294, 295f
stress in porous resin, 296
Sol-gel processing, applications, 145
Sol-gel transition
colloidal suspensions, 151f
rheology and diffusing wave
spectroscopy (DWS), 153f
Solubility, adhesion and, 367–368
Solvation, ions in liquids near
interfaces, 197, 199
Solvation potential, modeling, 201
Solvents
effects on phase boundaries, 75
percent variation of lattice parameter
at different polar organic solvent
contents, 76f
polar organic, and phase behavior of
$EO_{37}PO_{58}EO_{37}$ in water, 74f
poly(ethylene oxide)–poly(propylene
oxide) block copolymers, 71, 74
self-assembly of amphiphiles, 61–62
Spin relaxation
aggregate micro-viscosity, 37

alkyl chain order and dynamics in
aggregates, 36
molecular motion, 30
Spring constant, macroscopic
properties, 154
Stability
ethylene oxide/1,2-butylene oxide
(EO/BO) structures, 340–343
mesoscopic regions, 82
particle size distributions of EO/BO
after various treatments, 341, 342f
shear, of EO/BO, 340–342
thermal, of liquid crystalline
structures in EO/BO multi-
lamellar vesicles (MLVs), 342–
343
Static light scattering
modeled static intensity, 131–132
normalized scattering intensity
Rayleigh ratio, 130–131
simulation of static scattering
intensity for clusters, fibers, and
micelles, 132f
static intensity measured for pure
C_8E_4 and micellar contribution
from model, 131f
surfactant Triton X-100, 55f
total static intensity measured and
model curve, 131f
weighted form factors P(q) of three
aggregate types, 132
See also Dynamic light scattering;
Light scattering
Statistical mechanics, modeling and
simulation, 209f
Stejskal–Tanner relation, equation,
31–32
Stokes–Einstein–Debye (SED) model
breakdown with molecular size, 178
hydrodynamics, 178
SED breakdown and
poly(isobutylene) results, 186–187
See also Nanoscale vs. macroscale
friction
Stokes–Einstein equation,
hydrodynamic radius, 49–50

Stokes formula, ion mobility, 195
Storage modulus, suspension of destabilized latex particles, 150*f*
Stress, porous resin, 296
Structure, stable suspensions, 146–149
Structure factors
  dynamic, 149
  polystyrene (PS) suspension, 148*f*
  static, of deionized polystyrene suspensions, 147*f*
Supercooled liquids, charge stabilized particles, 146
Superwetting cleaners
  change in phase topology with temperature, 385, 387–388
  control of kinetic properties and cleaning, 383–385
  control of rheological behavior, 377–379
  dissipative particle dynamics (DPD) method, 378
  dodecane uptake capacity, 382*f*
  DPD simulation of mesoscale structure for three-component system, 388*f*
  DPD simulation of system with less repulsive head group interactions, 380*f*
  DPD simulations of three-component systems, 386*f*
  effect of micelle shape on oil uptake capacity, 381
  effect of oil molar volume on upper and lower temperatures for oils in water-oil-diethylene glycol monobutyl ether system, 389*f*
  effect of oil molecular weight and co-surfactant concentration on oil uptake of microemulsion, 383*f*
  effect of shearing on mesoscale structure, 387, 388*f*
  effect of tri-critical point on soil removal kinetics, 390–391
  evolution of viscosity with increasing salt concentration, 381*f*
  evolution of water-oil interfacial tension vs. temperature, 389*f*
  experimental contour plot of dodecane uptake, 381, 382*f*
  mesoscale structure in 3-component oil/water/amphiphile systems, 383–385
  optimizing three-component phase behavior, 385, 387–388
  predicting mesoscale structure in surfactant solutions, 377–379
  relationship between cleaning time and distance from tri-critical point, 391*f*
  tar removal from treated glass surface, 390*f*
  tar removal kinetics, 391
  three-component phase diagram and critical point, 383–385
  tri-critical points, 388–390
Surface roughness, silicon substrates, 5
Surfactant mixtures
  air-water interface, 103–104
  anionic and nonionic surfactants sodium dodecyl sulfate (SDS) and hexaethylene glycol monododecyl ether (C12E6), 98
  applications, 97
  average scattering length density profile, 99
  background and resurgence, 97
  C10E6/C14E6 mixtures, 111–112
  C12E3/C12E8 mixtures, 111
  comparing composition of interfaces and micelles, 107*f*
  composition in mixed SDS/C12E6 micelles vs. solution concentration, 109*f*
  cross-partial structure factors, 100
  experimental, 98–102
  investigation techniques, 97–98
  liquid-liquid interface, 105–108
  liquid-solid interface, 104–105
  micelle form factor, 101
  micelles, 108–110

modeling micelles with standard core + shell model, 101–102

mole% SDS and mole% C12E6 vs. surfactant concentration for 70/30 mixture, 103*f*

neutron reflectivity, 98–100

neutron reflexivity at hydrophilic liquid-solid interface, 105*f*

nonionic mixtures at air-water interface, 110–112

optical matrix method, 99

pseudo phase approximation, 97–98

Q, wave-vector transfer, 98

regular solution theory (RST) approach, 97-98

scattered intensity at small Q, 101

scattered intensity for 70/30 SDS/C12E6, 108*f*

scattered intensity for hexadecane in D$_2$O emulsion with SDS/C12E6, 107*f*

scattered intensity in limit of small Q, 102

scattering length density, 99

scattering length density profile, 100

SDS/C12E6 mixtures, 102–110

small angle neutron scattering (SANS), 101–102

specular reflectivity, 98–99

specular reflectivity for nonionic mixture C12E3/C12E8, 100, 101*f*

variation in micelle aggregation number with solution composition, 110*f*

variation in surface composition and adsorbed amount vs. solution composition, 106*f*

variation of mole% SDS with surfactant concentration at air-water interface for 70/30 and 50/50 mixture, 104*f*

volume fraction distribution for C10E6/C14E6 mixture, 113*f*

volume fraction distribution for C12E8 alkyl and ethylene oxide chains and solvent in C12E3/C12E8 mixture, 112*f*

Surfactants

attraction to oil-water interfaces, 207

dissipative particle dynamics (DPD) for dissolution, 247–248

dynamic and static light scattering, 55*f*

electrophoretic nuclear magnetic resonance (NMR) of polymer and, 35–36

ethylene oxide/1,2-butylene oxide (EO/BO) copolymers as possible vesicle-forming, 331–332

formulation of products, 372–374

inversion with addition of, 401–402

micelle formation, 207–208

micellization, 34–35

monolayers at water-oil interface, 82

multiphase mixture of Pluronic, water, and oil, 168, 169*f*

organization by X-ray and neutron scattering, 82

predicting mesoscale structure, 377–379

problems in nuclear magnetic resonance (NMR), 29

self-assembled structures, 61

See also Cetyltrimethylammonium bromide (CTAB) surfactant; Phase inversion; Trisiloxane surfactants

Suspensions. *See* Colloidal suspensions

T

Tar removal, effect of tri-critical point on kinetics, 390–391

Temperature

change in phase topology, 385, 387–388

concept of effective, 163

dependence of pluronic in water, 233, 234

evolution of water-oil interfacial tension, 389*f*

interfacial tension vs., for alkanols at water-hexane interface, 86*f*

phase diagram for oil, water, and surfactant, 395*f*

phase inversion, 396–397

simulation of film voltage vs., 195–196

variation of monolayers at water-hexane interface, 91–92

water solvent quality and poly(ethylene oxide)–poly(propylene oxide) [PEO–PPO] block copolymers, 70–71

X-ray reflectivity normalized to Fresnel reflectivity vs., 92*f*

Thermal stability, ethylene oxide/1,2-butylene oxide (EO/BO), 342–343

Thermodynamic models, mean fiber length and polymer concentration, 125–127

Three-dimensional cross correlation (3DDLS)

multiple scattering, 146, 148

static structure factors of deionized suspensions of polystyrene latex particles, 147*f*

See also Colloidal suspensions

Torsional relaxations, model, 187–189

Trap model, soft glasses, 163

Triacontanol monolayers

disorder in chain, 89

electron density profile at water-hexane interface, 87, 88*f*

fitting parameters, 89*t*

water-vapor interface, 90–91

See also *n*-Alkanol monolayers

Tri-critical point

description, 388–390

effect on soil removal kinetics, 390–391

relationship between cleaning time and distance from tri-critical point, 391*f*

See also Superwetting cleaners

Trisiloxane surfactants

aggregate structure and surface hydrophobicity, 12

atomic force microscopy (AFM) imaging and force measurements, 5

$(CH_3)_3SiO)_2Si(CH_3)(CH_2)_3(OCH_2)_{12}$OH [$M(D'E_{12}OH)M$], 3

characterization methods, 4–5

concentration of $M(D'E_{12}OH)M$, 5

experimental, 4–5

head group area, 14

intermediate structures of $M(D'E_{12}OH)M$, 12, 14

jump-in event, 7

materials, 4

molecular packing parameter, 14

n-octadecyltrichlorosilane (OTS) monolayer coverage, 5

phase behavior and microstructure, 3

single isotropic phase for $M(D'E_{12}OH)M$ in water, 14

soft contact AFM, 3

soft contact AFM image and force vs. distance curve on oxidized silicon wafer in $M(D'E_{12}OH)M$ solution, 6*f*

structure of $M(D'E_{12}OH)M$ on OTS-modified oxidized silicon substrate with 20° water contact angle, 7, 8*f*

structure of $M(D'E_{12}OH)M$ on OTS-modified oxidized silicon substrate with 40° water contact angle, 7, 9*f*

structure of $M(D'E_{12}OH)M$ on OTS-modified oxidized silicon substrate with 80° water contact angle, 7, 10*f*, 11*f*, 12

structure of $M(D'E_{12}OH)M$ on OTS-modified oxidized silicon substrate with 107° water contact angle, 12, 13*f*

substrate surface modification, 4

surface aggregates, 14

surface roughness, 5

Two-step model, surfactant systems, 30

U

Ultra-sonicated structures, ethylene oxide/1,2-butylene oxide (EO/BO) diblock copolymers, 337–338
Ultrasonic treatment, cubosomes, 348*f*
Uni-lamellar vesicle (ULV) formation, 329–330
See also Ethylene oxide/1,2-butylene oxide (EO/BO) diblock copolymers

V

Validation
mesoscale modeling, 232–233
nanoviscometer, 180–181
Vapor phase deposition, *n*-octadecyltrichlorosilane (OTS), 4
Vesicles
choice of size and density of bilayer initial configuration, 251
diagram of ULV and MLV, 329*f*
dissipative particle dynamics (DPD), 248–253
effects of polydispersity on formation, 338–340
equilibrated mono-lamellar vesicle, 252
ethylene oxide/1,2-butylene oxide (EO/BO) copolymers as possible vesicle-forming surfactants, 331–332
flat aggregate coexisting with larger vesicle, 252
formation, 329–330
formation and growth, 248–253
homogeneous solution of micelles, 252
multi-lamellar vesicles (MLVs), 329
snapshots of DPD simulation, 250
structure formation and detection, 332–335
two small micelles coexisting with larger vesicle, 252

uni-lamellar vesicles (ULVs), 329
See also Dissipative particle dynamics (DPD); Ethylene oxide/1,2-butylene oxide (EO/BO) diblock copolymers
Viruses, light scattering, 57
Viscosity
aggregate micro-viscosity from spin relaxation data, 37
change for 3-methylpentane (3MP) film, 198
corrections for characterizing mesoscale structure, 37-38
friction, 178
Vogel–Tamman–Fucher equation, 195
See also Nanoscale vs. macroscale friction
Vogel–Tamman–Fucher equation, viscosity, 195

W

Water
ion solvation, 200
motion of ions across oil-water-oil interface, 199
Water contact angle. *See* Trisiloxane surfactants
Water-hexane interface
n-alkanols at, 82–83
disorder in chain, 89
electron density gradient normal to surface, 84
electron density profile normal to interface for triacontanol monolayer, 87, 88*f*
experimental, 83–85
fitting parameters for triacontanol monolayers, 89*t*
interfacial tension, 83
interfacial tension vs. temperature for alkanols, 86*f*
kinematics of X-ray reflectivity from, 84*f*

monolayers at, 85–90

nomenclature for interfacial layers, 84*f*

oscillations in reflectivity, 86–87

sample preparation, 83

single kink defect, 89

specular reflectivity, 85

temperature variation of monolayers at, 91–92

X-ray methods, 83–85

X-ray reflectivity normalized to Fresnel reflectivity vs. temperature, 92*f*

X-ray reflectivity vs. wave vector normal to interface, 87*f*

Water-vapor interface

electron density in head-group region, 90–91

electron density profile for triacontanol monolayer, 88*f*, 90

triacontanol monolayers at, 90–91

Weighting function, definition, 301

Wetting. *See* Adhesion

Winsor phases

emulsion, 394–395

paraffinic oil, coco benzyl dimethyl ammonium chloride, and C14E4, 401*f*

X

Xanthan paste

creep compliance curves, 172*f*

creep measurements, 170

diffusing wave spectroscopy (DWS), 170–171

fluctuation-dissipation plot, 171, 173

relationship between rheology and DWS, 168, 170

X-ray diffraction, cholesterol monohydrate crystal, 20*f*

X-ray reflectivity

alkanols at water-hexane interface, 86, 87*f*

electron density gradient normal to surface, 84

function of wave vector transfer normal to interface, 87*f*

kinematics, 84*f*

method, 83–85

normalized to Fresnel reflectivity vs. temperature, 92*f*

specular reflectivity, 85

X-ray scattering, surfactant organization, 82

Y

Young's modulus, porous glasses, 294, 295*f*

Yukawa particles, interaction, 144*f*

Highlights from ACS Books

Desk Reference of Functional Polymers: Syntheses and Applications
Reza Arshady, Editor
832 pages, clothbound, ISBN 0–8412–3469–8

Chemical Engineering for Chemists
Richard G. Griskey
352 pages, clothbound, ISBN 0–8412–2215–0

Controlled Drug Delivery: Challenges and Strategies
Kinam Park, Editor
720 pages, clothbound, ISBN 0–8412–3470–1

A Practical Guide to Combinatorial Chemistry
Anthony W. Czarnik and Sheila H. DeWitt
462 pages, clothbound, ISBN 0–8412–3485–X

Chiral Separations: Applications and Technology
Satinder Ahuja, Editor
368 pages, clothbound, ISBN 0–8412–3407–8

Molecular Diversity and Combinatorial Chemistry: Libraries and Drug Discovery
Irwin M. Chaiken and Kim D. Janda, Editors
336 pages, clothbound, ISBN 0–8412–3450–7

A Lifetime of Synergy with Theory and Experiment
Andrew Streitwieser, Jr.
320 pages, clothbound, ISBN 0–8412–1836–6

For further information contact:
Order Department
Oxford University Press
2001 Evans Road
Cary, NC 27513
Phone: 1-800-445-9714 or 919-677-0977
Fax: 919-677-1303

More Best Sellers from ACS Books

Microwave-Enhanced Chemistry: Fundamentals, Sample Preparation, and Applications
Edited by H. M. (Skip) Kingston and Stephen J. Haswell
800 pp; clothbound ISBN 0–8412–3375–6

Designing Bioactive Molecules: Three-Dimensional Techniques and Applications
Edited by Yvonne Connolly Martin and Peter Willett
352 pp; clothbound ISBN 0–8412–3490–6

Principles of Environmental Toxicology, Second Edition
By Sigmund F. Zakrzewski
352 pp; clothbound ISBN 0–8412–3380–2

Controlled Radical Polymerization
Edited by Krzysztof Matyjaszewski
484 pp; clothbound ISBN 0–8412–3545–7

The Chemistry of Mind-Altering Drugs: History, Pharmacology, and Cultural Context
By Daniel M. Perrine
500 pp; casebound ISBN 0–8412–3253–9

Computational Thermochemistry: Prediction and Estimation of Molecular Thermodynamics
Edited by Karl K. Irikura and David J. Frurip
480 pp; clothbound ISBN 0–8412–3533–3

Organic Coatings for Corrosion Control
Edited by Gordon P. Bierwagen
468 pp; clothbound ISBN 0–8412–3549–X

Polymers in Sensors: Theory and Practice
Edited by Naim Akmal and Arthur M. Usmani
320 pp; clothbound ISBN 0–8412–3550–3

Phytomedicines of Europe: Chemistry and Biological Activity
Edited by Larry D. Lawson and Rudolph Bauer
336 pp; clothbound ISBN 0–8412–3559–7

For further information contact:
Order Department
Oxford University Press
2001 Evans Road
Cary, NC 27513
Phone: 1-800-445-9714 or 919-677-0977